Fluorescent Polymers for Sensing and Imaging

Fluorescent Polymers for Sensing and Imaging

Special Issue Editor

Seiichi Uchiyama

MDPI • Basel • Beijing • Wuhan • Barcelona • Belgrade • Manchester • Tokyo • Cluj • Tianjin

Special Issue Editor
Seiichi Uchiyama
University of Tokyo
Japan

Editorial Office
MDPI
St. Alban-Anlage 66
4052 Basel, Switzerland

This is a reprint of articles from the Special Issue published online in the open access journal *Polymers* (ISSN 2073-4360) (available at: https://www.mdpi.com/journal/polymers/special_issues/fluorescent_polymers_sensing_imaging).

For citation purposes, cite each article independently as indicated on the article page online and as indicated below:

LastName, A.A.; LastName, B.B.; LastName, C.C. Article Title. *Journal Name* **Year**, *Article Number*, Page Range.

ISBN 978-3-03936-940-9 (Pbk)
ISBN 978-3-03936-941-6 (PDF)

Cover image courtesy of Seiichi Uchiyama.

Contents

About the Special Issue Editor

Seiichi Uchiyama has been an Assistant Professor at the University of Tokyo since 2005. He received his B.Sc., M.Sc., and Ph.D. in Pharmacy from the University of Tokyo (supervisor: Prof. Kazuhiro Imai). Then, he spent three years as a postdoctoral researcher at Nara Women's University (supervisor: Prof. Kaoru Iwai) and Queen's University of Belfast (supervisor: Prof. A. Prasanna de Silva). After returning to the University of Tokyo in 2005, he started his career as an experimental researcher and continues to work as an independent leader of scientific projects funded by the Japanese government since 2008. His current interests include analytical and photophysical chemistry and the development of fluorescent sensors based on novel functional mechanisms.

Editorial

The Magical Combination of Polymer Science and Fluorometry

Seiichi Uchiyama

Graduate School of Pharmaceutical Sciences, The University of Tokyo, Tokyo 113-0033, Japan; seiichi@mol.f.u-tokyo.ac.jp; Tel.: +81-3-5841-4768

Received: 26 March 2020; Accepted: 8 April 2020; Published: 10 April 2020

I am very pleased to announce the publication of "Fluorescent Polymers for Sensing and Imaging". This Special Issue includes a review, thirteen articles, and one communication, which represent the contributions of seventy researchers in nine countries. When I received an invitation to serve as a guest editor from the editorial office of Polymers, I immediately recalled the first time I read the striking results published in 2000 by Swager et al., who reported a dramatic increase in pH sensitivity due to the amplifying effects of polymers [1]. Before obtaining my PhD in 2002, my research was focused on developing novel fluorescent sensing systems using only small organic molecules. The Swager paper provided fresh insight into polymeric architecture, which often afforded extremely high sensitivity that could not be achieved with small molecules. At the same time, a fluorescent polymeric thermometer with sub-degree temperature resolution was developed in my laboratory [2], which, amazingly, enabled intracellular thermometry [3]. Of course, the robustness and multiple functionality of a polymer motif are different advantages.

I have made two contributions to the Special Issue as a guest editor. In my review article [4], I summarize the findings of a long-term investigation performed in collaboration with Prof. de Silva on polymer-based sensing systems and polymer-specific microenvironments. In a research article [5], I present new intracellular thermometry results obtained using a cationic fluorescent nanogel thermometer. Measuring the temperature within living cells with polymeric sensing materials is a challenging target. The cationic fluorescent nanogel thermometer, prepared with a new cationic radical initiator [6], can be taken up by mammalian cells. Sensitive and noncytotoxic fluorescent polymeric thermometers, based on the combination of a thermo-responsive polymer and a fluorophore sensitive to polarity and hydrogen bonding, are described in this article. D'Souza et al. compare the fluorescence properties of BODIPY fluorophores and report a closely related temperature-sensing system [7]. A sharp response to temperature variation was observed after wrapping one of the BODIPY derivatives in Pluronic copolymers consisting of hydrophilic poly(ethylene oxide) and hydrophobic poly(propylene oxide) units.

Ions are the universal targets of fluorescent sensors. As I mentioned at the beginning of the introduction, polymeric architectures can provide unprecedented sensitivity and selectivity. Three elegant examples are reported in this Special Issue. Kyhm et al. describe a sensing system for potassium ion (K^+) in aqueous solution. This system combined the characteristics of fluorescent polyfluorene and a 15 base-aptamer that could pack K^+ [8]. The aptamer was fluorometrically labeled with fluorescein and rhodamine. Upon binding K^+, variable fluorescence outputs by the sensing system via Förster resonance energy transfer (FRET) were observed. Lin et al. synthesized novel silane-containing polythioethers that fluoresced in ethanol [9]. Selective quenching by ferric ions (Fe^{3+}) was achieved by functionalizing the polythioether end groups with suitable sulfhydryl compounds. Preliminary experimental data obtained using human epithelial carcinoma (HeLa) cells were also reported. Seitz et al. prepared poly(*N*-isopropylacrylamide)-based nanoparticles that contained fluorescein and anilinodiacetic acid units and utilized them as sensors for cupric ions (Cu^{2+}) in water [10]. While the nanoparticles intrinsically responded to changes in temperature, remarkable fluorescein quenching by Cu^{2+} was observed at a fixed temperature.

Molecular sensors require a special design approach that differs from that used for ion sensors, because the strong ionic interactions between the sensors and target ions and the coordination between ligands and target ions are absent. The oxygen sensing mechanism is unique, since quenching is due to collisions between molecular oxygen and the sensor. Peng et al. prepared ratiometric luminescent nanoparticles for oxygen sensing [11]. The nanoparticles consisted of polystyrene and polyacrylate block copolymers. They also contained oxygen-sensitive Ru^{2+} complexes and oxygen-insensitive Tb^{3+} complexes to afford a ratiometric emission signal. The polymeric structure around the complexes prevented interferences by other ions and molecules. The authors then demonstrated its application for sensing oxygen in model tumor spheroids. Zhao et al. developed a fluorescence sensor for adenosine triphosphate (ATP) in aqueous solution and HeLa cells using polythiophene that bore boronic acid and quaternary ammonium moieties [12]. The boronic acid and quaternary ammonium groups in the polymer bound to the diol and phosphate groups in ATP, respectively. ATP binding by polythiophene led to the formation of supramolecular aggregates, which quenched polythiophene emission. Tanaka et al. synthesized polyhedral oligomeric silsesquioxane that attached to Ru^{2+} complexes [13]. Electrochemiluminescence by the silsesquioxane was not significantly quenched by oxygen. In contrast, the water pollutant oxytetracycline markedly quenched silsesquioxane electrochemiluminescence after the oxidization of oxytetracycline at an electrode. Oxytetracycline sensing could thus be performed in a phosphate-buffered saline (PBS) without requiring a degassing procedure.

Molecular imprinting is an extraordinarily powerful polymer chemistry technique used to construct selective sensors. A polymeric receptor that is extremely specific to a target can be created via polymerization with the target molecule. Zhang et al. used this technique to prepare luminescent multilayered nanoparticles [14]. The cores of the nanoparticles contained iron oxides and luminescent quantum dots. The target molecule, bisphenol A (2,2-bis(4-hydroxyphenyl)propane), could be trapped in the 3-aminopropyltriethoxysilane-based shells. The nanoparticles were applied for the optical detection of bisphenol A in tap water and lake water. Bisphenol A bound to the nanoparticles quenched emission by the quantum dots in the cores in a linear fashion. The same technique can be extended beyond molecules to detect biological species. Wang et al. molecularly imprinted a polymer to detect the Gram-positive pathogenic bacterium, *Listeria monocytogenes* [15]. The polymer bound to *L. monocytogenes*, but it did not bind to *Escherichia coli*, *Staphylococcus aureus*, or *Salmonella*. *L. monocytogenes* was spiked into milk and pork, and detected through significant quenching of the luminescent polymer.

I am less familiar with the themes of the other contributions in this Special Issue, which make them more interesting to me. Matsumura and Iwai investigated variations in the microenvironment related to the complexation of poly(acrylic acid) and polyacrylamide using 9-(4-*N,N*-dimethylaminophenyl) phenanthrene, a polarity-sensitive fluorophore [16]. Complexation of the polymers was pH-dependent, and it was accompanied by significant changes in the local polarity. Warman et al. fabricated a poly(*t*-butyl acrylate) macrogel 24 mm in diameter, which contained fluorescent pyrene units to mimic the human eye [17]. The authors suggest that the poly(*t*-butyl acrylate) macrogel would be useful for controlling radiotherapy dosage, because the dose-related effects of the beam on the eye can be visualized in a three-dimensional fluorescence image of the macrogel. Ayesta et al. dissolved rhodamine B in methanol and inserted the fluorescent dye into poly(methyl methacrylate) optical fibers [18]. The authors showed that the penetration rate and fluorescence behavior of the rhodamine B-doped fibers were temperature-dependent. Their findings will be helpful for the development of new sensor materials. Timofeyev et al. coated fluorescent poly(allylamine hydrochloride)/poly(sodium 4-styrenesulfonate) nanocapsules with polyethylene glycol and monitored their circulation in a model amphipod, *Eulimnogammarus verrucosus* [19]. Knowledge about the distribution of the nanocapsules over time and their toxicity will hasten their application as sensors to monitor the physiological status of biological species.

Finally, I would like to express appreciation for the great editorial contributions of Liz Li and Zora Zhu at MDPI. I was kindly encouraged to work as a guest editor during a pleasant conversation with Ms. Hannah Guo, who attended a MDPI booth at the World Polymer Congress Macro2018 in Cairns, Australia. It is hoped that the information provided in this Special Issue will facilitate significant advances in polymer science in the future.

Conflicts of Interest: The authors declare no conflict of interest.

References

1. McQuade, D.T.; Hegedus, A.H.; Swager, T.M. Signal amplification of a "Turn-on" sensor: Harvesting the light captured by a conjugated polymer. *J. Am. Chem. Soc.* **2000**, *122*, 12389–12390. [CrossRef]
2. Uchiyama, S.; Matsumura, Y.; de Silva, A.P.; Iwai, K. Fluorescent molecular thermometers based on polymers showing temperature-induced phase transitions and labeled with polarity-responsive benzofurazans. *Anal. Chem.* **2003**, *75*, 5926–5935. [CrossRef] [PubMed]
3. Uchiyama, S.; Gota, C.; Tsuji, T.; Inada, N. Intracellular temperature measurements with fluorescent polymeric thermometers. *Chem. Commun.* **2017**, *53*, 10976–10992. [CrossRef] [PubMed]
4. Yao, C.-Y.; Uchiyama, S.; de Silva, A.P. A personal journey across fluorescent sensing and logic associated with polymers of various kinds. *Polymers* **2019**, *11*, 1351. [CrossRef] [PubMed]
5. Hayashi, T.; Kawamoto, K.; Inada, N.; Uchiyama, S. Cationic fluorescent nanogel thermometers based on thermoresponsive poly(*N*-isopropylacrylamide) and environment-sensitive benzofurazan. *Polymers* **2019**, *11*, 305. [CrossRef] [PubMed]
6. Uchiyama, S.; Tsuji, T.; Kawamoto, K.; Okano, K.; Fukatsu, E.; Noro, T.; Ikado, K.; Yamada, S.; Shibata, Y.; Hayashi, T.; et al. A cell-targeted non-cytotoxic fluorescent nanogel thermometer created with an imidazolium-containing cationic radical initiator. *Angew. Chem. Int. Ed.* **2018**, *57*, 5413–5417. [CrossRef] [PubMed]
7. Saremi, B.; Bandi, V.; Kazemi, S.; Hong, Y.; D'Souza, F.; Yuan, B. Exploring NIR aza-BODIPY-based polarity sensitive probes with ON-and-OFF fluorescence switching in Pluronic nanoparticles. *Polymers* **2020**, *12*, 540. [CrossRef] [PubMed]
8. Kim, I.; Jung, J.-E.; Lee, W.; Park, S.; Kim, H.; Jho, Y.-D.; Woo, H.Y.; Kyhm, K. Two-step energy transfer dynamics in conjugated polymer and dye-labeled aptamer-based potassium ion detection assay. *Polymers* **2019**, *11*, 1206. [CrossRef] [PubMed]
9. Gou, Z.; Zhang, X.; Zuo, Y.; Lin, W. Synthesis of silane-based poly(thioether) via successive click reaction and their applications in ion detection and cell imaging. *Polymers* **2019**, *11*, 1235. [CrossRef] [PubMed]
10. Wang, F.; Planalp, R.P.; Seitz, W.R. A Cu(II) indicator platform based on Cu(II) induced swelling that changes the extent of fluorescein self-quenching. *Polymers* **2019**, *11*, 1935. [CrossRef] [PubMed]
11. Zhao, W.-x.; Zhou, C.; Peng, H.-s. Ratiometric luminescent nanoprobes based on ruthenium and terbium-containing metallopolymers for intracellular oxygen sensing. *Polymers* **2019**, *11*, 1290. [CrossRef] [PubMed]
12. Liu, L.; Zhao, L.; Cheng, D.; Yao, X.; Lu, Y. Highly selective fluorescence sensing and imaging of ATP using a boronic acid groups-bearing polythiophene derivate. *Polymers* **2019**, *11*, 1139. [CrossRef] [PubMed]
13. Nakamura, R.; Narikiyo, H.; Gon, M.; Tanaka, K.; Chujo, Y. Oxygen-resistant electrochemiluminescence system with polyhedral oligomeric silsesquioxane. *Polymers* **2019**, *11*, 1170. [CrossRef] [PubMed]
14. Zhang, X.; Yang, S.; Chen, W.; Li, Y.; Wei, Y.; Luo, A. Magnetic fluorescence molecularly imprinted polymer based on FeOx/ZnS nanocomposites for highly selective sensing bisphenol A. *Polymers* **2019**, *11*, 1210. [CrossRef] [PubMed]
15. Zhao, X.; Cui, Y.; Wang, J.; Wang, J. Preparation of fluorescent molecularly imprinted polymers via Pickering emulsion interfaces and the application for visual sensing analysis of *Listeria monocytogenes*. *Polymers* **2019**, *11*, 984. [CrossRef] [PubMed]

16. Matsumura, Y.; Iwai, K. pH behavior of polymer complexes between poly(carboxylic acids) and poly(acrylamide derivatives) using a fluorescence label technique. *Polymers* **2019**, *11*, 1196. [CrossRef] [PubMed]

17. Luthjens, L.H.; Yao, T.; Warman, J.M. A polymer-gel eye-phantom for 3D fluorescent imaging of millimetre radiation beams. *Polymers* **2018**, *10*, 1195. [CrossRef] [PubMed]

18. Ayesta, I.; Azkune, M.; Arrospide, E.; Arrue, J.; Illarramendi, M.A.; Durana, G.; Zubia, J. Fabrication of active polymer optical fibers by solution doping and their characterization. *Polymers* **2019**, *11*, 52. [CrossRef] [PubMed]

19. Shchapova, E.; Nazarova, A.; Gurkov, A.; Borvinskaya, E.; Rzhechitskiy, Y.; Dmitriev, I.; Meglinski, I.; Timofeyev, M. Application of PEG-covered non-biodegradable polyelectrolyte microcapsules in the crustacean circulatory system on the example of the amphipod *Eulimnogammarus verrucosus*. *Polymers* **2019**, *11*, 1246. [CrossRef] [PubMed]

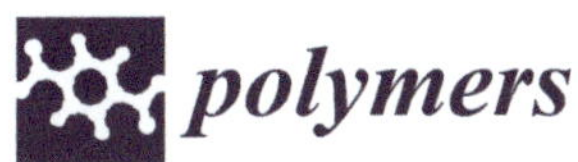 

Review

A Personal Journey across Fluorescent Sensing and Logic Associated with Polymers of Various Kinds

Chao-Yi Yao [1,*], Seiichi Uchiyama [2,*] and A. Prasanna de Silva [1,*]

[1] School of Chemistry and Chemical Engineering, Queen's University, BT9 5AG Belfast, Northern Ireland
[2] Graduate School of Pharmaceutical Sciences, The University of Tokyo, 7-3-1 Hongo Bunkyo-ku, Tokyo 113-0033, Japan
* Correspondence: cyao01@qub.ac.uk (C.-Y.Y.); seiichi@mol.f.u-tokyo.ac.jp (S.U.); a.desilva@qub.ac.uk (A.P.d.S.)

Received: 21 June 2019; Accepted: 12 August 2019; Published: 14 August 2019

Abstract: Our experiences concerning fluorescent molecular sensing and logic devices and their intersections with polymer science are the foci of this brief review. Proton-, metal ion- and polarity-responsive cases of these devices are placed in polymeric micro- or nano-environments, some of which involve phase separation. This leads to mapping of chemical species on the nanoscale. These devices also take advantage of thermal properties of some polymers in water in order to reincarnate themselves as thermometers. When the phase separation leads to particles, the latter can be labelled with identification tags based on molecular logic. Such particles also give rise to reusable sensors, although molecular-scale resolution is sacrificed in the process. Polymeric nano-environments also help to organize rather complex molecular logic systems from their simple components. Overall, our little experiences suggest that researchers in sensing and logic would benefit if they assimilate polymer concepts.

Keywords: fluorescence; polymer; particle; sensor; logic gate; pH; ion; temperature

---

1. Introduction

It has been our pleasure to investigate molecular-scale devices which communicate with us at the human-scale. Owing to their subnanometric dimensions, they operate across a range of larger size-scales and provide us with valuable information from these worlds. Fluorescence signals provide output while various chemical species serve as input signals. Excitation light powers these devices wirelessly. In order to carry information, some modulation is required in the fluorescence signal. Chemical responsiveness provides this by chemically biasing a competition [1–6] for the deactivation of the fluorophore excited state between fluorescence emission [7,8] and photoinduced electron transfer (PET) [9,10]. Because of the extreme nature of this responsiveness, it is easy to regard these devices as 'off-on' switches. This leads to the realization that molecular devices share many attributes with semiconductor logic counterparts [11], while differing in other features [12–38]. Chemical responsiveness of fluorescence signals can also be arranged via ionic/dipolar influences on internal charge transfer (ICT) excited states [2,8,39]. Extreme versions of this behavior can be seen in benzofurazan fluorophores which again lead to 'off-on' switchability. Having a binary digital basis in electronic engineering does not preclude analog operations for exquisitely fine measurements. Similarly, the Boolean character of molecular switching devices still allows for the accurate measurement of tiny changes in the input signal, whether it be a chemical concentration or a physical property, when substantial populations of molecules exert mass action. Accurate sensing is therefore available from digital molecular devices. A significant fraction of our research involves polymers of some kind, sometimes in crucial ways.

Although each of the authors had their research formation in photoscience of small molecules [40–43], it is clear to us that macromolecules have uniquely beneficial characteristics barred to small counterparts [44]. For instance, polymer molecules are large enough to possess their own environments at the nanometer-scale. Although objects as varied as proteins [45] and DNA origami [46] could be studied in this way, it would be more immediately productive to pay attention to simple symmetrical systems such as quasi-spherical detergent micelles in water. We can consider detergent micelles in water as supramolecular polymer systems held together by hydrophobic interactions and then examine the region bounded by their surfaces for H^+ distribution for instance. These are discussed in Section 2. Especially when cross-linked, polymer molecules are large enough to create their own phase-separated environments at the nanometer- to millimeter-scale. When solid particles are formed in this way, they serve as recyclable matrices to carry functional small molecules such as sensors. Section 3 represents these. Solid polymer particles can also be vehicles for functional small molecules such as drug candidates during their synthesis and their evaluation. These came to the fore during the combinatorial chemistry wave [47] and still have roles to play. It would be important therefore to be able to identify these vehicles individually within large populations. Section 4 presents a solution to this problem by tagging these vehicles with molecular logic gates. Linear macromolecules without cross-links can also create their own phase-separated environments in certain instances. Such a transition of extended linear to globular forms can occur as the temperature is ramped across a threshold value. Such transitions persist in some cross-linked gel versions as well. Fluorescence readout of these transitions is possible from polymer-linked probes. This opens the way to molecular thermometers, which are now throwing light on the foundations of biology (Section 5).

As indicated above, the polymer plays a variety of roles in these systems. These roles will depend on the chemical structures involved. Nano-environments will be set up by long hydrocarbon chain monomers carrying hydrophilic termini which aggregate in water. These micelles or membranes are non-covalent macromolecular (self-assembly) systems which are sisters of synthetic polymers. Some of these nano-environments will also be employed in an organizational role to assemble logic gates. Recyclable matrices will be created with diamondoid Si–O lattices. Vehicles for other molecules will be built from crosslinked polystyrene cores with oligoethyleneglycol shells. Sharp thermoresponsivity will be introduced with polyacrylamides carrying 2-propyl substituents and relatives.

2. Mapping Membrane-Bounded Species

Since compartmentalization is a key to the origin and maintenance of life, it is crucial to study membrane-bounded species, especially those which are key players in biology. H^+ is paramount in this capacity because of its vital role in bioenergetics [48]. Since fluorescent PET signalling began with H^+ sensing [49], Anthracenemethylamine derivative **1** (Scheme 1) is a straightforward adaptation of a 'fluorophore-spacer-receptor' system [50,51] with the addition of an anchoring module in the form of a hydrocarbon chain and a spatial tuning module in the form of amine substituents [52]. When H^+ is picked up by **1** from its neighbourhood, the amine receptor is no longer able to perform a PET operation to the anthracene fluorophore, and the fluorescence is switched 'on'. The neighbourhood being sampled is determined by the height/depth of the amine lone electron pair relative to the micelle-water interface, which in turn is controlled by the hydrophobicity of **1** as it gravitates to the appropriate point along the hydrophobicity/hydrophilicity continuum between polar water and the apolar micelle interior. The spatial tuning groups make fine adjustments to the positioning of the amine receptor. The local H^+ density relative to the value in bulk water is related to the difference in pK_a values determined by fluorescence-pH titrations for **1** in micellar media and for a very hydrophilic version of **1** in neat water [53]. Such ΔpK_a values obtained for structural variants of **1** can be correlated with the hydrophobicity of the spatial tuning module. These graphs provide a first glimpse into the spatial distribution of membrane-bounded H^+ and how it is controlled by electrostatic and dielectric effects [53].

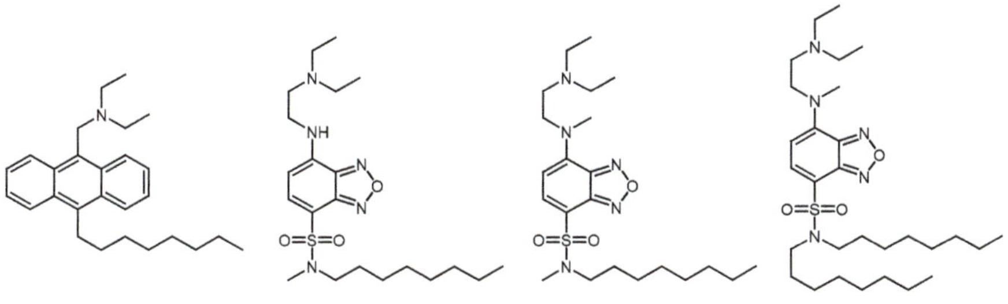

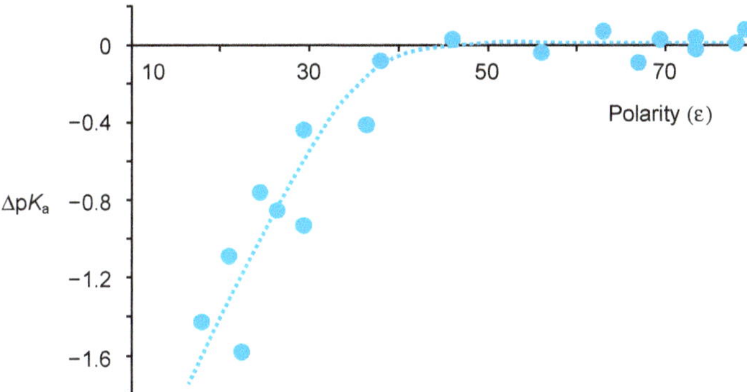

Scheme 1. Chemical structures of **1–4**.

A more proper mapping of H$^+$ in these micellar neighborhoods, in a cartographic sense, is achievable if the probe position can be determined at the same time as the ΔpK_a measurement. This is made possible by employing a variant of **1** outfitted with a fluorophore whose emission wavelength is dependent on environmental polarity. The position occupied by the probe on the hydrophobicity/hydrophilicity continuum between polar water and the apolar micelle interior will reflect the local polarity experienced by the probe, and hence its emission wavelength. ICT fluorophores fit the bill [54–57], and benzofurazans [58–63] are the best of all in our hands. **2** (Scheme 1) and its close derivatives produce rather educational maps of H$^+$ density near neutral Triton X-100 micelles [64] in water. In these fluorescent sensors, the effects of protonation at the terminal amino moiety (during H$^+$ sensing) on the fluorophore are dominantly observed in fluorescence efficiency but not in original absorption and emission wavelengths of the fluorophore, which enables accurate monitoring of both H$^+$ density and the environmental polarity simultaneously. As shown in Figure 1, the H$^+$ density near Triton X-100 micelles is hardly affected until we approach neighborhoods of an effective dielectric constant (ε) 40. As sensors go towards the micellar interior from the position of $\varepsilon = 40$ to that of $\varepsilon = 15$, H$^+$ density becomes suppressed to approximately 4% due to the dielectric repulsion (Figure 1). Our probes within the family represented by **2** are unable to get any closer to the micelle.

Figure 1. Local effective proton density (as measured by the shift of the acidity constant relative to bulk water, ΔpKa) near Triton X-100 micelles as a function of position (as measured by the local dielectric constant, ε). Adapted with permission from reference [64]. Copyright John Wiley and Sons 2018.

Further exploration of planet micelle is possible with **3** (Scheme 1) by providing useful mapping data from charged micelles. Although structurally close to **2**, **3** has no hydrogen-bond donor N-H group on the fluorophore. This is crucial because the N–H group at the anilino position is free from both protonation and deprotonation in a wide range of pH (e.g., $3 \leq pH \leq 12$) in water or aqueous micellar solutions and thus can engage in multiple hydrogen bondings with anionic head-groups of micelles (e.g., sulfate groups with considerable hydrogen bonding ability in sodium dodecyl sulfate (SDS) micelles [65]) to pin the probe to a narrow location, meaning that detailed mapping was not possible for anionic micelles via only hydrophobicity tuning. Once N–H is replaced by N–CH$_3$, this pinning effect disappears and a larger spatial distribution of probe positions opens up [66]. **2** does not fare much better with cationic micelles because of cation–pi interactions [67] between the micelle head-groups and the probe pi-system. **4** (Scheme 1) has stronger hydrophobic interactions due to the dioctyl chains so that the cation-pi interaction is relegated to a minor role. Better mapping is the result, though higher-resolution data remains our long-range goal.

Sentient beings depend on Na$^+$ near nerve membranes to convey and process environmental signals [68]. Membrane-bounded Na$^+$ is estimated by **5** (Scheme 2) [69], which takes a leaf out of **1**'s book by using a hydrocarbon chain for gross targeting and anchoring of the probe in the micelle. Owing to the relative structural complexity of the benzo-15-crown-5 ether receptor in **5** for Na$^+$ vis-à-vis the H$^+$ receptor amine in **1**, no spatial tuning module is available so far. Nevertheless, it is gratifying to find that Na$^+$ is concentrated a 100-fold near the surface of anionic micelles, whereas Na$^+$ is repelled so much from cationic micelles and even neutral micelles as to be immeasurable with **5**. The latter finding need not be a surprise because hydrophilic Na$^+$ would indeed be difficult to accommodate in a hydrophobic micelle neighbourhood when bulk water is available within travelling distance.

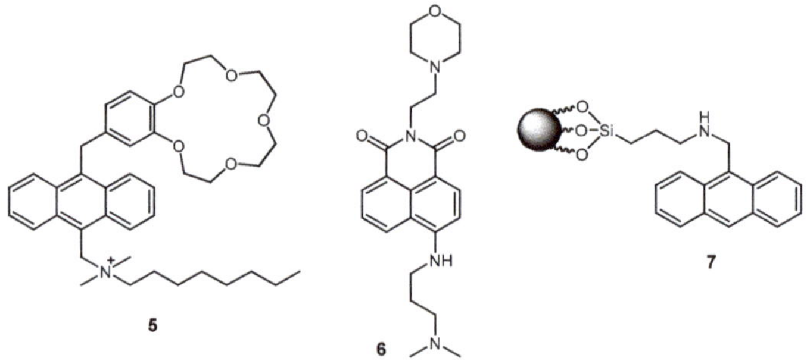

Scheme 2. Chemical structures of **5–7**.

Though nanometric in size, micelles are great containers which can organize sets of functional molecules. A pair of a fluorophore and a receptor is such an example of a self-assembled fluorescent PET sensor. Here, the role of the spacer in the fluorescent PET system is taken over by the micelle itself [70]. Inspired by this concept, we extend it to self-assembled AND logic systems with, e.g., a fluorophore and two different selective receptors [71]. The fact that various logic gates can be constructed by the step-by-step addition of components allows a 'plug-and-play' approach to some of the simpler molecular logic functions. Covalently bound AND logic gates operated within micelles represent computing at the smaller end of the nanoscale [72], which semiconductor devices still struggle to do.

We can cross from mapping and logic to the seemingly unrelated topic of photosynthetic reaction centre (PRC) mimics. In nature, the PRC is a marvel of supramolecular organization within a membrane in terms of structure and function [73]. This is a good thing too, since our origin and survival depend on it. We turn to micelles as a model membrane to contain PRC mimics of a receptor$_1$-spacer$_1$-fluorophore-spacer$_2$-receptor$_2$ format [74]. These have two PET pathways originating from the opposite termini of **6** (Scheme 2), of which one is favoured, somewhat similar to what is seen in the PRC. In its excited state, **6** has an internal electric field [75] to direct PET in one direction rather than the other.

3. Solid-Bound Sensors

Being modular, fluorescent sensors of the 'fluorophore-spacer-receptor' format are easily extended to 'fluorophore-spacer-receptor-spacer-particle' systems, e.g., **7** (Scheme 2) [76]. Beside the practical aspect of reusability, this SiO$_2$-bound amine receptor system shows the retardation of PET compared to homogeneous solution counterparts. Charge-separating processes of this kind are naturally slowed at solid surfaces because charge-stabilizing orientation polarization of water dipoles is less likely. Fortunately, this PET process, even after retardation, remains competitive with the radiative rate and so adequate H$^+$-sensing capability remains. This situation is maintained when H$^+$-sensing PET systems like **8** (Scheme 3) are embedded in polyvinylchloride, provided the polymer is suitably plasticized [77], and when Na$^+$-sensing PET systems like **9** (Scheme 3) are bonded to various fibres [78]. Solid-bound PET sensors continue to grow in number [79–82].

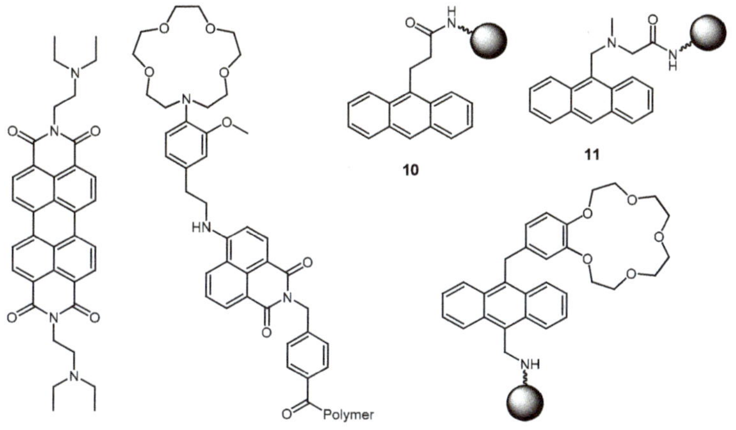

Scheme 3. Chemical structures of **8–12**.

4. Molecular Computational Identification (MCID)

In the previous section, the emphasis was on the fluorescent function with the particle being the new environment. Now the particle takes centre stage with the fluorescent function serving as an ID tag. As trailed above, polymer particles can be vehicles for various functional molecules but they can also be models for biological cells showing the way to cell diagnostics.

Whenever we encounter populations of objects, individual object identification is not a problem if they are at fixed locations. If they are not spatially addressable, some kind of tracking feature becomes necessary. Metamorphosing objects would also need tracking if some sense is to be made of their population. Modern information-based society is full of radiofrequency ID (RFID) tags which serve this tracking need [83], but these are limited to sizes above 10 μm because of the necessary antenna. Micrometric objects such as polymer particles and cells are therefore untouched by the RFID revolution and remain at large. Molecules would be capable of rectifying this situation if they possessed some easily detectable parameter which comes in a sufficiently large number of distinguishable values. For example, excitation and emission wavelengths of various fluorophores can encode up to 100 polymer beads, but not much more [84]. However, what about larger populations? It is possible to amplify these 100 codes many-fold by taking each fluorophore and making it conditionally switchable [85]. A given logic type represents this light output driven by chemical input. Many single-, double- and higher-logic types are available [26], e.g., PASS 1 (**10**, H$^+$-input), YES (**11**, H$^+$-input), and AND (**12**, Na$^+$-, H$^+$-inputs) (Scheme 3). Ternary logic types can be included as well [86]. Further amplification of diversity is made by attaching two or more tags to a given particle (Figure 2) [85,87,88]. The fluorescence wavelengths can extend into the near infrared to offer additional bandwidth [88]. MCID has recently been applied to populations, albeit rather small, so that object-to-object variations can be quantified (Figure 2). The method can then be used to unambiguously divide a population into sub-populations of a given logic type [88].

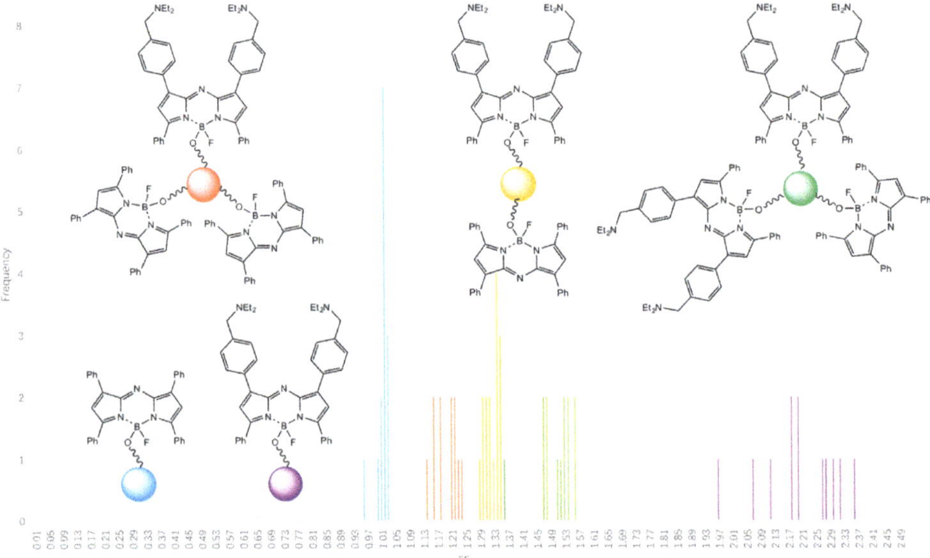

Figure 2. Histogram for the occurrence of various H^+-induced fluorescence enhancement factors (FE_{H+}) in logic-tagged beads for samples of identical copies. Adapted from reference [88] published by The Royal Society of Chemistry.

The mechanism of switching in YES gate **11**, AND gate **12** and the YES logic-based examples in Figure 2 are PET processes occurring within 'fluorophore-spacer-receptor-spacer-particle' and related type systems. Here, PET originates from an electron donor amine or benzocrown ether and terminates in an anthracene or azaBODIPY fluorophore. The PET rate is controlled by its thermodynamics as well as the length of the spacer. As usual, the PET process is arrested by protonation of the amine or by binding Na^+ to the benzocrown ether.

It is appropriate to mention some drawbacks, challenges and potential applications of MCID. The need to wash the samples with a chosen reagent can be considered as a drawback from some viewpoints, but chemical stimuli are common in biology. Applications of MCID can be imagined in tracking members of combinatorial chemistry libraries at the level of single polymer beads. The challenge will be to popularize this application. The road to application in cell diagnostics will be rockier, since MCID tags responding to suitably benign chemical stimuli would need to be found and validated.

5. Molecular Thermometers

As mentioned in the introduction section, some linear macromolecules switch between extended and collapsed forms in water as a response to varying temperature. Such cases with hydrophilic and hydrophobic groups which balance their effects display a lower critical solution temperature. The high degree of polymerization of these systems leads to strongly cooperative behaviour so that the transition occurs across a rather narrow temperature range. Outfitting of these polymers with a small amount of an environment (e.g., polarity and hydrogen bonding)-sensitive ICT fluorophore, 4-*N,N*-dimethylsulfamoyl-7-aminobenzofurazan (DBD), by means of copolymerization produces a fluorescent thermometer, **13** (Scheme 4), with greatly increased sensitivity (Figure 3) [89] over previous versions [90,91]. Other ICT fluorophores such as 7-aminocoumarin [92], BODIPY [93–96], dansylamine [97,98] and 4-amino-7-nitrobenzofurazan (NBD) [99,100] have also been applied in a similar way. At lower temperatures, **13** takes an open and extended form, and the solvent water molecules access the fluorophores in **13** to cause quenching. In contrast, **13** exists in a globular form at a higher temperature, where the fluorophores are surrounded by the hydrophobic backbone, and therefore emits strong fluorescence. In the similar system **14** (Scheme 4), fluorescence lifetime increases with temperature [101]. In contrast to fluorescence intensity, fluorescence lifetime is not influenced by the fluctuation of various experimental conditions (e.g., excitation light intensity and concentration of a sensor). Accordingly, the fluorescence lifetime can be a more reliable variable than the fluorescence intensity in some applications such as intracellular thermometry where the experimental conditions are relatively changeable. The downside is the need for more elaborate instrumentation for lifetime measurements.

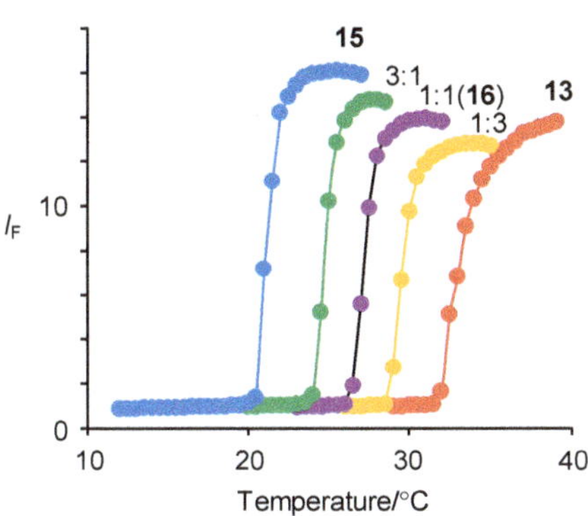

Scheme 4. Chemical structures of **13–17**.

Figure 3. Fluorescence intensity (IF) temperature profiles for **13** and **15** and for copolymers including **16** in the feed ratios given. Adapted with permission from reference [102]. Copyright American Chemical Society 2004.

In addition to high sensitivity, the polymeric design brings functional diversity to fluorescent thermometry. The functional temperature range can be easily tuned by using substituted monomers, e.g., **15** (Scheme 4) [89]. Interestingly, the functional temperature range of copolymer **16** (Scheme 4) with a 1:1 blend of the monomer units in **13** and **15** bisects the ranges of **13** and **15** [102]. So, the functional temperature range can be finely tuned by varying co-monomer feed ratios (Figure 3) [102]. The fluorescence wavelength of the polymeric thermometers can also be modified by using a different fluorophore, e.g., **17** (Scheme 4) [103] and **18** (Scheme 5) [104].

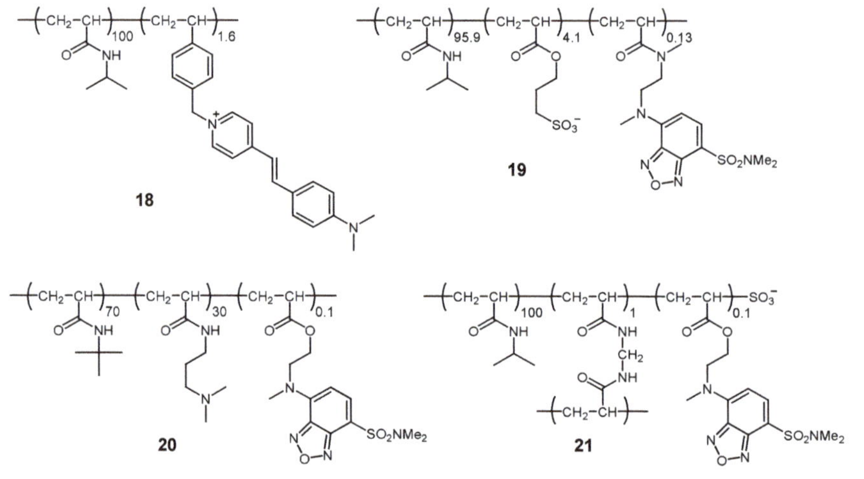

Scheme 5. Chemical structures of **18–21**.

Using an additional component in copolymers can improve physical and chemical features of fluorescent thermometers. **19** (Scheme 5) is a highly water-soluble thermometer because the ionic 3-sulfopropyl acrylate units prevent interpolymeric aggregation [105]. Such high solubility enables highly-resolved temperature measurements. In contrast, the incorporation of an H^+ receptor amine into a copolymer produces a molecular logic system with temperature and H^+ as multiple inputs [106–108]. For instance, the polymeric logic gate **20** (Scheme 5) fluoresces strongly only when environmental H^+ concentration is 'low' and temperature is 'high' to behave as an INHIBIT gate. High H^+ levels protonate the receptor, causing the copolymer to adopt the extended structure. So the fluorophore is surrounded by water to cause quenching whatever the temperature.

The gelation of solution-based fluorescent polymeric thermometers by adding a crosslinker results in robust nanogel particles [109]. The representative gel **21** (Scheme 5) exists as nanometric beads and shows nearly an order-of-magnitude fluorescence enhancement over a small temperature range when dispersed in water. The switching mechanism is essentially the same as that seen with the soluble version **13** [89]. The thermo-responsive fluorescent polymeric structure based on **13** has also been used as a shell structure of multifunctional magnetite nanoparticles [110]. These particles are expected to be useful in anticancer heat treatment where monitoring the temperature of target tumours is important.

While these fluorescent polymeric thermometers enabled temperature measurements of small objects such as an aqueous fluid in a heater-equipped microdevice [100,111] and turbid aqueous media heated with ultrasound-irradiation [112], the most attractive target is certainly biological cells. Highly water-soluble fluorescent gels (**22**, Scheme 6) microinjected into monkey's kidney COS7 cells showed a temperature-dependent fluorescence signal therein [113]. A linear polymeric thermometer (**23**, Scheme 6) unveiled temperature distribution of a COS7 cell with the aid of fluorescence lifetime imaging microscopy (FLIM), in which the inside of nuclei and neighbourhoods of mitochondria and centrosomes were remarkably hotter than the other cytoplasmic spaces (Figure 4) [114].

The above breakthrough in intracellular thermometry made biologists imagine the importance of the temperature at the single cell level and subsequently demand chemists to develop more user-friendly fluorescent thermometers. A cationic version of linear polymeric thermometers (**24**, Scheme 6) [115–117] now avoids microinjection procedures due to its ability to spontaneously enter cells. In addition, further labelling of this by a BODIPY structure (**25**, Scheme 6) enabled ratiometric thermometry, which offered high accuracy even without an expensive fluorescence lifetime imaging microscope [118]. Thermogenesis of brown adipocytes [119,120] and chemical stimulation of brain tissue [121] were successfully monitored with these thermometers.

Scheme 6. Chemical structures of **22–25**.

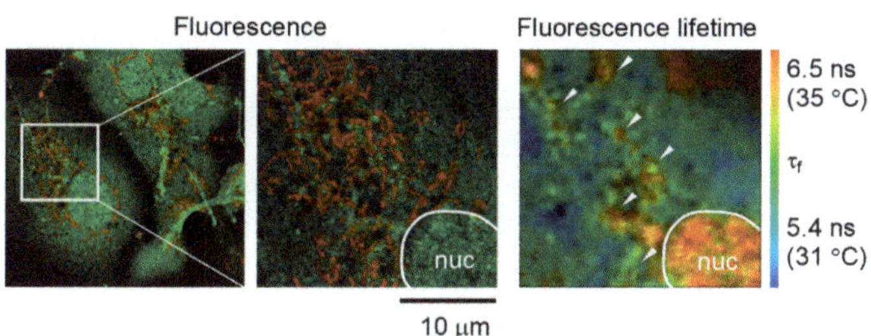

Figure 4. Intracellular temperature mapping with **23**. Confocal fluorescence images of **23** (green) and Mito Tracker Deep Red FM (red) (left and middle) and fluorescence lifetime image of **23** (right) in a live COS7 cell. 'nuc' indicate the nucleus. The region of interest shown in the white square in the left panel is enlarged in the middle and right panels. Arrowheads point to local heat production near mitochondria. Adopted from reference [114].

The future of fluorescent polymeric thermometers looks quite bright. The exemplified application of fluorescent polymeric thermometers in brain tissue will accelerate their use in in-vivo thermometry beyond a single cell level. In such cases, NIR (near infrared) ICT fluorophores will be preferred for the deep penetration of an excitation light and return of an emission signal to a detector. Another challenge of fluorescent polymeric thermometers is targeting to specific organelles of live cells by incorporating target signals into their chemical structures. It is expected that the localization of the fluorescent polymeric thermometer in heat-generating organelles, e.g., mitochondria, improves the detectability towards intracellular thermogenesis [122]. The cytotoxicity of fluorescent polymeric thermometers has also been concerning in biological and medical studies, and normal cell division and even differentiation were observed in a recent progress [123].

6. Conclusions

Whether constructed covalently from monomers or not, whether aqueous soluble or not, whether cross-linked or not, or whether solid or not, polymers offer unique environments and objects as playgrounds for sensors and logic designers. These efforts lead to reusable sensors, insights into the spatial distribution of chemical species near interfaces, membrane-assembled logic systems, temperature maps within living cells, and identification protocols for submillimetric objects.

Polymers **2019**, *11*, 1351

Author Contributions: Conceptualization, A.P.d.S.; writing—original draft preparation, C.-Y.Y., S.U., A.P.d.S.; writing—review and editing, C.-Y.Y., S.U., A.P.d.S.; funding acquisition, S.U.

Funding: This work was funded by China Scholarship Council and Japan Society for the Promotion of Science (Grant-in-Aid for Scientific Research (B) (17H03075)).

Acknowledgments: We thank Jean-Philippe Soumillion (Universite Louvain-la-Neuve, for collaboration on **7**), Otto Wolfbeis (Universitat Regensburg, for collaboration on **8**), Jim Tusa (Optimedical Inc, for collaboration on **9**), Mark James and Dave Pears (Avecia, for collaboration on **10–12**), Kaoru Iwai (Nara Women's University, for collaboration on **13**, **15**, **16**, **20** and **21**), Seiji Tobita and Toshitada Yoshihara (Gunma University, for collaboration on **14** and **17**), Kohki Okabe and Takashi Funatsu (The University of Tokyo, for collaboration on **22** and **23**), Yoshie Harada (Osaka University, for collaboration on **22** and **23**), Noriko Inada (Osaka Prefecture University, for collaboration on **23**) and Satoshi Yoshida and Aruto Yoshida (KIRIN Company Limited, for collaboration on **24** and **25**).

Conflicts of Interest: The authors declare no conflict of interest.

References

1. De Silva, A.P.; Gunaratne, H.Q.N.; Gunnlaugsson, T.; McCoy, C.P.; Maxwell, P.R.S.; Rademacher, J.T.; Rice, T.E. Photoionic devices with receptor-functionalized fluorophores. *Pure Appl. Chem.* **1996**, *68*, 1443–1448. [CrossRef]

2. De Silva, A.P.; Gunaratne, H.Q.N.; Gunnlaugsson, T.; Huxley, A.J.M.; McCoy, C.P.; Rademacher, J.T.; Rice, T.E. Signaling recognition events with fluorescent sensors and switches. *Chem. Rev.* **1997**, *97*, 1515–1566. [CrossRef] [PubMed]

3. De Silva, A.P.; Eilers, J.; Zlokarnik, G. Emerging fluorescence sensing technologies: From photophysical principles to cellular applications. *Proc. Natl. Acad. Sci. USA* **1999**, *96*, 8336–8337. [CrossRef] [PubMed]

4. De Silva, A.P.; Fox, D.B.; Huxley, A.J.M.; Moody, T.S. Combining luminescence, coordination and electron transfer for signalling purposes. *Coord. Chem. Rev.* **2000**, *205*, 41–57. [CrossRef]

5. De Silva, A.P.; Fox, D.B.; Moody, T.S.; Weir, S.M. Luminescent sensors and photonic switches. *Pure Appl. Chem.* **2001**, *73*, 503–511. [CrossRef]

6. Daly, B.; Ling, J.; de Silva, A.P. Current developments in fluorescent PET (photoinduced electron transfer) sensors and switches. *Chem. Soc. Rev.* **2015**, *44*, 4203–4211. [CrossRef] [PubMed]

7. Lakowicz, J.R. *Principles of Fluorescence Spectroscopy*, 3rd ed.; Springer: New York, NY, USA, 2006.

8. Valeur, B.; Berberan-Santos, M.N. *Molecular Fluorescence: Principles and Applications*, 2nd ed.; Wiley-VCH: Weinheim, Germany, 2012.

9. Weller, A. Electron-transfer and complex formation in the excited state. *Pure Appl. Chem.* **1968**, *16*, 115–123. [CrossRef]

10. Balzani, V. (Ed.) *Electron Transfer in Chemistry*; Wiley-VCH: Weinheim, Germany, 2001.

11. Malvino, A.P.; Brown, J.A. *Digital Computer Electronics*, 3rd ed.; Glencoe: Lake Forest, CA, USA, 1993.

12. De Silva, A.P.; Gunaratne, H.Q.N.; McCoy, C.P. A molecular photoionic AND gate based on fluorescent signalling. *Nature* **1993**, *364*, 42–44. [CrossRef]

13. Raymo, F.M. Digital processing and communication with molecular switches. *Adv. Mater.* **2002**, *14*, 401–414. [CrossRef]

14. Katz, E.; Privman, V. Enzyme-based logic systems for information processing. *Chem. Soc. Rev.* **2010**, *39*, 1835–1857. [CrossRef]

15. Amelia, M.; Zou, L.; Credi, A. Signal processing with multicomponent systems based on metal complexes. *Coord. Chem. Rev.* **2010**, *254*, 2267–2280. [CrossRef]

16. De Silva, A.P.; Uchiyama, S. Molecular logic gates and luminescent sensors based on photoinduced electron transfer. *Top. Curr. Chem.* **2011**, *300*, 1–28. [PubMed]

17. De Silva, A.P. Luminescent photoinduced electron transfer (PET) molecules for sensing and logic operations. *J. Phys. Chem. Lett.* **2011**, *2*, 2865–2871. [CrossRef]

18. Feringa, B.L.; Browne, W.R. (Eds.) *Molecular Switches*, 2nd ed.; Wiley-VCH: Weinheim, Germany, 2011.

19. De Ruiter, G.; van der Boom, M.E. Surface-confined assemblies and polymers for molecular logic. *Acc. Chem. Res.* **2011**, *44*, 563–573. [CrossRef] [PubMed]

20. De Silva, A.P. Molecular logic gate arrays. *Chem. Asian J.* **2011**, *6*, 750–766. [CrossRef] [PubMed]

21. Katz, E. (Ed.) *Molecular and Supramolecular Information Processing*; Wiley-VCH: Weinheim, Germany, 2012.

22. Katz, E. (Ed.) *Biomolecular Information Processing: From Logic Systems to Smart Sensors and Actuators*; Wiley-VCH: Weinheim, Germany, 2012.
23. Szaciłowski, K. *Infochemistry: Information Processing at the Nanoscale*; Wiley: Chichester, UK, 2012.
24. Balzani, V.; Credi, A.; Venturi, M. Molecular logic circuits. *ChemPhysChem* **2003**, *4*, 49–59. [CrossRef]
25. Gust, D.; Andréasson, J.; Pischel, U.; Moore, T.A.; Moore, A.L. Data and signal processing using photochromic molecules. *Chem. Commun.* **2012**, *48*, 1947–1957. [CrossRef] [PubMed]
26. De Silva, A.P. *Molecular Logic-Based Computation; Monographs in Supramolecular Chemistry No. 12*; The Royal Society of Chemistry: Cambridge, UK, 2013.
27. De Silva, A.P.; Uchiyama, S. *Molecular Logic Gates—Functional Molecules with the Ability of Information Processing*; Kodansha: Tokyo, Japan, 2014. (In Japanese)
28. Stojanovic, M.N.; Stefanovic, D.; Rudchenko, S. Exercises in molecular computing. *Acc. Chem. Res.* **2014**, *47*, 1845–1852. [CrossRef]
29. Ling, J.; Daly, B.; Silverson, V.A.D.; de Silva, A.P. Taking baby steps in molecular logic-based computation. *Chem. Commun.* **2015**, *51*, 8403–8409. [CrossRef]
30. Andréasson, J.; Pischel, U. Molecules with a sense of logic: A progress report. *Chem. Soc. Rev.* **2015**, *44*, 1053–1069. [CrossRef]
31. Andréasson, J.; Pischel, U. Molecules for security measures: From keypad locks to advanced communication protocols. *Chem. Soc. Rev.* **2018**, *47*, 2266–2279. [CrossRef] [PubMed]
32. De Silva, A.P.; McClenaghan, N.D. Molecular-scale logic gates. *Chem. Eur. J.* **2004**, *10*, 574–586. [CrossRef] [PubMed]
33. De Silva, A.P.; Leydet, Y.; Lincheneau, C.; McClenaghan, N.D. Chemical approaches to nanometre-scale logic gates. *J. Phys. Condens. Matter* **2006**, *18*, S1847–S1872. [CrossRef]
34. Magri, D.C.; Vance, T.P.; de Silva, A.P. From complexation to computation: Recent progress in molecular logic. *Inorg. Chim. Acta* **2007**, *360*, 751–764. [CrossRef]
35. De Silva, A.P.; Uchiyama, S.; Vance, T.P.; Wannalerse, B. A supramolecular chemistry basis for molecular logic and computation. *Coord. Chem. Rev.* **2007**, *251*, 1623–1632. [CrossRef]
36. De Silva, A.P.; Uchiyama, S. Molecular logic and computing. *Nat. Nanotechnol.* **2007**, *2*, 399–410. [CrossRef] [PubMed]
37. Balzani, V.; Credi, A.; Venturi, M. *Molecular Devices and Machines: Concepts and Perspectives for the Nanoworld*, 2nd ed.; Wiley-VCH: Weinheim, Germany, 2008.
38. Andréasson, J.; Pischel, U. Smart molecules at work—Mimicking advanced logic operations. *Chem. Soc. Rev.* **2010**, *39*, 174–188. [CrossRef] [PubMed]
39. Misra, R.; Bhattacharyya, S.P. *Intramolecular Charge Transfer: Theory and Applications*; Wiley-VCH: Weinheim, Germany, 2018.
40. Grimshaw, J.; de Silva, A.P. Photochemistry and photocyclization of aryl halides. *Chem. Soc. Rev.* **1981**, *10*, 181–203. [CrossRef]
41. De Silva, A.P.; de Silva, S.A. Fluorescent signalling crown ethers; 'switching on' of fluorescence by alkali metal ion recognition and binding in situ. *J. Chem. Soc. Chem. Commun.* **1986**, 1709–1710. [CrossRef]
42. Uchiyama, S.; Santa, T.; Okiyama, N.; Azuma, K.; Imai, K. Semi-empirical PM3 calculations predict the fluorescence quantum yields (Φ) of 4-monosubstituted benzofurazan compounds. *J. Chem. Soc. Perkin Trans. 2* **2000**, 1199–1207. [CrossRef]
43. Daly, B.; Moody, T.S.; Huxley, A.J.M.; Yao, C.; Schazmann, B.; Alves-Areias, A.; Malone, J.F.; Gunaratne, H.Q.N.; Nockemann, P.; de Silva, A.P. Molecular memory with downstream logic processing exemplified by switchable and self-indicating guest capture and release. *Nat. Commun.* **2019**, *10*, 49. [CrossRef] [PubMed]
44. Ringsdorf, H. Hermann Staudinger and the future of polymer research jubilees—Beloved occasions for cultural piety. *Angew. Chem. Int. Ed.* **2004**, *43*, 1064–1076. [CrossRef] [PubMed]
45. Berg, J.M.; Tymoczko, J.L.; Gatto, G.J., Jr.; Stryer, L. *Biochemistry*, 9th ed.; Freeman: New York, NY, USA, 2019.
46. Rothemund, P.W.K. Folding DNA to create nanoscale shapes and patterns. *Nature* **2006**, *440*, 297–302. [CrossRef] [PubMed]
47. Lam, K.S.; Salmon, S.E.; Hersh, E.M.; Hruby, V.J.; Kazmierski, W.M.; Knapp, R.J. A new type of synthetic peptide library for identifying ligand-binding activity. *Nature* **1991**, *354*, 82–84. [CrossRef] [PubMed]
48. Harold, F.M. *The Vital Force: A Study of Bioenergetics*; Freeman: New York, NY, USA, 1986.

49. De Silva, A.P.; Rupasinghe, R.A.D.D. A new class of fluorescent pH indicators based on photo-induced electron transfer. *J. Chem. Soc. Chem. Commun.* **1985**, 1669–1670. [CrossRef]

50. Bissell, R.A.; de Silva, A.P.; Gunaratne, H.Q.N.; Lynch, P.L.M.; Maguire, G.E.M.; Sandanayake, K.R.A.S. Molecular fluorescent signalling with 'fluor–spacer–receptor' systems: Approaches to sensing and switching devices via supramolecular photophysics. *Chem. Soc. Rev.* **1992**, *21*, 187–195. [CrossRef]

51. De Silva, A.P.; Vance, T.P.; West, M.E.S.; Wright, G.D. Bright molecules with sense, logic, numeracy and utility. *Org. Biomol. Chem.* **2008**, *6*, 2468–2481. [CrossRef]

52. Bissell, R.A.; Bryan, A.J.; de Silva, A.P.; McCoy, C.P. Fluorescent PET (photoinduced electron transfer) sensors with targeting/anchoring modules as molecular versions of submarine periscopes for mapping membrane-bounded protons. *J. Chem. Soc. Chem. Commun.* **1994**, 405–407. [CrossRef]

53. Fernández, M.S.; Fromherz, P. Lipoid pH indicators as probes of electrical potential and polarity in micelles. *J. Phys. Chem.* **1977**, *81*, 1755–1761. [CrossRef]

54. Yang, Z.; Cao, J.; He, Y.; Yang, J.H.; Kim, T.; Peng, X.; Kim, J.S. Macro-/micro-environment-sensitive chemosensing and biological imaging. *Chem. Soc. Rev.* **2014**, *43*, 4563–4601. [CrossRef]

55. Jiang, N.; Fan, J.; Xu, F.; Peng, X.; Mu, H.; Wang, J.; Xiong, X. Ratiometric fluorescence imaging of cellular polarity: Decrease in mitochondrial polarity in cancer cells. *Angew. Chem. Int. Ed.* **2015**, *54*, 2510–2514. [CrossRef] [PubMed]

56. Yu, C.; Miao, W.; Wang, J.; Hao, E.; Jiao, L. PyrrolyBODIPYs: Syntheses, properties, and application as environment-sensitive fluorescence probes. *ACS Omega* **2017**, *2*, 3551–3561. [CrossRef]

57. Collot, M.; Bou, S.; Fam, T.K.; Richert, L.; Mély, Y.; Danglot, L.; Klymchenko, A.S. Probing polarity and heterogeneity of lipid droplets in live cells using a push-pull fluorophore. *Anal. Chem.* **2019**, *91*, 1928–1935. [CrossRef] [PubMed]

58. Uchiyama, S.; Santa, T.; Imai, K. Fluorescence characteristics of six 4,7-disubstituted benzofurazan compounds: An experimental and semi-empirical MO study. *J. Chem. Soc. Perkin Trans. 2* **1999**, 2525–2532. [CrossRef]

59. Numasawa, Y.; Okabe, K.; Uchiyama, S.; Santa, T.; Imai, K. Fluorescence characteristics of ionic benzofurazans, 7-substituted-2,1,3-benzoxadiazole-4-sulfonates. *Dyes Pigment.* **2005**, *67*, 189–195. [CrossRef]

60. Uchiyama, S.; Kimura, K.; Gota, C.; Okabe, K.; Kawamoto, K.; Inada, N.; Yoshihara, T.; Tobita, S. Environment-sensitive fluorophores with benzothiadiazole and benzoselenadiazole structures as candidate components of a fluorescent polymeric thermometer. *Chem. Eur. J.* **2012**, *18*, 9552–9563. [CrossRef] [PubMed]

61. Zhuang, Y.-D.; Chiang, P.-Y.; Wang, C.-W.; Tan, K.-T. Environment-sensitive fluorescent turn-on probes targeting hydrophobic ligand-binding domains for selective protein detection. *Angew. Chem. Int. Ed.* **2013**, *52*, 8124–8128. [CrossRef] [PubMed]

62. Appelqvist, H.; Stranius, K.; Börjesson, K.; Nilsson, K.P.R.; Dyrager, C. Specific imaging of intracellular lipid droplets using a benzothiadiazole derivative with solvatochromic properties. *Bioconjugate Chem.* **2017**, *28*, 1363–1370. [CrossRef] [PubMed]

63. Thooft, A.M.; Cassaidy, K.; VanVeller, B. A small push-pull fluorophore for turn-on fluorescence. *J. Org. Chem.* **2017**, *82*, 8842–8847. [CrossRef] [PubMed]

64. Uchiyama, S.; Iwai, K.; de Silva, A.P. Multiplexing sensory molecules map protons near micellar membranes. *Angew. Chem. Int. Ed.* **2008**, *47*, 4667–4669. [CrossRef] [PubMed]

65. Vitha, M.F.; Weckwerth, J.D.; Odland, K.; Dema, V.; Carr, P.W. Study of the polarity and hydrogen bond ability of sodium dodecyl sulfate micelles by the Kamlet-Taft solvatochromic comparison method. *J. Phys. Chem.* **1996**, *100*, 18823–18828. [CrossRef]

66. Uchiyama, S.; Yano, K.; Fukatsu, E.; de Silva, A.P. Precise proton mapping near ionic micellar membranes with fluorescent photoinduced-electron-transfer sensors. *Chem. Eur. J.* **2019**, *25*, 8522–8527. [PubMed]

67. Dougherty, D.A. Cation-π interactions in chemistry and biology: A new view of benzene, Phe, Tyr, and Trp. *Science* **1996**, *271*, 163–168. [CrossRef] [PubMed]

68. Kandel, E.R.; Schwartz, J.H.; Jessell, T.M. *Principles of Neural Science*, 4th ed.; McGraw-Hill: New York, NY, USA, 2000.

69. Uchiyama, S.; Fukatsu, E.; McClean, G.D.; de Silva, A.P. Measurement of local sodium ion levels near micelle surfaces with fluorescent photoinduced-electron-transfer sensors. *Angew. Chem. Int. Ed.* **2016**, *55*, 768–771. [CrossRef] [PubMed]

70. Diaz-Fernandez, Y.; Foti, F.; Mangano, C.; Pallavicini, P.; Patroni, S.; Perez-Gramatges, A.; Rodriguez-Calvo, S. Micelles for the self-assembly of "off-on-off" fluorescent sensors for pH windows. *Chem. Eur. J.* **2006**, *12*, 921–930. [CrossRef] [PubMed]

71. De Silva, A.P.; Dobbin, C.M.; Vance, T.P.; Wannalerse, B. Multiply reconfigurable 'plug and play' molecular logic via self-assembly. *Chem. Commun.* **2009**, 1386–1388. [CrossRef] [PubMed]

72. Uchiyama, S.; McClean, G.D.; Iwai, K.; de Silva, A.P. Membrane media create small nanospaces for molecular computation. *J. Am. Chem. Soc.* **2005**, *127*, 8920–8921. [CrossRef]

73. Deisenhofer, J.; Norris, J.R. *The Photosynthetic Reaction Center*; Academic Press: San Diego, CA, USA, 1993.

74. Liu, J.; de Silva, A.P. Path-selective photoinduced electron transfer (PET) in a membrane-associated system studied by pH-dependent fluorescence. *Inorg. Chim. Acta* **2012**, *381*, 243–246. [CrossRef]

75. De Silva, A.P.; Gunaratne, H.Q.N.; Habib-Jiwan, J.-L.; McCoy, C.P.; Rice, T.E.; Soumillion, J.-P. New fluorescent model compounds for the study of photoinduced electron transfer: The influence of a molecular electric field in the excited state. *Angew. Chem. Int. Ed. Engl.* **1995**, *34*, 1728–1731. [CrossRef]

76. Ayadim, M.; Jiwan, J.L.H.; de Silva, A.P.; Soumillion, J.P. Photosensing by a fluorescing probe covalently attached to the silica. *Tetrahedron Lett.* **1996**, *37*, 7039–7042. [CrossRef]

77. Daffy, L.M.; de Silva, A.P.; Gunaratne, H.Q.N.; Huber, C.; Lynch, P.L.M.; Werner, T.; Wolfbeis, O.S. Arenedicarboximide building blocks for fluorescent photoinduced electron transfer pH sensors applicable with different media and communication wavelengths. *Chem. Eur. J.* **1998**, *4*, 1810–1815. [CrossRef]

78. Tusa, J.K.; He, H. Critical care analyzer with fluorescent optical chemosensors for blood analytes. *J. Mater. Chem.* **2005**, *15*, 2640–2647. [CrossRef]

79. Descalzo, A.B.; Marcos, M.D.; Martínez-Máñez, R.; Soto, J.; Beltrán, D.; Amorós, P. Anthrylmethylamine functionalised mesoporous silica-based materials as hybrid fluorescent chemosensors for ATP. *J. Mater. Chem.* **2005**, *15*, 2721–2731. [CrossRef]

80. Refalo, M.V.; Spiteri, J.C.; Magri, D.C. Covalent attachment of a fluorescent 'Pourbaix sensor' onto a polymer bead for sensing in water. *New J. Chem.* **2018**, *42*, 16474–16477. [CrossRef]

81. Fernández-Alonso, S.; Corrales, T.; Pablos, J.L.; Catalina, F. A switchable fluorescence solid sensor for Hg^{2+} detection in aqueous media based on a photocrosslinked membrane functionalized with (benzimidazolyl)methyl-piperazine derivative of 1,8-naphthalimide. *Sens. Actuators B-Chem.* **2018**, *270*, 256–262. [CrossRef]

82. Gui, B.; Meng, Y.; Xie, Y.; Tian, J.; Yu, G.; Zeng, W.; Zhang, G.; Gong, S.; Yang, C.; Zhang, D.; et al. Tuning the photoinduced electron transfer in a Zr-MOF: Toward solid-state fluorescent molecular switch and turn-on sensor. *Adv. Mater.* **2018**, *30*, 1802329. [CrossRef] [PubMed]

83. Shepard, S. *RFID: Radio Frequency Identification*; McGraw-Hill: New York, NY, USA, 2005.

84. Yamashita, D.S.; Weinstock, J. Method of Encoding a Series of Combinatorial Libraries and Developing Structure Activity Relationships. U.S. Patent 6,210,900 B1, 3 April 2001.

85. De Silva, A.P.; James, M.R.; McKinney, B.O.F.; Pears, D.A.; Weir, S.M. Molecular computational elements encode large populations of small objects. *Nat. Mater.* **2006**, *5*, 787–790. [CrossRef]

86. Brown, G.J.; de Silva, A.P.; James, M.R.; McKinney, B.O.F.; Pears, D.A.; Weir, S.M. Solid-bound, proton-driven, fluorescent 'off-on–off' switches based on PET (photoinduced electron transfer). *Tetrahedron* **2008**, *64*, 8301–8306. [CrossRef]

87. McKinney, B.O.F.; Daly, B.; Yao, C.; Schroeder, M.; de Silva, A.P. Consolidating molecular logic with new solid-bound YES and PASS 1 gates and their combinations. *ChemPhysChem* **2017**, *18*, 1760–1766. [CrossRef]

88. Yao, C.; Ling, J.; Chen, L.; de Silva, A.P. Population analysis to increase the robustness of molecular computational identification and its extension into the near-infrared for substantial numbers of small objects. *Chem. Sci.* **2019**, *10*, 2272–2279. [CrossRef]

89. Uchiyama, S.; Matsumura, Y.; de Silva, A.P.; Iwai, K. Fluorescent molecular thermometers based on polymers showing temperature-induced phase transitions and labeled with polarity-responsive benzofurazans. *Anal. Chem.* **2003**, *75*, 5926–5935. [CrossRef] [PubMed]

90. Chandrasekharan, N.; Kelly, L.A. Progress towards fluorescent molecular thermometers. In *Reviews in Fluorescence 2004*; Geddes, C.D., Lakowicz, J.R., Eds.; Kluwer Academic/Plenum Publishes: New York, NY, USA, 2004; pp. 21–40.

91. Uchiyama, S.; de Silva, A.P.; Iwai, K. Luminescent molecular thermometers. *J. Chem. Educ.* **2006**, *83*, 720–727. [CrossRef]

92. Inal, S.; Kölsch, J.D.; Sellrie, F.; Schenk, J.A.; Wischerhoff, E.; Laschewsky, A.; Neher, D. A water soluble fluorescent polymer as a dual colour sensor for temperature and a specific protein. *J. Mater. Chem. B* **2013**, *1*, 6373–6381. [CrossRef]

93. Alemdaroglu, F.E.; Alexander, S.C.; Ji, D.; Prusty, D.K.; Börsch, M.; Herrmann, A. Poly(BODIPY)s: A new class of tunable polymeric dyes. *Macromolecules* **2009**, *42*, 6529–6536. [CrossRef]

94. Sen, C.P.; Goud, V.D.; Shrestha, R.G.; Shrestha, L.K.; Ariga, K.; Valiyaveettil, S. BODIPY based hyperbranched conjugated polymers for detecting organic vapors. *Polym. Chem.* **2016**, *7*, 4213–4225. [CrossRef]

95. Squeo, B.M.; Gregoriou, V.G.; Avgeropoulos, A.; Baysec, S.; Allard, S.; Scherf, U.; Chochos, C.L. BODIPY-based polymeric dyes as emerging horizon materials for biological sensing and organic electronic applications. *Progr. Polym. Sci.* **2017**, *71*, 26–52.

96. Gong, D.; Cao, T.; Han, S.-C.; Zhu, X.; Iqbal, A.; Liu, W.; Qin, W.; Guo, H. Fluorescence enhancement thermoresponsive polymer luminescent sensors based on BODIPY for intracellular temperature. *Sens. Actuator B Chem.* **2017**, *252*, 577–583. [CrossRef]

97. Hattori, Y.; Nagase, K.; Kobayashi, J.; Kikuchi, A.; Akiyama, Y.; Kanazawa, H.; Okano, T. Hydration of poly (*N*-isopropylacrylamide) brushes on micro-silica beads measured by a fluorescent probe. *Chem. Phys. Lett.* **2010**, *491*, 193–198. [CrossRef]

98. Yamada, A.; Hiruta, Y.; Wang, J.; Ayano, E.; Kanazawa, H. Design of environmentally responsive fluorescent polymer probes for cellular imaging. *Biomacromolecules* **2015**, *16*, 2356–2362. [CrossRef]

99. Qiao, J.; Chen, C.; Qi, L.; Liu, M.; Dong, P.; Jiang, Q.; Yang, X.; Mu, X.; Mao, L. Intracellular temperature sensing by a ratiometric fluorescent polymer thermometer. *J. Mater. Chem. B* **2014**, *2*, 7544–7550. [CrossRef]

100. Cellini, F.; Peteron, S.D.; Porfiri, M. Flow velocity and temperature sensing using thermosensitive fluorescent polymer seed particles in water. *Int. J. Smart Nano Mater.* **2017**, *8*, 232–252. [CrossRef]

101. Gota, C.; Uchiyama, S.; Yoshihara, T.; Tobita, S.; Ohwada, T. Temperature-dependent fluorescence lifetime of a fluorescent polymeric thermometer, poly(*N*-isopropylacrylamide), labeled by polarity and hydrogen bonding sensitive 4-sulfamoyl-7-aminobenzofurazan. *J. Phys. Chem. B* **2008**, *112*, 2829–2836. [CrossRef] [PubMed]

102. Uchiyama, S.; Matsumura, Y.; de Silva, A.P.; Iwai, K. Modulation of the sensitive temperature range of fluorescent molecular thermometers based on thermoresponsive polymers. *Anal. Chem.* **2004**, *76*, 1793–1798. [CrossRef] [PubMed]

103. Uchiyama, S.; Takehira, K.; Yoshihara, T.; Tobita, S.; Ohwada, T. Environment-sensitive fluorophore emitting in protic environments. *Org. Lett.* **2006**, *8*, 5869–5872. [CrossRef] [PubMed]

104. Shiraishi, Y.; Miyamoto, R.; Hirai, T. A hemicyanine-conjugated copolymer as a highly sensitive fluorescent thermometer. *Langmuir* **2008**, *24*, 4273–4279. [CrossRef] [PubMed]

105. Gota, C.; Uchiyama, S.; Ohwada, T. Accurate fluorescent polymeric thermometers containing an ionic component. *Analyst* **2007**, *132*, 121–126. [CrossRef] [PubMed]

106. Uchiyama, S.; Kawai, N.; de Silva, A.P.; Iwai, K. Fluorescent polymeric AND logic gate with temperature and pH as inputs. *J. Am. Chem. Soc.* **2004**, *126*, 3032–3033. [CrossRef] [PubMed]

107. Shiraishi, Y.; Miyamoto, R.; Hirai, T. Temperature-driven on/off fluorescent indicator of pH window: An anthracene-conjugated thermoresponsive polymer. *Tetrahedron Lett.* **2007**, *48*, 6660–6664. [CrossRef]

108. Shiraishi, Y.; Miyamoto, R.; Zhang, X.; Hirai, T. Rhodamine-based fluorescent thermometer exhibiting selective emission enhancement at a specific temperature range. *Org. Lett.* **2007**, *9*, 3921–3924. [CrossRef]

109. Iwai, K.; Matsumura, Y.; Uchiyama, S.; de Silva, A.P. Development of fluorescent microgel thermometers based on thermo-responsive polymers and their modulation of sensitivity range. *J. Mater. Chem.* **2005**, *15*, 2796–2800. [CrossRef]

110. Herrera, A.P.; Rodríguez, M.; Torres-Lugo, M.; Rinaldi, C. Multifunctional magnetite nanoparticles coated with fluorescent thermo-responsive polymeric shells. *J. Mater. Chem.* **2008**, *18*, 855–858. [CrossRef]

111. Graham, E.M.; Iwai, K.; Uchiyama, S.; de Silva, A.P.; Magennis, S.W.; Jones, A.C. Quantitative mapping of aqueous microfluidic temperature with sub-degree resolution using fluorescence lifetime imaging microscopy. *Lab Chip* **2010**, *10*, 1267–1273. [CrossRef] [PubMed]

112. Yuan, B.; Uchiyama, S.; Liu, Y.; Nguyen, K.T.; Alexandrakis, G. High-resolution imaging in a deep turbid medium based on an ultrasound-switchable fluorescence technique. *Appl. Phys. Lett.* **2012**, *101*, 033703. [CrossRef] [PubMed]

113. Gota, C.; Okabe, K.; Funatsu, T.; Harada, Y.; Uchiyama, S. Hydrophilic fluorescent nanogel thermometer for intracellular thermometry. *J. Am. Chem. Soc.* **2009**, *131*, 2766–2767. [CrossRef]
114. Okabe, K.; Inada, N.; Gota, C.; Harada, Y.; Funatsu, T.; Uchiyama, S. Intracellular temperature mapping with a fluorescent polymeric thermometer and fluorescence lifetime imaging microscopy. *Nat. Commun.* **2012**, *3*, 705. [CrossRef] [PubMed]
115. Tsuji, T.; Yoshida, S.; Yoshida, A.; Uchiyama, S. Cationic fluorescent polymeric thermometers with the ability to enter yeast and mammalian cells for practical intracellular temperature measurements. *Anal. Chem.* **2013**, *85*, 9815–9823. [CrossRef] [PubMed]
116. Hayashi, T.; Fukuda, N.; Uchiyama, S.; Inada, N. A cell-permeable fluorescent polymeric thermometer for intracellular temperature mapping in mammalian cell lines. *PLoS ONE* **2015**, *10*, e0117677. [CrossRef]
117. Inada, N.; Fukuda, N.; Hayashi, T.; Uchiyama, S. Temperature imaging using a cationic linear fluorescent polymeric thermometer and fluorescence lifetime imaging microscopy. *Nat. Protoc.* **2019**, *14*, 1293–1321. [CrossRef]
118. Uchiyama, S.; Tsuji, T.; Ikado, K.; Yoshida, A.; Kawamoto, K.; Hayashi, T.; Inada, N. A cationic fluorescent polymeric thermometer for the ratiometric sensing of intracellular temperature. *Analyst* **2015**, *140*, 4498–4506. [CrossRef]
119. Tsuji, T.; Ikado, K.; Koizumi, H.; Uchiyama, S.; Kajimoto, K. Difference in intracellular temperature rise between matured and precursor brown adipocytes in response to uncoupler and β-adrenergic agonist stimuli. *Sci. Rep.* **2017**, *7*, 12889. [CrossRef]
120. Kimura, H.; Nagoshi, T.; Yoshii, A.; Kashiwagi, Y.; Tanaka, Y.; Ito, K.; Yoshino, T.; Tanaka, T.D.; Yoshimura, M. The thermogenic actions of natriuretic peptide in brown adipocytes: The direct measurement of the intracellular temperature using a fluorescent thermoprobe. *Sci. Rep.* **2017**, *7*, 12978. [CrossRef] [PubMed]
121. Hoshi, Y.; Okabe, K.; Shibasaki, K.; Funatsu, T.; Matsuki, N.; Ikegaya, Y.; Koyama, R. Ischemic brain injury leads to brain edema via hyperthermia-induced TRPV4 activation. *J. Neurosci.* **2018**, *38*, 5700–5709. [CrossRef] [PubMed]
122. Ano, T.; Kishimoto, F.; Sasaki, R.; Tsubaki, S.; Maitani, M.M.; Suzuki, E.; Wada, Y. In situ temperature measurements of reaction spaces under microwave irradiation using photoluminescent probes. *Phys. Chem. Chem. Phys.* **2016**, *18*, 13173–13179. [CrossRef] [PubMed]
123. Uchiyama, S.; Tsuji, T.; Kawamoto, K.; Okano, K.; Fukatsu, E.; Noro, T.; Ikado, K.; Yamada, S.; Shibata, Y.; Hayashi, T.; et al. A cell-targeted non-cytotoxic fluorescent nanogel thermometer created with an imidazolium-containing cationic radical initiator. *Angew. Chem. Int. Ed.* **2018**, *57*, 5413–5417. [CrossRef] [PubMed]

Communication

A Polymer-Gel Eye-Phantom for 3D Fluorescent Imaging of Millimetre Radiation Beams

Leonard H. Luthjens, Tiantian Yao and John M. Warman *

Delft University of Technology, Faculty of Applied Sciences, Department of Radiation Science and Technology, Section Radiation and Isotopes for Health, Mekelweg 15, 2629 JB Delft, The Netherlands; l.h.luthjens@tudelft.nl (L.H.L.); t.yao@tudelft.nl (T.Y.)
* Correspondence: j.m.warman@tudelft.nl

Received: 30 September 2018; Accepted: 23 October 2018; Published: 26 October 2018

Abstract: We have filled a 24 mm diameter glass sphere with a transparent polymer-gel that is radio-fluorogenic, i.e., it becomes (permanently) fluorescent when irradiated, with an intensity proportional to the local dose deposited. The gel consists of >99.9% tertiary-butyl acrylate (TBA), pre-polymerized to ~15% conversion, and ~100 ppm maleimido-pyrene (MPy). Its dimensions and physical properties are close to those of the vitreous body of the human eye. We have irradiated the gel with a 3 mm diameter, 200 kVp X-ray beam with a dose rate of ~1 Gy/min. A three-dimensional (3D) (video) view of the beam within the gel has been constructed from tomographic images obtained by scanning the sample through a thin sheet of UV light. To minimize optical artefacts, the cell was immersed in a square tank containing a refractive-index-matching medium. The 20–80% penumbra of the beam was determined to be ~0.4 mm. This research was a preparatory investigation of the possibility of using this method to monitor the millimetre diameter proton pencil beams used in ocular radiotherapy.

Keywords: radiotherapy eye-phantom; radio-fluorogenic gel; X-ray beam imaging; 3D radiation imaging; polymer gel dosimetry

---

1. Introduction

The treatment of ocular tumours is fraught with difficulties due to the proximity of important healthy tissues involved in sight and/or brain function. Radiotherapy has been an important method of treatment using either brachytherapy or external beam procedures [1–4]. The small dimension of the organ makes the use of highly collimated beams of millimetre dimensions imperative. With the recent proliferation of clinical proton beam sources [5], the incorporation of proton pencil beams in cancer protocols has become a worldwide reality [6–8]. Because of this, there is a current need for a medium capable of providing three-dimensional (3D) images with sub-millimetre spatial resolution of dose deposition in phantoms for confirmation of computer derived treatment protocols, equipment functioning, and for training clinical personnel [9].

A variety of methods involving 3D matrixes of one- and two-dimensional (2D) individual detectors are commercially available and presently used in the clinic. However, they are incapable of achieving the sub-millimetre spatial resolution required for small beam applications. For this, molecular media that undergo a measureable physico-chemical change on a microscopic scale when irradiated are required. Since the first suggestion in 1950 by Day and Stein [10] of using a quasi-rigid gel medium, several methods based on this basic formula have been developed and reviewed [11–17]. Unfortunately, none have become generally accepted in the radiotherapy clinic [18–20]. The reasons for this lack of adoption can be ascribed to a combination of: the complexity of the gel formulations, often involving up to five components, and the corresponding difficulty of universal reproducibility;

the lack of off-the-shelf availability compared with gaseous (IC) or solid-state devices; the delay in post-irradiation data analysis; and, possibly, the cost.

In this report, we show how a radio-fluorogenic (RFG) polymer-gel [21,22] can be used to provide a 3D image of the track of a narrow (3 mm diameter) X-ray beam in a volume as small as the human eye. An RFG gel is a two-component mixture whose components are commercially available, inexpensive, and whose composition can be accurately determined by optical spectroscopy. The gel, which is tissue equivalent, becomes permanently fluorescent on irradiation with intensity proportional to the local dose deposited. The fluorescence can be tomographically scanned on-site immediately after irradiation and processed within minutes to yield 3D images of the radiation field [23]. The work reported here is a pilot study investigating the feasibility of applying RFG gel phantoms to the control of radiotherapy protocols and equipment for proton pencil beam treatment of eye tumours. It was driven by the recent construction of a proton radiotherapy clinic within the grounds of the authors' institute. The capability of making 2D bulk images of proton pencil beams in an RFG gel was demonstrated previously by one of the authors using a high-energy physics particle accelerator [24].

2. Materials and Methods

2.1. RFG Gel Formation

The gel container used is a spherical borosilicate glass bulb 24 mm in outer diameter with a wall thickness of 1.7 mm; similar dimensions to those of the adult human eye. The bulb contains 5 mL of a transparent radiofluorogenic (RFG) gel with a gravimetric density of 0.91 kgL^{-1} and an electron density of 3.00×10^{26} L^{-1}, values that are close to the 1.01 kgL^{-1} and 3.35×10^{26} L^{-1} of the vitreous body of the eye, and the 1.18 kgL^{-1} and 3.84×10^{26} L^{-1} of chemically-related PMMA (often denoted as "solid water"). The RFG gel consists of 15% pre-polymerised and inhibitor-free tertiary-butyl acrylate (TBA, Sigma Aldrich purum #01775, Sigma Aldrich company, Zwijndrecht, The Netherlands), to which is added approximately 100 ppm of the fluorogenic compound maleimido-pyrene (MPy, Sigma Aldrich P7908, Sigma Aldrich company, Zwijndrecht, The Netherlands).

The RFG gel preparation procedure has been described in full previously [21,22]. In summary, the gel is preformed in the cell by ^{60}Co γ-irradiation (~15 Gy at ~1 Gy/min) of pure, de-aerated, and inhibitor-free TBA to ~15% monomer conversion. Excess monomer is then pumped off, leaving the polymer network deposited on the cell wall, as shown in Figure 1. The gel is reformed by adding a dilute solution of MPy in TBA equal to the volume of monomer pumped off and allowing time (~2 weeks) for reformation of the gel.

Figure 1. Left: The "eye-phantom" cell after radiation-induced polymerization of tertiary-butyl acrylate (TBA) to ~15% conversion and pumping-off the residual monomer leaving the polymer network. **Right**: The clear gel reformed on addition of a dilute solution of MPy in TBA and swelling of the polymer network.

2.2. X-ray Beam Irradiation

The gel, contained in the eye-phantom cell, was irradiated with 200 kVp X-rays from a Philips MCN 321 X-ray tube. The conical beam was restricted by a brass collimator with a circular channel 3.0 mm in diameter and 30 mm long inserted in a 120×120 mm^2 square, 27 mm thick Al/Pb/Al attenuator. The dose rate at the exit of the collimator, 460 mm from the source target, was ~1.0 Gy/min for a beam current of 15 mA. The cell was irradiated for 20 min resulting in a total incident dose of ~20 Gy. The basic irradiation set-up with optical rail, collimator and outline laser has been shown in a previous publication [23].

MD-V3 Gafchromic film (Lot Nr 02101602, Promis Electo-Optics B.V., Wijchen, The Netherlands) was used as a secondary, 2D dosimeter for monitoring the incident and exit beam dose rates and cross-sections. The dose sensitivity of the film was calibrated against an X-ray Exposure Meter (Vinten Instruments) that was routinely calibrated at the Netherlands Metrology Institute. The derivation of the dose from the measured intensity reflected by the scanned films has been described in full in reference [25]. The film was attached to the front and rear faces of a 24 mm square borosilicate glass cell placed at the same position as the 24 mm diameter eye-phantom cell. These control measurements were made with the square cell containing either air or an identical gel to that used in the eye-phantom.

2.3. Fluorescence Imaging

Fluorescence imaging was carried out using the ultraviolet (UV) slit scanning method previously reported [23]. This yields a series of tomographic images that can be used to generate a 3D image of the fluorescence within the gel. The sheet of UV light at 385 nm was 2 mm thick; the cell was transported in 1 mm steps through the sheet for a total distance of 30 mm. The cell was immersed in a glycerol bath contained in a 40×40 mm^2 square borosilicate glass container with optically flat sides. This resulted in close matching of the refractive indexes of the surroundings (glycerol $n = 1.47$) with that of the borosilicate glass cell walls ($n = 1.47$) and the gel ($n = 1.42$). This diminished optical artefacts caused by lensing and reflections at dielectric interfaces.

3. Results and Discussion

3.1. Test Measurement with a Standard Fluorescent Solution

In Figure 2 we show an image taken of a cell containing a micromolar concentration of the fluorescence-standard diphenyl anthracene (DPA) in cyclohexane. The image was taken with the UV excitation sheet positioned at the centre of the cell.

The pixel profile shown in Figure 2 displays a close to uniform fluorescence within the volume of the cell. The slight (~8%) upward curvature was most likely caused by a lensing effect due to the difference in refractive index between cyclohexane and glass (1.43 versus 1.47). The black "shadow" surrounding the liquid was due to the non-fluorescent glass cell wall. The slight increase in fluorescence outside the glass wall resulted from the fact that, in this measurement, the sample of glycerol used contained a fluorescent impurity. The presence of the impurity in the glycerol had a positive side to it, since it allowed us to determine the number of pixels per mm in the images from the accurately measureable outer diameter of the cell of 24.0 mm. The value determined was 44 pixels per mm, or 0.023 mm per pixel. This represents the ultimate practical limit to the spatial resolution of the measurements in the x/y plane. Additionally, thickness of the cell wall and the diameter of the gel could be accurately determined from the scan in Figure 2 to be 1.7 and 20.6 mm, respectively.

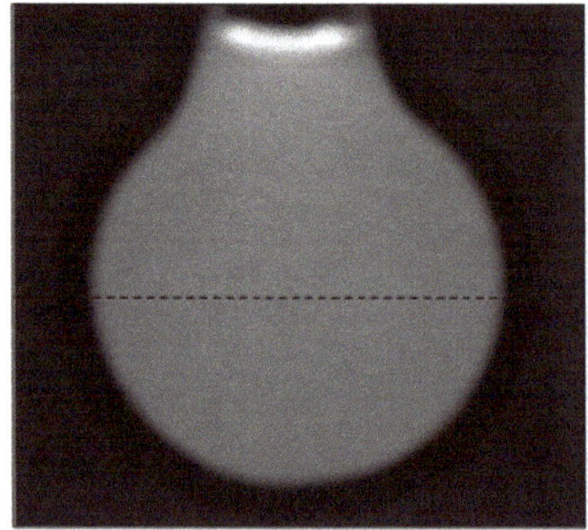

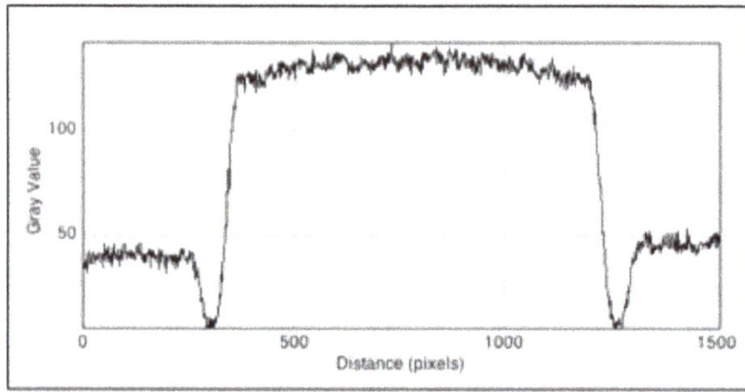

Figure 2. The fluorescence of a dilute solution of the fluorescence-standard diphenyl anthracene in the "eye-phantom" cell. The black "shadow" surrounding the cell, and corresponding intensity dips in the lower pixel profile, result from the non-fluorescent glass wall of the cell surrounded by the slightly fluorescent glycerol dielectric-matching bath. The position of the pixel profile scan is shown by the dashed line in the upper image. The pixel to pixel distance represents 0.023 mm in the gel.

3.2. Radiochromic Film Measurements

An MD-V3 radiochromic film image of the X-ray beam incident on the cell is shown in Figure 3. As can be seen, the beam was slightly oval with a long axis 14% larger than the short axis. This phenomenon was attributed to the fact that the 3.0 mm (perfectly) circular collimator was smaller than the 4 mm dimension of the tungsten target of the source resulting in partial "pinhole imaging" of the X-ray emission from the target.

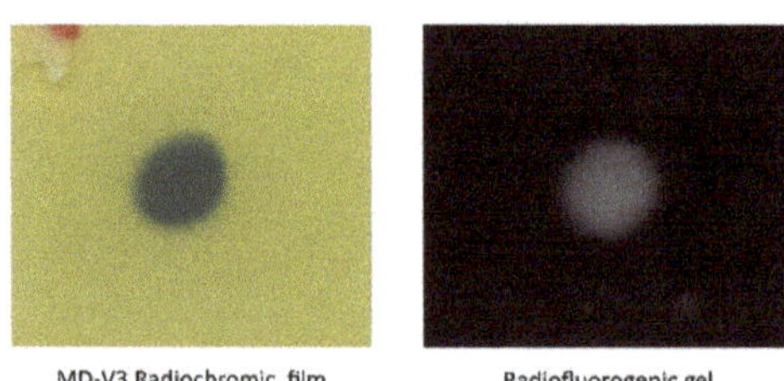

Figure 3. Left: The cross-sectional beam geometry as recorded using MD-V3 radiochromic film at the front face of the cell. **Right**: The fluorescent image as recorded in the radio-fluorogenic (RFG) gel looking along the propagation (z) axis of the beam.

From a pixel profile line scan across the RC beam image in Figure 3 the 20–80% penumbra of the beam was found to be 0.41 and 0.54 mm for the incident and exit beams, respectively. The spatial resolution of the film images is 300 dpi or 0.085 mm/pixel. The much larger penumbra value was, therefore, attributed to edge-dispersion in the intensity of the X-ray beam.

The central optical absorption in the profile scan was used together with Equation (1) to determine the incident and exit dose rates. For a cell containing air, the values were 0.94 and 0.67 Gy/min; a decrease by 0.713. For a cell containing an RFG gel, the values were 0.89 and 0.47 Gy/min; a decrease by 0.528. The decrease in dose rate between the incident and exit films was due to the conical expansion of the beam from z_1 (47.6 cm) to z_2 (50.0 cm) and photon attenuation in the intervening media.

$$D'(z_2)/D'(z_1) = (z_1/z_2)^2 \exp - \{2[M\rho\delta]_{glass} + [M\rho\delta]_{gel}\} \tag{1}$$

In Equation (1) M, ρ, and δ are the mass attenuation coefficient, the gravimetric density, and the length of the attenuating medium, respectively. For the air-containing cell, the attenuation by the cell contents was negligible; the overall decrease by a factor of 0.71 lead to a value of $2\,[M\rho\delta]_{glass} = 0.244$. Taking a borosilicate glass density of 2.3 g/cm^3 and a wall thickness of 0.17 cm resulted in a mass attenuation coefficient for the cell wall of 0.312 cm^2/g. From the dose rate decrease for the gel-containing cell we determined a value for $(2\,[M\rho\delta]_{glass} + [M\rho\delta]_{gel})$ of 0.540, or $[M\rho\delta]_{gel} = 0.296$ after subtracting the glass wall contribution determined above. Taking for the gel $\delta = 2.0$ cm and $\rho = 0.91$ g/cm^3 resulted in $M_{gel} = 0.163$ cm^2/g, which is close to the value of 0.154 cm^2/g previously found [26] and within the range of 0.21 to 0.13 cm^2/g determined for 50 to 200 keV photons in PMMA, a compound of similar chemical composition [27].

3.3. Fluorescent Images of the Irradiated Gel

The irradiated cell was scanned in 1 mm steps with the UV sheet orthogonal to the propagation (z) axis of the X-ray beam. The resulting 30 tomographic images of the fluorescence were used to construct a 3D translucent image of the beam within the gel as described in reference [23,26]. This 3D representation can be viewed here in the form of a video either in full colour or in blue-pixel grey-scale: Video file links can be found in the Supplementary Materials ("Movie full color.MP4" and "Movie gray.MP4").

An end-on view of the fluorescent image is shown in Figure 3 where it is compared with the incident beam cross-section found using the radiochromic film. The same oval form of the beam cross section is apparent in the RFG gel. A pixel profile scan across the beam is shown in Figure 4. From this, a penumbra value of 0.44 mm was found. This lies between the entrance and exit values of 0.41 and 0.54 mm found with the RC film. The spatial resolution of the profile for images taken in the x-y plane is 0.023 mm/pixel, which is considerably smaller than the penumbra value of 0.44 mm. As for the film measurements, the penumbra was ascribed to edge-dispersion in the X-ray beam, rather than spatial resolution of the measurements.

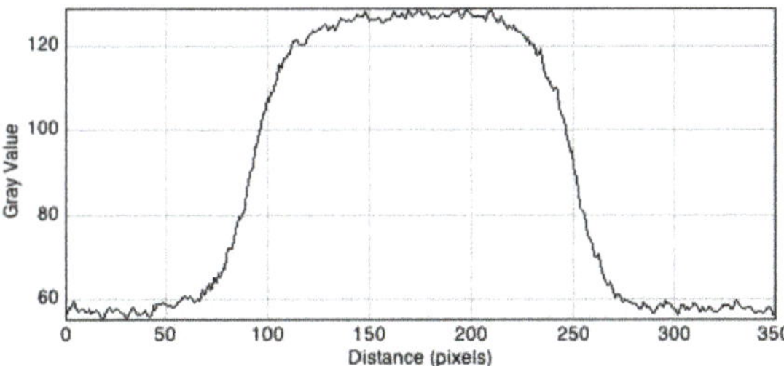

Figure 4. A pixel profile line scan across the center of the RFG gel image in Figure 3. The spatial resolution is 0.023 mm per pixel; the average 20–80% rise and fall value (the "penumbra") is 0.44 mm.

In Figure 5 (lower) a side view (y/z plane) of the fluorescence reconstruction is shown. This displays a gradual decrease in intensity with increasing depth in the gel as expected.

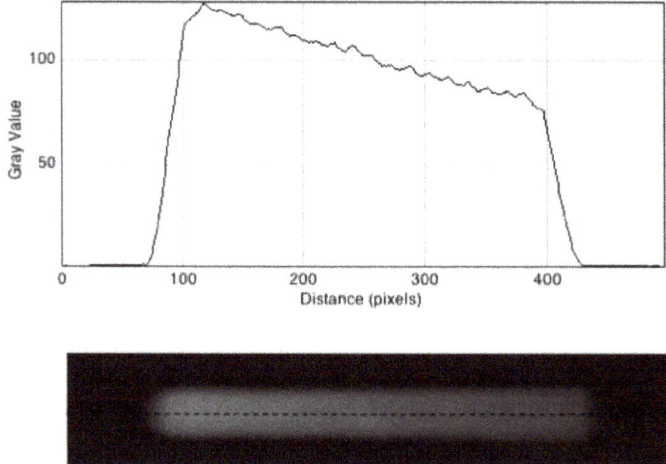

Figure 5. Lower: A still image of the reconstructed X-ray beam fluorescence side-on. **Upper**: A pixel intensity profile of the image taken along the dashed line shown.

The decrease is quantified in the pixel profile scan in Figure 5 (Upper) which decreases by a factor of 0.63 over the length of the gel. This is close to the value of 0.67, which would be expected for the decrease in dose rate across the gel based on the value of $[M\rho\delta]_{gel} = 0.296$, determined in Section 3.2, and a conical expansion by 10%.

Precise agreement might not be expected because of the superlinear dependence found for the fluorescence intensity on dose and the sublinear dependence found on dose rate [22]. These compensatory effects are the object of ongoing research into methods of making corrections to the experimental data required for dosimetry applications.

4. Conclusions

We have shown that it is possible to produce a three-dimensional, fluorescent image of the energy deposited by a 3 mm beam of high-energy radiation in a gel medium of physical properties and dimensions close to those of the vitreous body of the human eye. A 3D video image of the beam, with submillimetre spatial resolution, can be produced in-house within minutes of irradiation using a portable tomographic fluorescence-scanning apparatus with refractive index matching. We intend to apply the method to the study of energy deposition by proton pencil beams and to the control of radiotherapy protocols and equipment when the Holland Proton Therapy Clinic is eventually commissioned (www.hollandptc.nl/en/).

Supplementary Materials: The following are available online at http://www.mdpi.com/2073-4360/10/11/1195/s1, Video S1: "Movie full color.MP4" and "Movie gray.MP4".

Author Contributions: Conceptualization, L.H.L., T.Y. and J.M.W.; Methodology, L.H.L., T.Y. and J.M.W; Validation, L.H.L., T.Y. and J.M.W.; Formal Analysis, L.H.L., T.Y. and J.M.W.; Writing-Review & Editing, J.M.W.

Acknowledgments: The authors wish to acknowledge Gertjan Bon of the glassblowing department of the University of Amsterdam who realized the form of the eye phantom cell.

Conflicts of Interest: The authors declare no conflict of interest.

References

1. Stannard, C.; Sauerwein, W.; Maree, G.; Lecuona, K. Radiotherapy for ocular tumours. *Eye* **2013**, *27*, 119–127. [CrossRef] [PubMed]
2. Poder, J.; Corde, S. I-125 ROPES eye plaque dosimetry: Validation of a commercial 3D opthalmic brachytherapy treatment planning system and independent dose calculation software with GafChromic EBT3 films. *Med. Phys.* **2013**, *40*, 121709. [CrossRef] [PubMed]
3. Eye Cancer: Treatment Options. Available online: https://www.cancer.net/cancer-types/eye-cancer/treatment-options (accessed on 6 June 2018).
4. Radiation Therapy for Eye Cancer. Available online: https://www.cancer.org/cancer/eye-cancer/treating/radiation-therapy.html (accessed on 6 June 2018).
5. The Changing Landscape of Cancer Therapy. Available online: https://physicsworld.com/a/the-changing-landscape-of-cancer-therapy/ (accessed on 6 June 2018).
6. McAuley, G.A.; Heczko, S.L.; Nguyen, T.; Slater, J.M.; Slater, J.D.; Wroe, A.J. Monte Carlo evaluation of magnetically focused proton beams for radiosurgery. *Phys. Med. Biol.* **2018**, *63*, 055010. [CrossRef] [PubMed]
7. Allen, A.M.; Pawlicki, T.; Dong, L.; Fourkal, E.; Buyyounousli, M.; Cengel, K.; Plastaras, J.; Bucci, M.K.; Yock, Y.I.; Bonilla, L.; et al. An evidence based review of proton beam therapy: The report of ASTRO's emerging technology committee. *Radiother. Oncol.* **2012**, *103*, 8–11. [CrossRef] [PubMed]
8. Goitein, G. Proton radiation therapy of ocular melonoma. *PTCOG 49 Teaching Course.* 2010. Available online: http://www.ptcog.ch/archive/conference_p&t&v/PTCOG49/presentationsEW/18-4-3_Eye.pdf (accessed on 4 September 2018).
9. Pasler, M.; Hernandez, V.; Jornet, N.; Clark, C.H. Novel methodologies for dosimetry audits: Adapting to advanced radiotherapy techniques. *Phys. Imaging Radiat. Oncol.* **2018**, *5*, 76–84. [CrossRef]
10. Day, M.J.; Stein, G. Chemical effects of ionizing radiation in some gels. *Nature* **1950**, *166*, 146–147. [CrossRef] [PubMed]
11. Baldock, C.; De Deene, Y.; Doran, S.; Ibbott, G.; Jirasek, A.; Lepage, M.; McAuley, K.B.; Oldham, M.; Schreiner, L.J. Polymer gel dosimetry. *Phys. Med. Biol.* **2010**, *55*, R1. [CrossRef] [PubMed]
12. Watanabe, Y.; Warmington, L.; Gopishankar, N. Three-dimensional radiation dosimetry using polymer gel and solid radiochromic polymer: From basics to clinical applications. *World J. Radiol.* **2017**, *9*, 112–125. [CrossRef] [PubMed]
13. Vandecasteele, J.; De Deene, Y. Evaluation of radiochromic gel dosimetry and polymer gel dosimetry in a clinical dose verification. *Phys. Med. Biol.* **2013**, *58*, 6241–6252. [CrossRef] [PubMed]
14. Schreiner, L.J. True 3D chemical dosimetry (gels, plastics): Development and clinical role. *J. Phys. Conf. Ser.* **2015**, *573*, 012003. [CrossRef]
15. Oldham, M. Radiochromic 3D Detectors. *J. Phys. Conf. Ser.* **2015**, *573*, 012006. [CrossRef]
16. Maryanski, M.J.; Schulz, R.J.; Ibbott, G.S.; Gatenby, J.C.; Xie, J.; Horton, D.; Gore, J.C. Magnetic resonance imaging of radiation dose distributions using a polymer gel dosimeter. *Phys. Med. Biol.* **1994**, *39*, 1437–1455. [CrossRef] [PubMed]
17. Gore, J.C.; Ranade, M.; Maryanski, M.J.; Schulz, R.J. Radiation dose distributions in three dimensions from tomographic optical density scanning of polymer gels: I. Development of an optical scanner. *Phys. Med. Biol.* **1996**, *41*, 2695–2704. [CrossRef] [PubMed]
18. Cameron, M.; Cornelius, I.; Cutajar, D.L.; Davis, J.A.; Rosenfeld, A.B.; Lerch, M.; Guatelli, S. Comparison of phantom materials for use in quality assurance of microbeam radiation therapy. *J. Synchrotron Radiat.* **2017**, *24*, 866–876. [CrossRef] [PubMed]
19. Kron, T.; Lehmann, J.; Greer, P. Dosimetry of ionising radiation in modern radiation oncology. *Phys. Med. Biol.* **2016**, *61*, R167–R205. [CrossRef] [PubMed]
20. Newton, J.; Oldham, M.; Thomas, A.; Li, Y.; Adamovics, J.; Kirsch, D.G.; Das, S. Commissioning a small-field biological irradiator using point, 2D, and 3D dosimetry techniques. *Med. Phys.* **2011**, *38*, 6754–6762. [CrossRef] [PubMed]
21. Warman, J.M.; Luthjens, L.H.; de Haas, M.P. High-energy radiation monitoring based on radio-fluorogenic co-polymerization II: Fixed fluorescent images of collimated X-ray beams using an RFCP gel. *Phys. Med. Biol.* **2011**, *56*, 1487–1508. [CrossRef] [PubMed]

22. Warman, J.M.; de Haas, M.P.; Luthjens, L.H.; Denkova, A.G.; Yao, T. A radio-fluorogenic polymer-gel makes fixed fluorescent images of complex radiation fields. *Polymers* **2018**, *10*, 685. [CrossRef]
23. Yao, T.; Gasparini, A.; de Haas, M.P.; Luthjens, L.H.; Denkova, A.G.; Warman, J.M. A tomographic UV-sheet scanning technique for producing 3D fluorescence images of x-ray beams in a radio-fluorogenic gel. *Biomed. Phys. Eng. Express* **2017**, *3*, 027004. [CrossRef]
24. Warman, J.M.; de Haas, M.P.; Luthjens, L.H.; Denkova, A.G.; Kavatsyuk, O.; van Gothem, M.-J.; Kiewiet, H.H.; Brandenburg, S. Fixed fluorescent images of an 80 MeV proton pencil beam. *Rad. Phys. Chem.* **2013**, *85*, 179–183. [CrossRef]
25. Yao, T.; Luthjens, L.H.; Gasparini, A.; Warman, J.M. A study of four radiochromic films currently used for (2D) radiation dosimetry. *Rad. Phys. Chem.* **2017**, *133*, 37–44. [CrossRef]
26. Yao, T. 3D Radiation Dosimetry using a Radio-Fluorogenic Gel. Ph.D. Thesis, Technische Universiteit Delft, Mekelweg, Delft, The Netherlands, January 2017.
27. Hubbell, J.H.; Seltzer, S.M. *Tables of X-ray Mass Attenuation Coefficients and Mass Energy-Absorption Coefficients from 1 keV to 20 MeV for Elements Z = 1 to 92 and 48 Additional Substances of Dosimetric Interest*; PLM: Gaithersburg, MD, USA, 1995; NISTIR-5632.

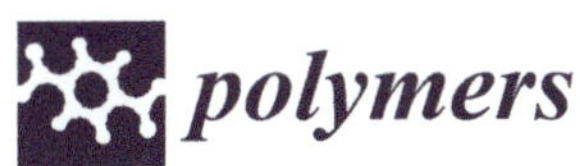 

Article

Fabrication of Active Polymer Optical Fibers by Solution Doping and Their Characterization

Igor Ayesta [1,*], Mikel Azkune [2], Eneko Arrospide [1], Jon Arrue [2], María Asunción Illarramendi [3], Gaizka Durana [2] and Joseba Zubia [2]

[1] Department of Applied Mathematics, Engineering School of Bilbao, University of the Basque Country (UPV/EHU), Plaza Ingeniero Torres Quevedo, 1, E-48013 Bilbao, Spain; eneko.arrospide@ehu.eus

[2] Department of Communications Engineering, Engineering School of Bilbao, University of the Basque Country (UPV/EHU), Plaza Ingeniero Torres Quevedo, 1, E-48013 Bilbao, Spain; mikel.azkune@ehu.eus (M.A.); jon.arrue@ehu.eus (J.A.); gaizka.durana@ehu.eus (G.D.); joseba.zubia@ehu.eus (J.Z.)

[3] Department of Applied Physics I, Engineering School of Bilbao, University of the Basque Country (UPV/EHU), Plaza Ingeniero Torres Quevedo, 1, E-48013 Bilbao, Spain; ma.illarramendi@ehu.eus

[*] Correspondence: igor.ayesta@ehu.eus; Tel.: +34-946-014-153

Received: 27 November 2018; Accepted: 26 December 2018; Published: 31 December 2018

Abstract: This paper employs the solution-doping technique for the fabrication of active polymer optical fibers (POFs), in which the dopant molecules are directly incorporated into the core of non-doped uncladded fibers. Firstly, we characterize the insertion of a solution of rhodamine B and methanol into the core of the fiber samples at different temperatures, and we show that better optical characteristics, especially in the attenuation coefficient, are achieved at lower temperatures. Moreover, we also analyze the dependence of the emission features of doped fibers on both the propagation distance and the excitation time. Some of these features and the corresponding ones reported in the literature for typical active POFs doped with the same dopant are quantitatively similar among them. This applies to the spectral location of the absorption and the emission bands, the spectral displacement with propagation distance, and the linear attenuation coefficient. The samples prepared in the way described in this work present higher photostability than typical samples reported in the literature, which are prepared in different ways.

Keywords: polymer optical fibers; rhodamine B; solution-doping technique; photostability; optical characterization; penetration of the dopant solution

---

1. Introduction

In the last few years, interest in the field of photonics has increased as a result of the incorporation of functional materials into solid-state organic hosts, especially into polymer optical fibers (POFs) [1,2]. As compared to glass fibers, POFs are easier and more economical to manufacture, safer to handle, and much more flexible [3,4]. As the manufacturing temperature of POFs is much lower than that of glass fibers, it is possible to embed a wide range of dopants in the fiber core, from organic dyes and conjugated polymers to other kinds of materials, such as rare-earth ions and quantum dots. Some of these dopants can generate or amplify visible light at the low-attenuation windows of POFs [5,6]. The emission and absorption features of these dopants can be suitable for achieving luminescence in the visible region of the spectrum, which is interesting for a wide range of applications [7–9]. Consequently, research on doped POFs is on an upward trend. The results obtained are being applied for the development of optical sensors [7,8], of solar concentrators that collect and transport solar light [9], and of superluminiscent speckle-free light sources [5,10]. Additionally, the use of optical fibers allows for symmetrical output beams, due to their circular symmetry, and for longer interaction distances between light and dopant, owing to the guidance of light. Moreover, the confinement of light as it propagates facilitates connectivity, and the high ratio between surface area and volume in the fiber improves the heat-dissipation efficiency.

There are different approaches to incorporate the dopant molecules into the Poly(methyl methacrylate) (PMMA) core. Commonly, the doping is carried out at the pre-polymerization stage. The dopant molecules are added to the monomer mixture before carrying out the polymerization process [11,12]. Alternatively, the dopant molecules, together with the melted PMMA, can be mixed during the extrusion process carried out to fabricate the POF preform [13]. The third approach consists in introducing the dopant molecules into the microstructured fiber preform by using the solution-doping technique [14]. However, all these approaches are performed at the fiber preform level, which means that a fiber-drawing tower is needed to obtain the doped POFs. These towers are very expensive, so not all researchers can afford to have one. In contrast, this work describes a doping procedure, based on the solution-doping technique, in which the incorporation of dopant molecules into the core material is carried out directly into the fiber, instead of into the preform. This procedure also opens up the opportunity for any research group to prepare their own ad-hoc doped POFs for any specific application requiring precise characteristics such as distinct dopant molecules or fiber diameters. To date, few works employing the solution-doping technique in POFs have been reported, and only the preform or the fiber cladding were doped [14,15]. In [15], cladded POFs were employed. The dopant molecules could not thereby penetrate into the core, because the interface between the core and the cladding acted as a barrier for them, while the solvent (methanol) could penetrate. Therefore, the cladding was the only doped region. However, following the procedure described in this paper, we can obtain POFs whose core is completely and uniformly doped with active molecules. The dopant utilized for this work is the organic dye rhodamine B. As far as we know, this is the first time that this procedure and its optical characterization have been reported.

The paper is organized as follows. First, the method for preparing the samples is explained. Then, the experimental setup is described, together with the penetration characteristics of the solution of methanol and dopant into the fiber samples. Afterwards, the optical characterization is shown and discussed.

2. Materials and Methods

2.1. Sample Preparation

The POF samples employed in this work were fabricated by our research group from Plexiglass® extrusion rods purchased from Evonik (Essen, Germany). These were 20 mm in diameter and had a weight-average molecular weight (M_w) of 110301. They were annealed in a C-70/200 climate chamber (Controltecnia-CTS, Hechingen, Germany) for 7 days at a temperature of 90 °C under low relative humidity. Afterwards, they were drawn into only core fibers of 500 µm in diameter using our own POF drawing tower, which enabled us to control the fiber diameter with an accuracy of ±1% [16]. The refractive-index profile of the fibers was step-index in all cases. Alternatively, there are only core POFs commercially available, which could also be employed for the preparation of doped fibers.

To incorporate rhodamine B dopant molecules into the pristine only core fibers, we cut samples of around 5 cm and put them in an oven at a temperature of 60 °C in order to eliminate any possible residual stress [17]. The solution employed in this work was prepared by dissolving 500 mg of rhodamine B (Merck kGaA, Darmstadt, Germany) in 200 ml of methanol (Thermo Fisher Scientific Inc., Madrid, Spain). As will be shown in this paper, the resultant dopant concentration was high enough for us to visualize the penetration of the solution into the samples. Rhodamine B is a well-known and widely used dopant for optical-communication purposes because its emission spectrum is located in one of the low-attenuation windows of the PMMA material [18]. In any case, other dopants of similar size, such as different rhodamines or fluorescein, could also be employed in our solution-doping technique. The reason for utilizing methanol as solvent is that rhodamine B can be easily dissolved in it, but not the PMMA. This property prevents the formation of dopant aggregates, which has detrimental effects on the light-emission features of doped fibers [19]. In addition, the maximum immersion time of the host PMMA material in order for it not to be damaged by the solvent is larger with methanol than with other solvents [17]. Moreover, methanol is also quite volatile, which allows it to be removed from the PMMA below its glass transition temperature [14]. The solution prepared in this way was employed for different samples by filling multiple small laboratory glass bottles with the solution and introducing samples that were to be treated at the same doping temperature in each bottle. Afterwards, these bottles were kept at different temperatures, depending on the desired doping temperature: in a fridge (10 °C), in a chamber at room temperature (20 °C), or in a laboratory water bath (from 35 °C to 50 °C). After the desired time of thermal treatment for each bottle, the samples were taken out of the solution, thoroughly rinsed and dried at room temperature. Then, a close look at the cross-sections of these samples through an optical microscope allowed us to monitor the different lateral penetrations of the solution of solvent and dye. To acquire images of these cross-sections, each sample was cut at a distance of around 3 mm from one of its ends, which was polished by hand using polishing paper. Each sample was then connectorized with fiber-chuck connectors and placed on the sample holder of the optical microscope in an upright position, in order for the image of its cross-section to be easily taken.

Some of the doped samples obtained in this way were then cladded with a layer of around 10 microns. For this purpose, the only core doped samples were immersed once for 30 s at room temperature in a solution containing the material of the cladding. This consisted of the polymer PC-404PF (Luvantix ADM, Daejeon, Korea) in isopropyl-alcohol at the concentration of 1/3 v/v. Finally, the cladded samples were cured with UV radiation.

2.2. Experimental Setup

Figure 1 sketches the experimental setup employed to measure the optical spectra by means of the side-illumination technique [20,21]. The POF sample is excited laterally and the light emitted from the excitation point propagates along the sample over a distance that can be adjusted by moving the linear stage employed to hold the sample. The light is finally collected by an optical spectrometer placed at the end of the sample. This is the procedure followed for all the measurements carried out as functions of either the propagation distance or the excitation time. The excitation light source was a Mai Tai HP laser system (Spectra-Physics, Newport, Santa Clara, CA, USA), which emitted Gaussian light pulses of around 100 fs at a repetition rate of 80 MHz, with peak powers of 300 kW and an average power of 2.5 W. The spot of the laser beam was 1.2 mm in diameter. The wavelength of the original light was reduced by using a second harmonic generator (Inspire Blue, Radiantis, Barcelona, Spain) in order to be able to excite from 345 nm to 520 nm. The excitation irradiance was controlled by placing a variable attenuator after the laser output and adjusting it by hand.

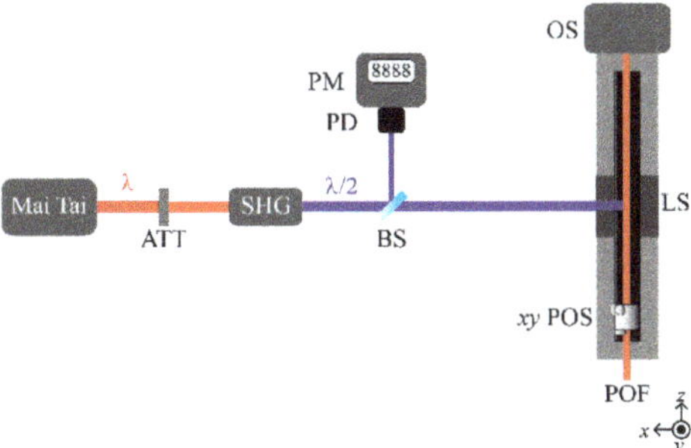

Figure 1. Experimental setup employed to measure the emission spectra of our doped POFs. Legend: ATT: variable attenuator; SHG: Inspire Blue second harmonic generator; BS: beam splitter; PD: photodetector; PM: power meter; LS: linear stage; *xy*-POS: xy-micropositioner; OS: optical spectrometer.

Each of the doped POFs to be measured was held in place with xy-micropositioners standing on a linear stage, thus allowing for the fiber to be maintained completely horizontal while centering the incident laser beam on the fiber symmetry axis. For all of the fiber samples, the spectrum emitted from the polished fiber end was measured by means of a USB4000-UV-Vis spectrometer (Ocean Optics, Largo, FL, USA) with an optical resolution of 1.5 nm for the full width at half-maximum (FWHM). All the obtained data were corrected for the response of the detection system. Additionally, a reference signal was taken using a beam splitter in order to cancel out any power fluctuation of the laser. By changing the position of the laser-excitation point, the variation of the emission spectrum as a function of the propagation distance through the doped fiber was also analyzed. For this purpose, an ILS250CC (Newport, Santa Clara, CA, USA) linear stage driven by an ESP300 (Newport, Santa Clara, CA, USA) motion controller was utilized. The whole acquisition system was automated by means of a LabVIEW program elaborated by us.

Additionally, the absorption spectra of the doped fibers were measured by a Cary 50 UV-Vis spectrophotometer (Agilent Technologies, Santa Clara, CA, USA). In these measurements, the length of the fibers was short enough (1 cm) for the absorption bands of the dopant to be detected.

3. Results and Discussion

3.1. Penetration of the Solution

Figure 2 shows cross-sectional images of fiber samples that were immersed in a solution of rhodamine B and methanol at 10 °C for different time intervals. The lateral penetration of the solution is clearly visible, because of the clear boundary between the outer ring, which is swollen and orange, and the intact transparent inner region. As the immersion time increased, the solution penetrated deeper into the fiber until full penetration was reached. Although Figure 2 only shows the progress in the penetration process corresponding to 10 °C, the process of formation of an outer ring is analogous at other temperatures, but the dynamics are different in relation to its evolution with time.

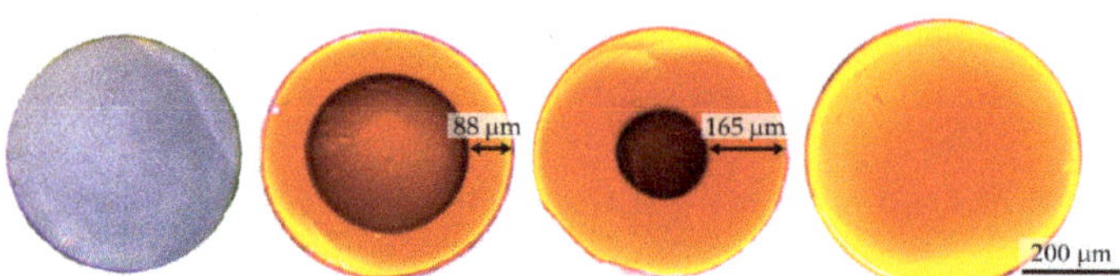

Figure 2. Microscope images of the cross-sections of POF samples immersed in a solution of rhodamine B and methanol at 10 °C for different time intervals, namely (from left to right): 0 h, 30 h, 54 h and 96 h. The image of the pristine fiber was taken in transmitting mode. The images of the doped fibers were obtained under lateral UV excitation.

To assess the penetration of the solution into the fibers, the ring's front thickness and the relative dopant penetration were considered. The latter is defined as the quotient $(\varphi_{out} - \varphi_{in})/\varphi_{out}$, where φ_{out} is the outer diameter of the fiber and φ_{in} is the diameter of the inner unpenetrated region [13]. Both parameters were calculated for three samples doped under the same conditions, and each experimental point of Figure 3 shows the corresponding average value.

According to the relative values of penetrant mobility and segment relaxation of the polymer, the diffusion behavior can be categorized into three different cases [21,22]. These are denoted as follows: Case I, or Fickian diffusion, in which the penetrant mobility is much lower than the segment relaxation of the polymer; Anomalous Case, in which the penetrant mobility is similar to the segment relaxation; and Case II, or non-Fickian diffusion, in which the penetrant mobility is much larger than the segment relaxation. The latter case features a sharp boundary between the outer swollen layer and the inner core material, which advances with uniform velocity. In all these cases, the front penetration thickness (d) is related to the immersion time (t) through the following expression [23]:

$$d = k \cdot t^n, \tag{1}$$

where k is a constant and n is a real number that varies from case to case: usually 0.5 for Case I, 1 for Case II, and any intermediate value for the Anomalous Case.

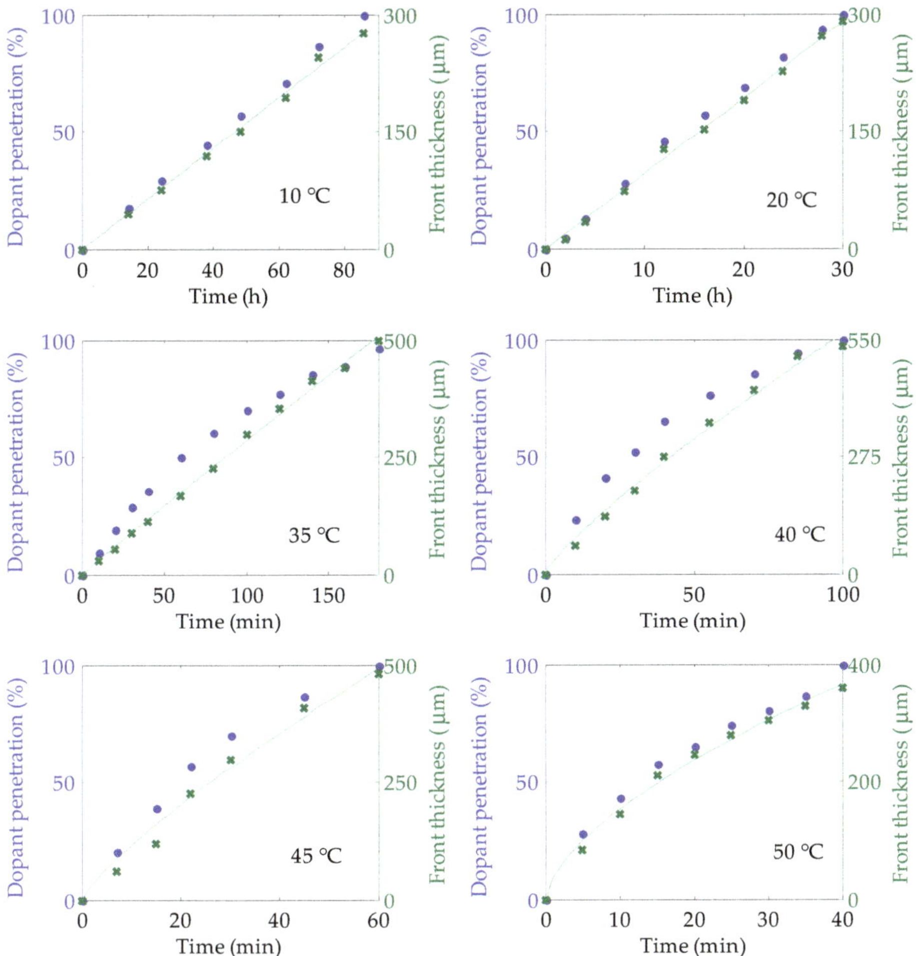

Figure 3. Temporal evolutions of the relative dopant penetration (blue circles) and of the ring's front thickness (green crosses) for six different doping temperatures: 10 °C, 20 °C, 35 °C, 40 °C, 45 °C and 50 °C. The green lines are the fittings of the front thicknesses at each temperature to Equation (1).

The results presented in Figure 3 and in Table 1 demonstrate that the penetration behavior of our solution of rhodamine B and methanol into our POF samples gradually changed from Case II to Case I as the temperature increased. For low temperatures (10 °C and 20 °C), both the relative dopant penetration and the front thickness advanced at constant speed. Moreover, the exponent n was equal to 1, which means that the category of the diffusion behavior was clearly Case II. For higher temperatures (above 20 °C), the advance of the front thickness was much faster than before. However, the diffusion of the solution through the swollen layer could no longer keep a constant concentration at the front, thus producing a slight deceleration in the advance of the front as the penetration approached the center of the fiber considered. Consequently, the value of n shifted towards 0.5, which is characteristic of Case I.

Table 1. Values of the fittings of the experimental front thicknesses of Figure 3 to Equation (1), and the corresponding coefficients of determination.

Temperature (°C)	k	n	R^2
10	3.2 ± 0.1	1 ± 0.05	0.9961
20	9.6 ± 0.2	1 ± 0.05	0.9972
35	3.3 ± 0.9	0.97 ± 0.06	0.9972
40	12 ± 5	0.83 ± 0.09	0.9922
45	18 ± 12	0.81 ± 0.09	0.9835
50	37 ± 9	0.62 ± 0.07	0.9935

At first glance, the use of high temperatures seems to be more interesting from the point of view of the time required to complete the doping procedure. However, as will be shown in the next section, the optical characteristics would be negatively affected at doping temperatures that are close to or above room temperature.

Figure 4 shows the Neperian logarithm of the penetration rates obtained from Figure 3 plotted against the inverse of the absolute temperature (i.e., the Arrhenius plot), where T is measured in kelvins. The slope of the fitting of the points to a straight line is proportional to the activation energy for the occurrence of the diffusion. In our case, this energy is 27.64 kcal/mol, which is close to the values of some PMMA films immersed in methanol (27 kcal/mol was reported in [24]). This means that the activation energy is not significantly affected by the presence or not of the dopant or by the geometry of the PMMA system.

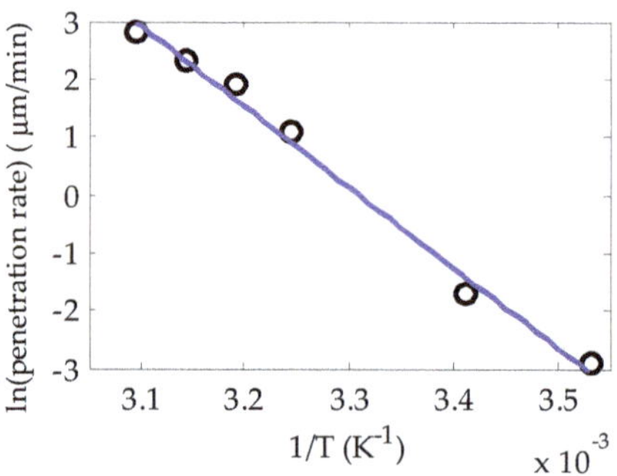

Figure 4. Arrhenius plot for the solution of rhodamine B and methanol penetrating in PMMA POFs. The coefficient of determination of the fitting is 0.9901.

3.2. Optical Characterization

The analysis of the doping diffusion at different temperatures revealed that the translucency of the doped material was much lower when the doping process was carried out above room temperature (i.e., at any of the temperatures equal to or greater than 35 °C in Figure 3), foreshadowing bad light-transmission features. This effect is clearly noticeable in Figure 5, and it is a direct consequence of the interaction between the methanol solvent and the PMMA of the fiber core. It can be explained as follows: the glass transition temperature (T_g) of a polymeric material can change when it is immersed in a solvent. In fact, the T_g value of the polymer can be reduced down to room temperature if it is immersed in methanol [25]. Specifically, the T_g for a methanol-PMMA system can range between 20 °C and 30 °C, depending on the weight-average molecular weight (M_w) of PMMA ($T_g = 20$ °C for $M_w = 23,500$ g/mol, and $T_g = 30$ °C for $M_w = 550,000$ g/mol) [26]. As we mentioned in the section about the sample preparation, the value of M_w for our fibers is 110301, which means that the T_g of our samples immersed in methanol should be only slightly higher than 20 °C. When the temperature of the doping process is higher than the T_g, as was the case in some of our experiments, small holes are formed in the PMMA material. Moreover, as the doping temperature is increased, the size and the population of these holes increases, which raises the optical attenuation of the fibers dramatically [27]. This effect can be clearly seen in Figure 6, in which the attenuation coefficients were measured by the spectrophotometer for different fiber samples immersed in methanol at the temperatures mentioned in Figure 3. A complete methanol penetration was achieved in all cases. For fibers doped at 10 °C, the optical attenuation was not much affected by the immersion, obtaining almost the same attenuation values as those of pristine fibers. This suggests that holes could not grow because the immersion temperature was still below the corresponding T_g value. However, at 20 °C, the attenuation began to be affected, especially in the visible part of the spectrum. As for the rest of the temperatures, i.e., above 35 °C, measurements could not be carried out, because the light attenuation was too high.

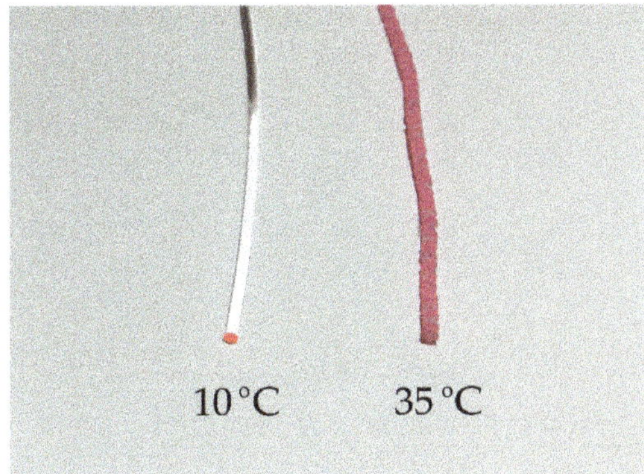

Figure 5. External appearance of two doped samples prepared at 10 °C and at 35 °C.

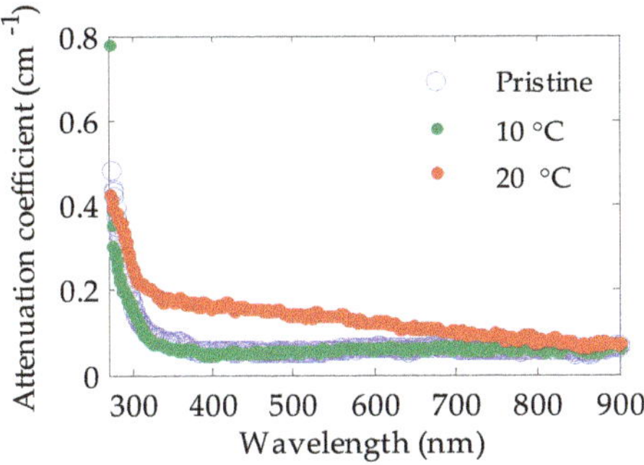

Figure 6. Attenuation coefficients measured for pristine PMMA POFs and for samples immersed in methanol at two different temperatures. At the temperature of 35 °C or above it the attenuation was too high for the measurements.

Heating the immersed fiber above its T_g value also has consequences on the relaxation of internal stresses in the PMMA material. As a matter of fact, the relaxation is usually followed by changes in the physical dimensions, such as shrinkages in the axial direction as well as growths in the diameter [28,29]. We observed these effects even at 20 °C, so we can conclude that the most appropriate doping temperature is around 10 °C for the fibers employed in this work.

Taking the aforementioned results into account, samples of 12 cm were fabricated at the suitable temperature of 10 °C. Moreover, in a second stage we incorporated a cladding layer of around 10 μm of a polymeric material to these optimum samples, in order to improve the light transmission features. Additionally, the core material is less prone to being affected by external unwanted agents if a cladding is added. This process was carried out in the way described in the section about the preparation of the samples. The absorption coefficient and the emission spectrum of these samples in the near-ultraviolet and visible regions are shown in Figure 7. The presence of the dopant is apparent in the prominent absorption band peaking at 558 nm. The emission spectrum was measured by exciting the sample at 520 nm with 22 nJ/cm^2 of irradiance. Its maximum value is located at 592 nm. Actually, the peak wavelength is affected by the red shift produced by reabsorption and reemission processes along the connectorized initial length of the fiber [30], so shorter emission wavelengths would be expected if the spectrum were measured in a thin piece of bulk material. In any case, the obtained absorption and emission peaks are very similar to those reported for PMMA fibers that were also doped with rhodamine B, but utilizing different doping techniques, such as the interfacial gel polymerization [31,32]. We have estimated that the dopant concentration of our doped fibers is in the range between 1 and 10 ppm, by using the method reported in [33], which is based on the red shift of the emission as the fiber concentration is varied.

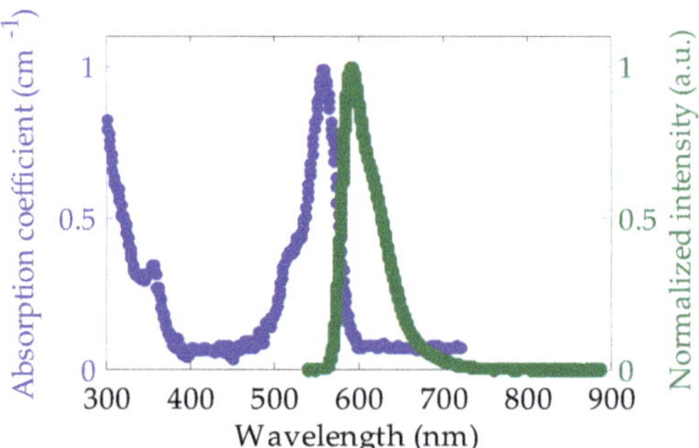

Figure 7. Absorption coefficient and emission spectrum corresponding to our POF doped with rhodamine B by using the solution-doping technique at 10 °C. The emission spectrum corresponds to an excitation wavelength of 520 nm and to a propagation distance of 4 cm.

The evolution of the emission spectra obtained for each propagation distance z along the fiber is illustrated in Figure 8, which corresponds to our POF doped with rhodamine B at 10 °C. These spectra were not symmetrical, so their evolution is characterized by means of their first and second moments, N_1 and N_2, respectively. The first moment represents the average emission wavelength, and the square root of the second moment is proportional to the spectral width. As is well known, separating the excitation point on the fiber away from the detector causes a reduction in the emitted irradiance, while the spectrum is shifted toward longer wavelengths, with no significant changes in the spectral widths [19]. The spectral behavior as a function of the propagation distance shown in Figure 8 for our fibers is the same as that reported for other doped fibers manufactured using other techniques such as the interfacial gel polymerization. Please note that these kinds of spectral shifts could be employed for applications such as the manufacture of tunable light sources or the design of displacement sensors based on doped fibers [34].

The attenuation of each of our doped fibers was calculated by measuring the decrease in the irradiance of the fluorescence spectra as the light propagation distance z increased. Assuming that the illuminated fiber section behaves as a plane-wave source, the light irradiance propagating towards the photodetector at any wavelength λ decays exponentially with z as follows [20,21]:

$$I(\lambda, z) = I_0(\lambda) \cdot \exp(-\alpha(\lambda) \cdot z), \tag{2}$$

where $I_0(\lambda)$ represents the light irradiance measured at $z = 0$ at the wavelength λ, and $\alpha(\lambda)$ is the linear attenuation coefficient at the considered wavelength. Figure 9 shows the linear-attenuation coefficients calculated for our doped fibers by fitting the experimental data to Equation (2) for several wavelengths [35]. The attenuation values obtained in the way described above are very similar to those reported in the literature for other doped PMMA POFs, including thermoplastic ones and even thermosetting ones doped with rhodamine 6G [36]. This implies that the doping procedure employed in this work did not add extra losses to the fibers with respect to the typical attenuation values of fibers fabricated by means of more conventional techniques, such as the interfacial gel polymerization. This holds true as long as the doping temperature is maintained below the corresponding T_g. This result, together with the previous ones, serves to validate the doping technique employed in this work.

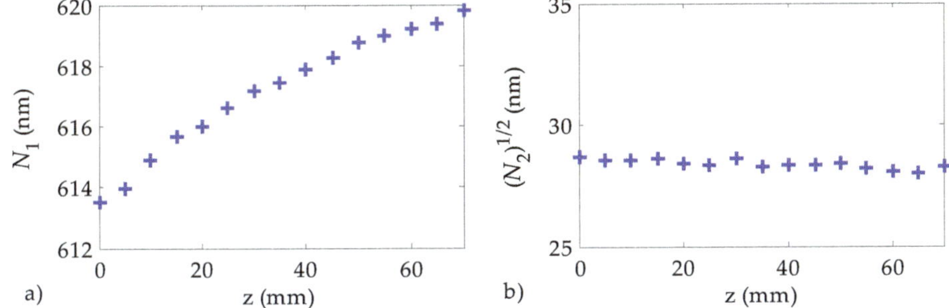

Figure 8. Evolution of (**a**) the first moment (N_1) and of (**b**) the square root of the second moment ($\sqrt{N_2}$) as a function of the excitation wavelength for our POFs doped with rhodamine B at 10 °C. The fiber was pumped at 520 nm with an irradiance of 22 nJ/cm^2. The z of the point of the fiber closest to the detector is normalized to 0.

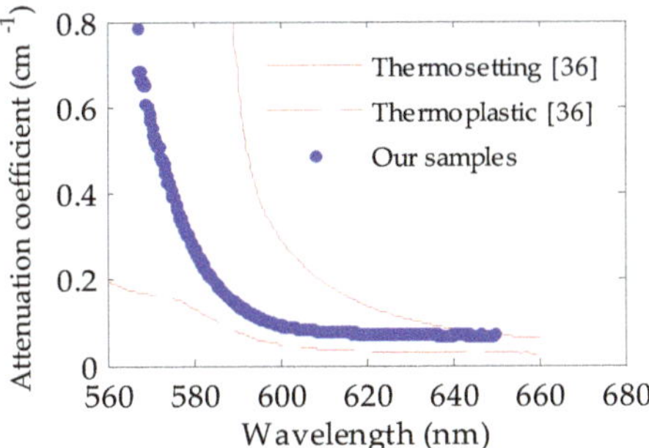

Figure 9. Linear attenuation coefficients of our POFs doped with rhodamine B at 10 °C (thick blue dots), together with those reported in [36] for thermosetting (thin solid line) and thermoplastic (thin dashed line) PMMA POFs doped with rhodamine 6G. The excitation wavelength and irradiance were 520 nm and 22 nJ/cm^2, respectively.

Finally, the photostability of the emitted spectra was also studied for our doped samples by pumping them at 520 nm with our femtosecond laser at an irradiance of 22 nJ/cm^2. The fibers utilized for this purpose had not been previously exposed to laser light, in order to avoid previous degradation that would have affected the results. The relative fluorescence intensities were normalized to 100% at the start of the measurements. As shown in Figure 10, the fluorescence capacity of the doped fibers was reduced by 32% after 60 min of exposure, which, in this case, was equivalent to 2.88×10^{11} laser shots. This reduction is smaller than those reported for other POF samples doped by means of the polymerization technique when subjected to the same exposure conditions [19,36]. Specifically, the photostability of our doped fibers was, respectively, 7% and 18% higher than those measured in the same conditions for thermoplastic and thermosetting POFs doped with rhodamine 6G. Moreover, as Table 2 shows, our doped fibers also presented a greater photostability than that reported for POFs doped with some conjugated polymers, such as PFO, F8T2 or PF3T [19]. The improvement in the photostability can be explained as follows. While the gel-polymerization technique tends to form dopant aggregates in the fiber core during the fabrication process [37], the doping procedure employed in this paper is less prone to the formation of such aggregates. The low segmental-relaxation rates of the PMMA, at least when the temperature is low enough, does not allow the settling of many dopant molecules together in the fiber core, so individual dopant molecules tend to be more spaced. The consequent absence of aggregates reduces the number of non-radiative relaxations of the dye molecules and, therefore, prevents them from optical bleaching [38]. Moreover, as the dopant molecules are mainly surrounded by the PMMA material, these are partially protected from the thermally induced degradation effects, so the fibers tend to conserve their emission capacity for a longer time [39]. The fact that the improvement in photostability is larger in the case of the comparison with the result reported for the thermosetting fiber could also be related to the much greater dopant concentration employed in their case. Improvements in photostability were also observed in the case of rhodamine-6G PMMA films pumped with visible or ultraviolet radiation when they were doped by means of a diffusion procedure instead of the typical gel-polymerization technique [40]. However, this is the first time that this effect is observed in POFs, as far as we know, which could be very useful in the design and development of all-optical devices based on this kind of fibers with longer lifespans.

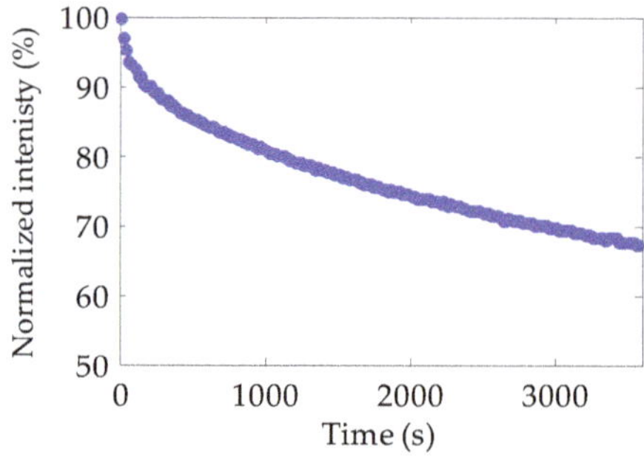

Figure 10. Fluorescence intensity as a function of the excitation time in our POF samples doped at 10 °C. The excitation wavelength was 520 nm, the excitation irradiance was 22 nJ/cm^2, and the light propagation distance was 4 cm.

Table 2. Percentage rate of degradation of the fluorescence intensity for several doped fibers under the same excitation conditions: 60 min of excitation with 22 nJ/cm^2 of irradiance.

Fiber Sample	Degradation (%)	Reference
Our samples	32	-
Thermoplastic fiber doped with rhodamine 6G	39	[36]
Thermosetting fiber doped with rhodamine 6G	50	[36]
PMMA POF doped with PFO	81	[19]
PMMA POF doped with F8T2	81	[19]
PMMA POF doped with PF3T	45	[19]

4. Conclusions

This paper shows a method of fabrication of active POFs starting from only core PMMA fibers, by immersing them in a solution of methanol and the organic dye rhodamine B for an appropriate period of time at a suitable temperature. We demonstrate that the penetration behavior gradually changes from the ideal diffusion category known as Case II to the one known as Case I or Fickian diffusion as the temperature is increased from low values (10 °C) to high ones (50 °C). Although the front penetrates more slowly at lower temperatures, the corresponding linear-attenuation coefficients are lower, due to the absence of hole formation in the core material if, and only if, the temperature of the doping process is lower than the corresponding glass-transition temperature. Moreover, we also measure and analyze the spectral characteristics and the intensity of the emission of doped fibers as functions of both the propagation distance and the excitation time. The corresponding results are in good agreement with those reported for similar active POFs doped by means of other fabrication techniques, such as the gel-polymerization technique. Furthermore, our fibers present a higher photostability, owing to the lower probability of formation of dopant aggregates when using our doping technique. This improvement, together with the rest of optical features and ease of fabrication, could be very interesting in the design process of all-optical devices based on active POFs, such as switches, lasers and amplifiers, optical sensors and solar concentrators.

Author Contributions: I.A. and J.Z. conceived and designed the experiments; E.A. and G.D. fabricated the PMMA optical fibers; I.A., M.A.I. and M.A. performed the experiments; I.A., J.Z., and J.A. analyzed the data; all authors contributed to the scientific discussion and wrote the paper.

Funding: This research was funded by European Regional Development Fund (ERDF), by Ministerio de Economía y Competitividad (MINECO) (TEC2015-638263-C03-1-R) and by Eusko Jaurlaritza (ELKARTEK KK-2016/0030, ELKARTEK KK-2016/0059, ELKARTEK KK-2017/00033, ELKARTEK KK-2017/00089, IT933-16). The work of Mikel Azkune was supported in part by a research fellowship from the Universidad del País Vasco/Euskal Herriko Unibertsitatea (UPV/EHU), Vicerrectorado de Euskera y Formación Continua, while working on a Ph.D. degree.

Acknowledgments: We would like to thank Ester Zuza for her insightful and constructive comments during the development of this work.

Conflicts of Interest: The authors declare no conflict of interest.

References

1. Clark, J.; Lanzani, G. Organic photonics for communications. *Nat. Photonics* **2010**, *7*, 438–446. [CrossRef]
2. Arrue, J.; Jiménez, F.; Ayesta, I.; Illarramendi, M.A.; Zubia, J. Polymer-optical-fiber lasers and amplifiers doped with organic dyes. *Polymers* **2011**, *3*, 1162–1180. [CrossRef]
3. Koike, Y. *Fundamentals of Plastic Optical Fibers*; Wiley-VCH: Weinheim, Germany, 2015; ISBN 978-3-527-41006-4.
4. Ziemann, O.; Krauser, J.; Zamzow, P.E.; Daum, W. *POF-Polymer Optical Fibers for Data Communication*; Springer: Berlin/Heidelberg, Germany, 2002; ISBN 978-3-662-04863-4.
5. Ayesta, I.; Arrue, J.; Jiménez, F.; Illarramendi, M.A.; Zubia, J. Analysis of the emission features in graded-index polymer optical fiber amplifiers. *J. Lightwave Technol.* **2011**, *29*, 2629–2635. [CrossRef]
6. Parola, I.; Illarramendi, M.A.; Arrue, J.; Ayesta, I.; Jiménez, F.; Zubia, J.; Tagaya, A.; Koike, Y. Characterization of the optical gain in doped polymer optical fibres. *J. Lumin.* **2016**, *8*, 1–8. [CrossRef]
7. Peters, K. Polymer optical fiber sensors-A review. *Smart Mater. Struct.* **2011**, *20*, 1–17. [CrossRef]
8. Kamimura, S.; Furukawa, R. Strain sensing based on radiative emission-absorption mechanism using dye-doped polymer optical fiber. *Appl. Phys. Lett.* **2017**, *111*, 1–5. [CrossRef]
9. Parola, I.; Zaremba, D.; Evert, R.; Kielhorn, J.; Jakobs, F.; Illarramendi, M.A.; Zubia, J.; Kowalsky, W.; Johanes, H.-H. High performance fluorescent fiber solar concentrators employing double-doped polymer optical fibers. *Sol. Energy Mater. Sol. Cells.* **2018**, *178*, 20–28. [CrossRef]
10. He, J.; Chan, W.-K.E.; Cheng, X.; Tse, M.-L.V.; Lu, C.; Wai, P.-K.A.; Savovic, S.; Tam, H.-Y. Experimental and theoretical investigation of the polymer optical fiber random laser with resonant feedback. *Adv. Opt. Mater.* **2018**, *6*, 1701187. [CrossRef]
11. Liang, H.; Zheng, Z.; Li, Z.; Xu, J.; Chen, B.; Zhao, H.; Zhang, Q.; Ming, H. Fabrication and amplification of Rhodamine B-doped step-index polymer optical fiber. *J. Appl. Polym. Sci.* **2004**, *93*, 681–685. [CrossRef]
12. Sheeba, M.; Rajesh, M.; Nampoori, V.P.N.; Radhakrishnan, P. Fabrication and characterization of dye mixture doped polymer optical fiber as a broad wavelength optical amplifier. *Appl. Optics* **2008**, *47*, 884–889. [CrossRef]
13. Parola, I.; Arrospide, E.; Recart, F.; Illarramendi, M.A.; Durana, G.; Guarrotxena, N.; García, O.; Zubia, J. Fabrication and characterization of polymer optical fibers doped with perylene-derivatives for fluorescent lighting applications. *Fibers* **2017**, *5*, 28. [CrossRef]
14. Large, M.C.J.; Ponrathnam, S.; Argyros, A.; Pujari, N.S.; Cox, F. Solution doping of microstructured polymer optical fibres. *Opt. Express* **2004**, *12*, 1966–1971. [CrossRef] [PubMed]

15. Stajanca, P.; Topolniak, L.; Pötschke, S.; Krebber, K. Solution-mediated cladding doping of commercial polymer optical fibers. *Opt. Fiber Technol.* **2018**, *41*, 227–234. [CrossRef]

16. Arrospide, E.; Durana, G.; Azkune, M.; Aldabaldetreku, G.; Bikandi, I.; Ruiz-Rubio, L.; Zubia, J. Polymers beyond standard optical fibres—Fabrication of microstructured polymer optical fibres. *Polym. Int.* **2018**, *67*, 1155–1163. [CrossRef]

17. Argyros, A.; Van Eijkelenborg, M.A.; Jackson, S.D.; Mildren, R.P. Microstructured polymer fiber laser. *Opt. Express* **2004**, 1882–1884. [CrossRef]

18. Zubia, J.; Arrue, J. Plastic optical fibers: An introduction to their technological processes and applications. *Opt. Fiber Technol.* **2001**, *7*, 101–140. [CrossRef]

19. Ayesta, I.; Illarramendi, M.A.; Arrue, J.; Jiménez, F.; Zubia, J.; Bikandi, I.; Ugartemendia, J.M.; Sarasua, J.R. Luminescence study of polymer optical fibers doped with conjugated polymers. *J. Lightwave Technol.* **2012**, *30*, 3367–3375. [CrossRef]

20. Kruhlak, R.; Kuzyk, M. Side-illumination fluorescence spectroscopy. I. principles. *J. Opt. Soc. Am. B-Opt. Phys.* **1999**, *16*, 1749–1755. [CrossRef]

21. Kruhlak, R.; Kuzyk, M. Side-illumination fluorescence spectroscopy. II. Applications to squaraine-dye-doped polymer optical fibers. *J. Opt. Soc. Am. B-Opt. Phys.* **1999**, *16*, 1756–1767. [CrossRef]

22. Alfrey, T.; Gurnee, E.F.; Lloyd, W.G. Diffusion in glassy polymers. *J. Polym. Sci.* **1966**, *12*, 249–261. [CrossRef]

23. Thomas, N.; Windle, A.H. Transport of methanol in poly(methyl methacrylate). *Polymer* **1978**, *19*, 255–265. [CrossRef]

24. Thomas, N.; Windle, A.H. A deformation model for case II diffusion. *Polymer* **1980**, *21*, 613–619. [CrossRef]

25. Fasano, A.; Woyessa, G.; Janting, J.; Rasmussen, H.K.; Bang, O. Solution-mediated annealing of polymer optical fiber bragg gratings at room temperature. *IEEE Photonics Technol. Lett.* **2017**, *29*, 687–690. [CrossRef]

26. Williams, D.R.G.; Allen, P.E.M.; Truong, V.T. Glass transition temperature and stress relaxation of methanol equilibrated poly (methyl methacrylate). *Eur. Polym. J.* **1986**, *22*, 911–919. [CrossRef]

27. Lin, C.B.; Liu, K.S.; Lee, S. Methanol-induced opacity in poly(methyl methacrylate). *J. Polym. Sci. Pt. B-Polym. Phys.* **1991**, *29*, 1457–1466. [CrossRef]

28. Jiang, C.; Kuzyk, M.G.; Ding, J.-L.; Johns, W.E.; Welker, D.J. Fabrication and mechanical behavior of dye-doped polymer optical fiber. *J. Appl. Phys.* **2002**, *92*, 4–12. [CrossRef]

29. Stajanca, P.; Cetinkaya, O.; Schukar, M.; Krebber, K. Molecular alignment relaxation in polymer optical fibers for sensing applications. *Opt. Fiber Technol.* **2016**, *28*, 11–17. [CrossRef]

30. Arrue, J.; Jiménez, F.; Illarramendi, M.A.; Zubia, J.; Ayesta, I.; Bikandi, I.; Berganza, A. Computational analysis of the power spectral shifts and widths along dye-doped polymer optical fibers. *IEEE Photonics J.* **2010**, *2*, 521–531. [CrossRef]

31. Tagaya, A.; Teramoto, S.; Nihei, E.; Sasaki, K.; Koike, Y. High-power and high-gain organic dye-doped polymer optical fiber amplifiers: Novel techniques for preparation and spectral investigation. *Appl. Optics* **1997**, *36*, 572–578. [CrossRef]

32. Li, Z.-C.; Liang, H.; Zheng, Z.-Q.; Zhang, Q.-J.; Ming, H. Amplified spontaneous emission of rhodamine B-doped step-index polymer optical fibre. *Chin. Phys. Lett.* **2005**, *22*, 618–620.

33. Kurian, A.; George, M.A.; Paul, B.; Nampoori, V.P.N.; Vallabhan, C.P.G. Studies on fluorescence efficiency and photodegradation of rhodamine 6g doped pmma using a dual beam thermal lens technique. *Laser Chem.* **2002**, *20*, 99–110. [CrossRef]

34. Aiestaran, P.; Dominguez, V.; Arrue, J.; Zubia, J. A fluorescent linear optical fiber position sensor. *Opt. Mat.* **2009**, *31*, 1101–1104. [CrossRef]

35. Illarramendi, M.A.; Zubia, J.; Bazzana, L.; Durana, G.; Aldabaldetreku, G.; Sarasua, J.R. Spectroscopic characterization of plastic optical fibers doped with fluorene oligomers. *J. Lightwave Technol.* **2009**, *27*, 3220–3226. [CrossRef]

36. Ayesta, I.; Illarramendi, M.A.; Arrue, J.; Parola, I.; Jiménez, F.; Zubia, J.; Tagaya, A.; Koike, Y. Optical characterization of doped thermoplastic and thermosetting polymer-Optical-Fibers. *Polymers* **2017**, *9*, 90. [CrossRef]

37. Peng, G.D.; Xiong, Z.; Chu, P.L. Fluorescence decay and recovery in organic dye-doped polymer optical fibers. *J. Lightwave Technol.* **1998**, *16*, 2365–2369. [CrossRef]

38. De la Rosa-Cruz, E.; Dirk, C.W.; Rodríguez, A.; Castaño, V.M. Characterization of fluorescence induced by side illumination of rhodamine B doped plastic optical fibers. *Fiber Integr. Opt.* **2001**, *20*, 457–464. [CrossRef]

39. Faloss, M.; Canva, M.; Georges, P.; Brun, A.; Chaput, F.; Boilot, J.P. Toward millions of laser pulses with pyrromethene-and perylene-doped xerogels. *Appl. Opt.* **1997**, *36*, 6760–6763. [CrossRef]
40. Muto, J.; Kobayashi, K. Photobleaching behaviours of rhodamine 6G diffused in polymethyl methacrylate and in the copolymer of methyl methacrylate with methacrylic acid. *Phys. Status. Solidi (a)* **1984**, *84*, K29–K33. [CrossRef]

Article

Preparation of Fluorescent Molecularly Imprinted Polymers via Pickering Emulsion Interfaces and the Application for Visual Sensing Analysis of *Listeria Monocytogenes*

Xiaolei Zhao [1,†], Yan Cui [1,†], Junping Wang [1,*] and Junying Wang [2,*]

[1] Tianjin University of Science and Technology, No. 29 The Thirteenth Road, Tianjin Economy and Technology, Development Area, Tianjin 300457, China; zxl1989330@163.com (X.Z.); yancuitust@163.com (Y.C.)

[2] The Biotechnology Research Institute of Chinese Academy of Agricultural Sciences, No 12, Zhongguancun South Street, Beijing 100081, China

* Correspondence: wangjp@tust.edu.cn (Junp.W.); wangjunying@caas.cn (Juny.W.);
Tel.: +86-022-60912487 (Junp.W.); +86-010-68519687 (Juny.W.)

† These authors contributed equally to this work.

Received: 2 April 2019; Accepted: 22 May 2019; Published: 4 June 2019

Abstract: In this work, a novel molecularly imprinted polymer (MIP) with water-soluble CdTe quantum dots (QDs) was synthesized by oil-in-water Pickering emulsion polymerization using whole *Listeria monocytogenes* as the template. *Listeria monocytogenes* was first treated by acryloyl-functionalized chitosan with QDs to form a bacteria–chitosan network as the water phase. This was then stabilized in an oil-in-water emulsion comprising a cross-linker, monomer, and initiator, causing recognition sites on the surface of microspheres embedded with CdTe QDs. The resulting MIP microspheres enabled selective capture of the target bacteria via recognition cavities. The target bacteria *Listeria monocytogenes* was detected. Scanning electron microscopy (SEM) characterization showed that the MIPs had a rough spherical shape. There was visual fluorescence detection via quenching in the presence of the target molecule, which offered qualitative detection of *Listeria monocytogenes* in milk and pork samples. The developed method simplified the analysis process and did not require any sample pretreatment. In addition, the fluorescence sensor provided an effective, fast, and convenient method for *Listeria monocytogenes* detection in food samples.

Keywords: pickering emulsion; quantum dots; *Listeria monocytogenes*

---

1. Introduction

Foodborne pathogens are an emerging global public health problem [1]. Even with increased awareness in food safety and quality control, there are still about 48 million cases of foodborne diseases in the United States every year [2,3]. *Salmonella*, *Escherichia coli O157:H7*, *Staphylococcus aureus*, *Listeria monocytogenes*, and *Bacillus cereus* are the most common foodborne pathogens [4,5]. Of these, *Listeria monocytogenes* (*L. monocytogenes*) is more likely to cause death and is associated with listeriosis. Aside from its high fatality rate (20–30%), listeriosis can cause many diseases such as sepsis and meningitis. It readily infects immunocompromised people and in particular neonates, pregnant women, and people over age 65 [6,7]. Meats, fruits, vegetables, seafood, milk, and dairy products are common foods associated with *Listeria monocytogenes* [8,9].

Culture and colony counting are most often used for bacterial identification, but these are time-consuming and laborious processes and require several steps (sample pre-enrichment, selective enrichment, and confirmation) [10,11]. To shorten the analysis time, various rapid detection methods have been presented, including enzyme-linked immunosorbent assay (ELISA) [12], polymerase chain reaction (PCR) [13], and surface plasmon resonance (SPR) [14]. These strategies offer good specificity and sensitivity but require expensive instruments, multiple steps, and well-trained technicians. For example, the high cost of antibodies and the multiple steps involved in manual applications limit immunoassays methods [15]. In some PCR-based assays, DNA extraction and false-positive results can lead to cross-contamination of samples. This is a major drawback of PCR [11]. Hence, a fast, real-time, and effective detection method is urgently needed for *L. monocytogenes* in food.

Molecular imprinting technology (MIT) is an attractive strategy for designing a matrix using customized materials with high selectivity for template molecules or which are related to analogous compounds. Through continuing development of this technology, molecularly imprinted polymers (MIPs) have wide potential application in the fields of biosensors [16], separation [17], drug delivery [18], and catalysis [19]. To date, the synthesis of MIPs of small organic molecules has been straightforward, but MIT for larger templates such as biomacromolecules (e.g., proteins) and bacteria remains highly challenging [20].

Pan et al. prepared MIPs via an inverse-phase suspension and bulk polymerization to detect *Staphylococcus aureus* using *S. aureus* protein A (SpA) as a template protein. However, because of the fragility and complexity of bacteria, it is difficult to generate cavities of a specific size and shape during the imprinting reaction. To address this problem, Shen et al. presented a novel Pickering emulsion polymerization strategy for the preparation of a series of small molecules and protein imprinted polymers [21,22] in which the dispersed liquid droplets are stabilized by solid particles instead of surfactants. Recently, they further proposed the synthesis of bacteria recognition polymers to detect *Escherichia coli* and *Micrococcus luteus* by exploiting the capability of bacteria to self-assemble at an oil–water interface [23].

The Pickering emulsion is a solid particle-stabilized emulsion which is either oil-in-water (O/W) or water-in-oil (W/O) [24]. Versus traditional emulsion polymerization, dispersed liquid droplets are stabilized by solid particles instead of conventional surfactants [25]. Because of the lower toxicity, well controlled size, and high mechanical strength, Pickering emulsion polymerization has been widely applied to synthesize MIPs for the specific recognition of small molecules, such as bifenthrin, malachite green, and bisphenols [26–28]. However, the application of Pickering emulsion polymerization for bacteria-based MIPs is still relatively unexplored.

Quantum dots (QDs) are semiconductor nanocrystals with broad absorption spectra, narrow and tunable emission, and high photoluminescence [29]. Combining the unique optical ability of QDs and specific recognition of MIPs, the common applications of imprinted polymers in optical sensors have been reported for the detection of different molecules. For example, Wang et al. developed molecularly imprinted silica layers coated with QDs for diethylstilbestrol [30]. Feng et al. obtained molecularly imprinted sensor-coated QDs to determine tetrabromobisphenol-A (TBBPA) by sol-gel method [31]. Huang et al. have described a novel fluorescent sensing platform by employing inorganic perovskite quantum dots as a fluorescence signal for the detection of omethoate [32]. However, there have been no fluorescence-imprinted sensor studies used to detect *Listeria monocytogenes* in food samples.

The objective of this study was, therefore, to fabricate stable, rapid, low-cost, and convenient fluorescence-imprinted polymers for the visual qualitative identification of *L. monocytogenes* via a fluorescence microscope. The fluorescence MIPs were prepared by Pickering emulsion polymerization, in which whole *L. monocytogenes* were directly used as the template, and the bacteria-chitosan-QDs network stabilized the particles. In addition, a novel MIP-based sensor based on the response of fluorescence intensity was developed for the qualitative detection of *L. monocytogenes*.

2. Materials and Methods

2.1. Materials

N-(3-dimethylaminopropyl)-*N′*-ethylcarbodiimide hydrochloride (EDC), chitosan, and divinylbenzene (DVB) were purchased from Sigma-Aldrich (Shanghai, China). *N,N*-dimethylacetamide

(DMAC), triethylamine, NaBH$_4$, sodium dodecyl sulphate (SDS), and potassium tellurite (K$_2$TeO$_3$) were purchased from Sinopharm Chemical Reagent Co., Ltd. (Tianjin, China). Trimethylolpropane trimethacrylate (TRIM), *N,N*-dimethylaniline (DMA), and benzoyl peroxide (BPO) were purchased from Aladdin Industrial Co., Ltd. (Shanghai, China). Thioglycolic acid (TGA) and acryloyl chloride were purchased from Alfa Aesar (Tianjin, China), and Cd(CH$_3$COO)$_2$·2H$_2$O was purchased from TianJin Kemel Chemical Reagent Co., Ltd. (Tianjin, China). Before use, the DVB was passed through an aluminum oxide column to remove the stabilizer.

2.2. Instruments

Scanning electron microscopy (SEM, SU1510, Hitachi, Tokyo, Japan), transmission electron microscopy (TEM, JEOL-2010 FEF, Tokyo, Japan), Fourier transform infrared (FT-IR) spectrophotometry (Tensor 27, Bruker, Karlsruhe, Germany), and fluorescence spectrometry (Lumina, Thermo Scientific, Waltham, MT, USA) were used to characterize the polymers.

2.3. Bacterial Strains and Cultivation of Strains

Listeria monocytogenes strain ATCC 19111, *S. aureus* strain ATCC 25923, *E. coli* O157:H7 strain ATCC 35150, and *Salmonella* strain ATCC 14028 were obtained from the American Type Culture Collection (ATCC). All bacteria stains were cultivated in Luria-Bertani (LB) broth at 37 °C with shaking overnight. The bacterial cells were suspended in phosphate buffer solution (PBS buffer) under gentle vortex mixing [33].

For the Pickering emulsion polymerization, the *L. monocytogenes* were cultivated to an OD$_{600}$ (the optical density was measured at 600 nm by a UV spectrophotometer) of about 0.6 to 0.8. After centrifugation at 5000 rpm for 5 min, the bacteria cells were collected and washed three times with PBS buffer. Finally, the bacteria were resuspended in PBS and adjusted to a QD$_{600}$ of 2, this being the template in Pickering emulsion polymerization.

2.4. Synthesis of CdTe QDs

The CdTe QDs were synthesized following the method given in [34]. Briefly, 53.2 mg of Cd(CH$_3$COO)$_2$·2H$_2$O was dissolved into 50 mL of deionized water in a 100 mL flask. Then, 18 µL of TGA was added and the pH value was adjusted to 10.5 with a 1 M NaOH solution. After stirring for 5 min, 50 mL of 0.2 mg ml^{-1} K$_2$TeO$_3$ solution and 80 mg of NaBH$_4$ were successively added to the solution. Next, the reaction proceeded for another 5 min, and the flask was attached to a condenser and refluxed for 1 h at 100 °C. After cooling to room temperature, the CdTe QDs with an emission peak at 556 nm were obtained and stored at −4 °C for subsequent experiments.

2.5. Preparation of N-Acrylchitosan (NAC) and NAC-QD Complex

A NAC preparation method was followed according to the literature with slight modifications [23]. Here, 1.61 g of chitosan was dispersed in 40 mL of DMAC and named solution A. Solution A was stirred for 12 h at room temperature, purged with argon gas for 10 min at 0 °C, and treated with 600 µL of triethylamine under continuous stirring. Solution B was prepared by adding 322 µL of acryloyl chloride to 5 mL of DMAC. This was then added dropwise into solution A. The mixture was stirred at 0 °C for 4 h followed by stirring at 25 °C for 20 h, and successively washed in DMAC, dichloromethane, and methanol. Finally, the NAC powder was collected by filtration and dried in a vacuum chamber.

The NAC-QD complex was prepared as follows: 1.5 mg of NAC was dissolved in 5 mL of acetic acid solution (0.03%). Then, 5 mL of CdTe QDs and 5 mg of EDC were separately added dropwise to the above solution and stirred overnight at room temperature.

2.6. Synthesis of MIPs by Pickering Emulsion Polymerization

Bacterial-stabilized Pickering emulsions were prepared using a method similar to that described by Shen et al., except that the NAC-QD complex was used to replace NAC to form a bacteria-pre-polymer complex [23]. First, 900 µL of the NAC-QD solution and 300 µL of the *L. monocytogenes* PBS suspension (OD_{600} = 2) were mixed; the mixed solution was set aside for 30 min to form the NAC-QD bacteria network, which was used as the water phase in the Pickering polymerization. Second, 0.6 mL of TRIM, 0.6 mL of DVB, 6.2 mg of BPO, and 31.3 µL of DMA were added in another 5 mL tube as the oil phase. Subsequently, the two phases were mixed by vigorous hand shaking for 10 min, such that a stable Pickering emulsion was established. The Pickering emulsion was stabilized via the NAC-QD-bacteria network and was kept still for 24 h at room temperature without agitation. To remove the template, the resulting polymer beads were successively washed with 10% acetic acid, 1% SDS, water, and methanol. After drying in a vacuum chamber, MIPs with specific binding sites for *L. monocytogenes* were obtained.

As a control, non-imprinted polymer (NIP) beads were also prepared using the same procedure, with the exception of the absence of *L. monocytogenes*.

2.7. Analysis of Bacterial Binding Properties

Adsorption kinetic data were tested as follows: 5 mg of the MIPs was added to 1 mL *L. monocytogenes* PBS solution (1.0×10^5 colony forming unit (CFU) mL^{-1}). The supernatant was removed after gently shaking for different periods of time (0.5, 1, 1.5, 2, 2.5, 3, 4, 5, 6, and 7 h) and allowing the sample to settle for 3 min. The target bacteria absorbed on the polymer beads were eluted with 1 mL PBST (PBS containing 5 mL L^{-1} Tween 20) solution, and the number of bacteria in elution was analyzed by viable cell counting. The binding amount was calculated as followed:

$$Q = \frac{C_s}{M} \times V$$

Here, Q (CFU mg^{-1}) is the adsorption capacity and C_s is the bacteria concentration in the elution. V (mL) is the volume of bacterial suspension and M (mg) is the adsorption weight.

For the static adsorption testing, 5 mg of MIPs/NIPs were weighed into 1 mL PBS buffer with different concentrations of bacterial suspension (7.5×10^1, 1.6×10^2, 7.6×10^2, 1.5×10^3, 7.5×10^3, 1.5×10^4, 3.8×10^4, 7.5×10^4, 1.6×10^5, and 7.5×10^5 CFU mL^{-1}). After shaking for 2 h, the supernatant was removed and the binding amount of *L. monocytogenes* was recorded.

The selectivity study was conducted using two strategies depending on purpose. For visual studies, 5 mg of MIPs were incubated in 1 mL *L. monocytogenes* and *S. aureus* for 2 h under the same conditions, respectively. After removing the supernatant and Gram staining the MIPs, *L. monocytogenes* and *S. aureus* assembled on the polymer beads were directly observed by optical microscope. For a detailed evaluation, 5 mg of MIPs or NIPs were suspended in a 2 mL mixture containing *L. monocytogenes*, *E. coli*, *Salmonella*, and *S. aureus* (each 4.0×10^4 CFU mL^{-1}). The amounts of *L. monocytogenes*, *E. coli*, *Salmonella*, and *S. aureus* binding on the polymer beads were recorded via the plate-coating method.

2.8. Aplication to Real Samples

Here, 1 mL of milk was directly inoculated with *L. monocytogenes* at final concentrations of 1.0×10^3 and 1.0×10^5 CFU mL^{-1}. After adding 5 mg of MIPs, the mixture was thoroughly shaken for 2 h and sedimented for 3 min. After removing the supernatant, the residual MIPs were dried in a vacuum chamber and the fluorescence was directly observed via a fluorescence microscope.

One gram of pork was inoculated with 5 mL of 1.0×10^3 and 1.0×10^5 CFU mL^{-1} of *L. monocytogenes* and set aside overnight at 4 °C. Afterwards, 5 mL of the sample solution was collected and mixed with 5 mg of the MIPs. After thoroughly shaking for 2 h and sedimenting for 3 min, the fluorescence color of the residual MIPs was observed using a fluorescence microscope.

3. Results

3.1. Design and Preparation of MIPs

The design and preparation of the MIPs is shown in Scheme 1. MIPs were prepared via a Pickering emulsion polymerization, which was composed of a water phase and an oil phase. During imprinting, an NAC monomer containing enough free amino groups was first prepared by reacting acrylotl chloride with the amino groups in chitosan. Meanwhile, the CdTe QDs were functionalized with the carboxyl group from the TGA. NAC-QDs were obtained through the amide bond between the amino groups from the NAC and the carboxyl group from the CdTe QDs. The positively charged NAC-QD complex was easily bound with the negatively charged template *L. monocytogenes* via electrostatic interactions in the water phase [35]. The oil phase consisted of TRIM and DVB as co-cross-linkers and BPO as an initiator. After mixing the two phases, a stable emulation was obtained after vigorous hand shaking (see Figure 1), indicating the high efficiency of the self-assembled bacteria-NAC-QD network to construct a stable Pickering emulsion as a surfactant.

When polymerization was induced, the NAC-QD complex at the oil-water interface and the co-cross-linkers in the oil phase (TRIM and DVB) polymerized to form solid polymer beads; the template bacteria were located on the surface of the polymer beads. After removing the template bacteria, the imprinted sites were generated and were completely fitted with the template. Furthermore, the fluorescence intensity of MIPs decreased when the template was bound to the MIP beads; the intensity could recover after removing the template.

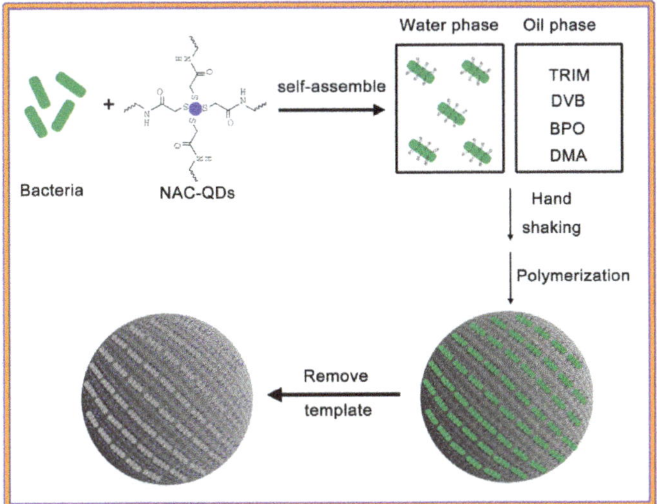

Scheme 1. Molecularly imprinted polymer (MIP) synthesis process (NAC, TRIM, DVB, BPO, and DMA relate to N-Acrylchitosan, trimethylolpropane trimethacrylate, divinylbenzene, benzoyl peroxide, and *N,N*-dimethylaniline).

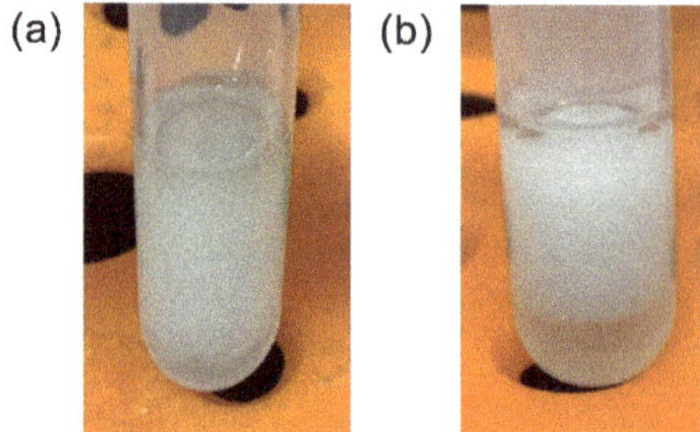

Figure 1. Optical images of emulsions obtained by shaking the oil-water mixture of (**a**) MIPs and (**b**) non-imprinted polymers (NIPs).

3.2. Characterization of Polymer Beads

The polymer beads were carefully characterized to validate this preparation concept. Figure 2 displays TEM images of the CdTe QDs and the NAC-QD complex. The diameters of the NAC-QDs were much larger than those of the original QDs, indicating a successful introduction of the NAC and formation of the NAC-QD complex. The morphology and size of the MIPs and NIPs were characterized by SEM (see Figure 3a,b). Both the MIPs and NIPs exhibited a uniform spherical structure with a particle diameter of about 200 μm. Furthermore, the surface of the MIPs was rough and irregular due to their imprinted cavity; the NIPs were relatively smooth. These results were confirmed via magnified SEM images, which can be seen in Figure 3c,d. A large number of effective and tailor-made imprinted sites were found on the surface of the MIPs; there were none on the NIP surfaces.

FT-IR spectra were used to analyze the surface groups of the MIPs and NIPs. The remarkable peaks at 1633 cm^{-1} and 1463 cm^{-1} were attributed to C=C stretching vibrations and the benzene ring vibration of DVB, respectively [36,37]. Strong bands were observed near 1160 cm^{-1} and 1720 cm^{-1} which corresponded to the C–O–C stretch and C=O vibration from cross-linking TRIM [38]. Furthermore, there was no significant difference between the FT-IR spectra of the MIPs and NIPs, confirming that the template was removed completely.

Figure 3f shows the optical properties of the QDs, NAC-QDs, MIPs, and NIPs. Compared with the emission peak of the pure QDs, the spectra peak position of the NAC-QDs had a slight red-shift to ~340 nm due to the aggregation of QDs during the self-assembly of the NAC-QDs [39]. The fluorescence intensity of the NAC-QDs was much higher than those of the pure QDs because of the surface passivation of the QDs [40]. However, the comparative blue shift in the emission peaks of the MIPs and NIPs was obvious. This might be because of the reduction in the surface charge of the QDs because of electrostatic interactions, leading to a smaller Stokes shift.

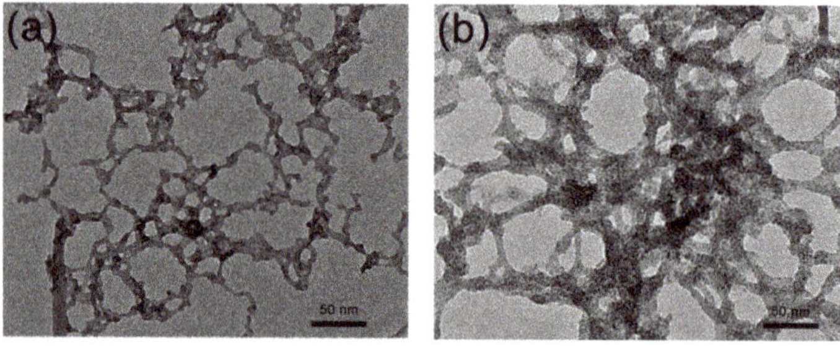

Figure 2. Transmission electron microscopy (TEM) images of (**a**) quantum dots (QDs) and (**b**) the NAC-QD complex.

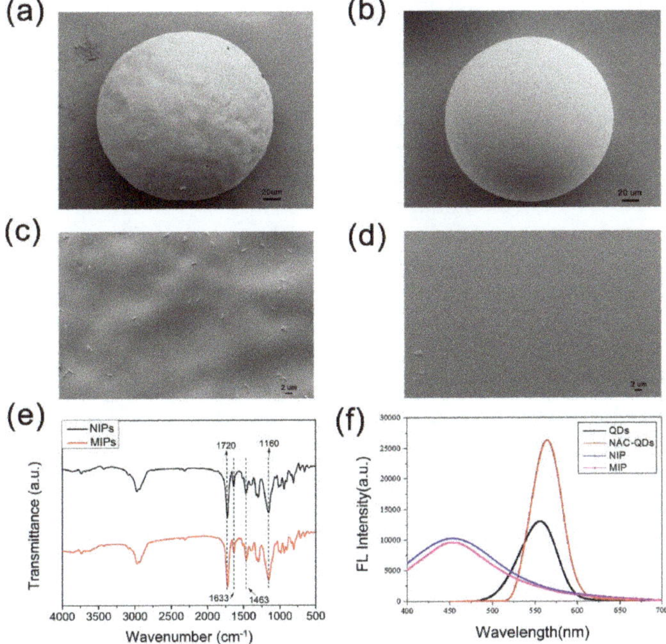

Figure 3. Characterization of polymers. Scanning electron microscopy (SEM) images of (**a**) MIPs and (**b**) NIPs which correspond to magnified SEM images of (**c**) MIPs and (**d**) NIPs, respectively; (**e**) Fourier transform infrared (FT-IR) spectra of MIPs and NIPs; and (**f**) fluorescence emission spectra of the QDs, NAC-QD complex, MIPs, and NIPs.

3.3. Adsorption Performance of MIPs and NIPs

3.3.1. Bacterial Binding under Overloading Conditions

Equal amounts of MIPs and NIPs (5 mg) were incubated with 1 mL of *L. monocytogenes* suspension (OD_{600} = 0.1, almost 1.0×10^8 CFU mL^{-1}) for 3 h. After removing the supernatant, the polymer beads were directly observed via SEM. Figure 4a,b show the surface morphology of the treated MIPs and NIPs with *L. monocytogenes*; the target bacteria are clearly visible on the surface of the polymer beads and the density of the cells absorbed on the MIPs was significantly greater than that of the NIPs due to the high adsorption efficiency of MIPs for *L. monocytogenes*.

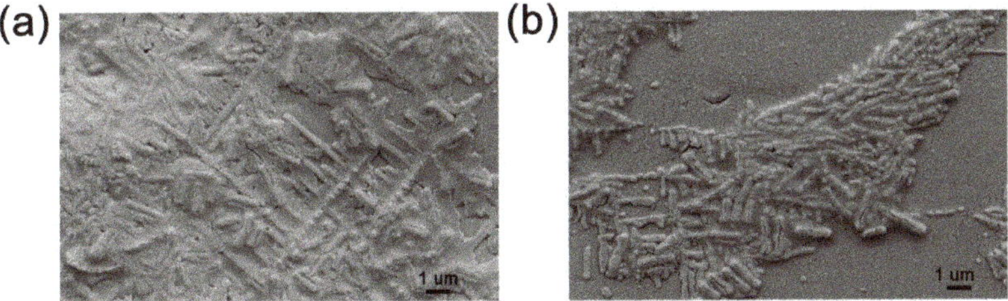

Figure 4. SEM images of (**a**) MIPs and (**b**) NIPs after the adsorption of *L. monocytogenes* (1.0×10^8 CFU mL^{-1}). CFU means the colony forming unit.

To further investigate the adsorption performance, equilibrium binding analysis, adsorption isotherms, and selectivity adsorption were also studied, as discussed below.

3.3.2. Kinetics Adsorption of MIPs

The kinetic adsorption curves of the MIPs and NIPs were all evaluated to explore the adsorption rate (see Figure 5a). A high adsorption rate was performed, and the binding amount linearly increased with increasing contact time before reaching an adsorption equilibrium. At bacteria concentrations of 1.0×10^5 CFU mL^{-1}, the equilibrium time of the polymer beads was almost 2 h, which could be shortened upon reducing the bacterial concentration. The fast adsorption rate could have occurred because the bacterial-NAC-QD complex in the water phase was used as the surfactant to construct the stable O/W Pickering emulsion during the imprinting process; thus, the template bacteria were located on the surface of the polymer beads. After removing the template, several effective and tailor-made imprinted sites remained on the MIP surface. These were easily accessible for the rebinding of template bacteria. In addition, a small reduction of adsorption amounts on the polymer beads occurred over an extended time. One possible reason for this observation is that the bacteria died through a nutrient deficiency in the PBS, resulting in a gradual decrease in *L. monocytogenes* viability.

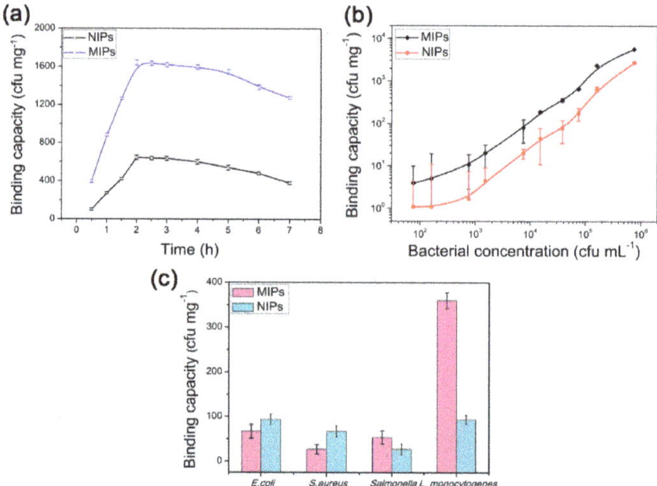

Figure 5. (**a**) Kinetic data of MIPs and NIPs, (**b**) static isotherm curve of MIPs and NIPs, and (**c**) the selectivity adsorption of MIPs and NIPs on four bacteria mixing solutions.

3.3.3. Adsorption Isotherm Analysis

To further investigate the binding performance of MIPs and NIPs, static adsorption data were collected at initial concentrations of *L. monocytogenes* ranging from 7.5×10^1 to 7.5×10^5 CFU mL^{-1}. Figure 5b shows the adsorption isotherm curve: the binding capacity increased with increasing initial bacteria concentrations and the MIPs exhibited a significantly higher adsorption capacity than that of the NIPs across the tested concentration range ($p < 0.05$). These results were mainly due to the specific adsorption of imprinted sites, leading to superior adsorption of *L. monocytogenes*. A trace of bacteria was absorbed on the NIPs simply based on non-specific adsorption. At an initial concentration of 3.8×10^4 CFU mL^{-1}, the adsorption capacities of the MIPs (355.6 CFU mg^{-1}) were almost 4.57-fold that of the NIPs (77.8 CFU mg^{-1}), while the imprinting factor (IF) value reduced to 2.04 when the initial concentration increased to 7.6×10^5 CFU mL^{-1}. At high concentrations, the surface-bound bacteria were able to attract more bacteria cells to self-assemble and aggregated as a membrane-bound cluster on the surface polymer beads, leading to an increase in non-specific adsorption.

3.3.4. Selectivity Study

For visual observation of the selectivity of MIPs, *L. monocytogenes* and *S. aureus* were separately absorbed on MIPs and were directly observed via an optical microscope (Figure 6). Many *L. monocytogenes* were observed, indicating the high binding uptake for template bacteria. By contrast, only a few *S. aureus* were absorbed onto the MIPs, showing that the spherical *S. aureus* did not fit into the imprinted sites. Those that were generated were completely fitted with the rod *L. monocytogenes*. The low binding capacity for *S. aureus* was simply dependent on the non-specific adsorption.

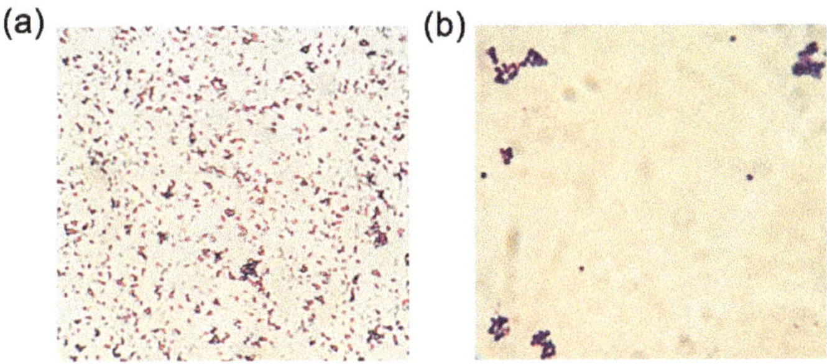

Figure 6. The optical images of MIPs absorbing (**a**) *L. monocytogenes* and (**b**) *S. aureus*.

The interference experiment evaluated selective recognition of the template bacteria. Some commonly foodborne pathogens were selected as competitors, including *E. coli*, *Salmonella*, and *S. aureus*. Figure 5c shows the adsorption capacities of MIPs and NIPs on different bacteria. Under the same conditions, the binding capacities of the MIPs on *L. monocytogenes* and the other three bacteria were significantly different ($p < 0.05$). The imprinted polymers expressed the highest specific adsorption for *L. monocytogenes* with significantly lower adsorption of other bacteria. By contrast, the NIPs absorbed fewer template bacteria than the MIP beads because of the absence of selective imprinted recognition sites. Therefore, other than the electrostatic interactions, the recognition mechanism of the MIP beads was closely related to the complementary shape and size of binding sites that were produced by the template bacteria. The results indicate that the high recognition specificity of the imprinted polymer enabled the MIP beads to bind the target bacteria.

3.4. Establishment of a Detection Method

During MIP preparation, several imprinted sites remained on the surface of the MIPs containing QDs. As a result, the template bacteria were easily bound to the polymer beads and caused a significant change in the fluorescence intensity. To confirm the utility of these materials, the MIPs and NIPs were first incubated with 10^3 and 10^5 CFU mL^{-1} concentrations of bacteria and the fluorescence intensity of the resulting polymer beads was directly observed using a fluorescence microscope under an ultraviolet lamp. Images are presented in Figure 7. The beads differed in brightness at different concentrations of *L. monocytogenes*, and the original MIPs and NIPs exhibited the brightest blue luminescence. With increasing bacteria concentrations ranging from 10^3 to 10^5 CFU mL^{-1}, the color of the polymer beads became increasingly darker with a more pronounced change in the MIPs. Considering that microbial imprinting has not reached the level of precision that can be achieved by imprinting small molecules, the fluorescence MIPs sensor was applied for fast and qualitative detection of *L. monocytogenes* with a limit of detection (LOD) of 10^3 CFU mL^{-1}.

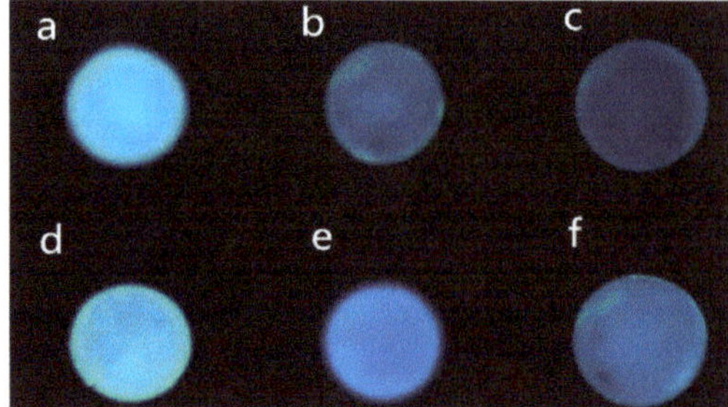

Figure 7. Images of MIPs (for **a–c**, the concentrations of bacteria were 0, 10^3, and 10^5 CFU mL^{-1}) and NIPs (for **d–f**, the concentrations of bacteria were 0, 10^3, and 10^5 CFU mL^{-1}) adsorbing different concentrations of bacteria.

3.5. Analysis in Real Samples

Milk and pork samples were purchased from a local supermarket and were verified to be free of *L. monocytogenes* according to the National Standard GB/4789.30-2016. Thus, to evaluate the binding performance of the MIP beads in real samples, the milk and pork samples were analyzed by spiking them with two levels (10^3 CFU mL^{-1} and 10^5 CFU mL^{-1}). Figure 8 shows MIP beads irradiated under an ultraviolet lamp. Due to matrix effects, the original MIPs exhibited bright blue-green fluorescence. The color of the MIP beads clearly became much darker with increasing concentrations of *L. monocytogenes*, indicating that this method could be applied for the rapid detection of *L. monocytogenes* in real samples.

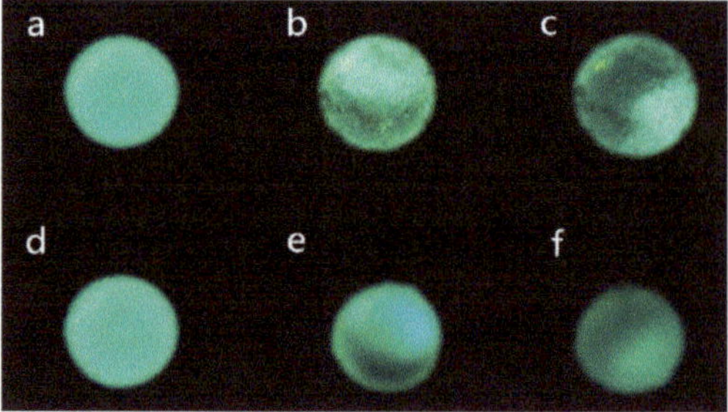

Figure 8. Images of MIPs in (**a–c**) milk and (**d–f**) pork. (for **a–c** and **d–f**, the concentrations of bacteria were 0, 10^3, and 10^5 CFU mL^{-1}).

4. Conclusions

In this work, novel fluorescence imprinted polymers were synthesized via an oil-in-water Pickering emulsion polymerization, in which the whole bacteria was used as the template, and a bacteria-NAC-QD complex was used as the stabilization particle at the oil-water interface. The MIPs provided a fast mass transfer rate and excellent specific adsorption for *L. monocytogenes*. In addition, the obtained MIPs offered sensitive and visual detection of *L. monocytogenes* in milk and pork samples based on changes in their fluorescence color under UV lamp excitation. We believe this strategy will promote the development of fluorescence bacteria imprinted technology and have the potential to be extended to the analysis of other foodborne pathogens.

Author Contributions: X.Z., Y.C.; investigation, Y.C.; writing—original draft preparation, X.Z.; writing—review and editing, J.W. (Junying Wang); supervision, J.W. (Junying Wang), J.W. (Junping Wang); funding acquisition.

Funding: This research was funded by the National Key R&D Program of China (grant no. 2016YFD0401202) and the Special Project of the Tianjin Innovation Platform (grant no. 17PTGCCX00230).

Acknowledgments: The authors are grateful for language help provided by LetPub.

Conflicts of Interest: The authors declare no conflict of interest.

References

1. Zhang, L.S.; Huang, R.; Liu, W.P.; Liu, H.X.; Zhou, X.M.; Xing, D. Rapid and visual detection of Listeria monocytogenes based on nanoparticle cluster catalyzed signal amplification. *Biosens. Bioelectron.* **2016**, *86*, 1–7. [CrossRef] [PubMed]

2. Bu, T.; Huang, Q.; Yan, L.Z.; Zhang, W.T.; Dou, L.N.; Huang, L.J.; Yang, Q.F.; Zhao, B.X.; Yang, B.W.; Li, T.; et al. Applicability of biological dye tracer in strip biosensor for ultrasensitive detection of pathogenic bacteria. *Food Chem.* **2019**, *274*, 816–821. [CrossRef] [PubMed]

3. Chiriaco, M.S.; Parlangeli, I.; Sirsi, F.; Poltronieri, P.; Primiceri, E. Impedance Sensing Platform for Detection of the Food Pathogen Listeria monocytogenes. *Electronics* **2018**, *7*, 11. [CrossRef]

4. Zhao, X.; Lin, C.W.; Wang, J.; Oh, D.H. Advances in Rapid Detection Methods for Foodborne Pathogens. *J. Microbiol. Biotechnol.* **2014**, *24*, 297–312. [CrossRef] [PubMed]

5. Alhogail, S.; Suaifan, G.; Zourob, M. Rapid colorimetric sensing platform for the detection of Listeria monocytogenes foodborne pathogen. *Biosens. Bioelectron.* **2016**, *86*, 1061–1066. [CrossRef] [PubMed]

6. Labrador, M.; Rota, M.C.; Perez-Arquillue, C.; Herrera, A.; Bayarri, S. Comparative evaluation of impedanciometry combined with chromogenic agars or RNA hybridization and real-time PCR methods for the detection of L. monocytogenes in dry-cured ham. *Food Control* **2018**, *94*, 108–115. [CrossRef]

7. Lee, S.; Rakic-Martinez, M.; Graves, L.M.; Ward, T.J.; Siletzky, R.M.; Kathariou, S. Genetic Determinants for Cadmium and Arsenic Resistance among Listeria monocytogenes Serotype 4b Isolates from Sporadic Human Listeriosis Patients. *Appl. Environ. Microbiol.* **2013**, *79*, 2471–2476. [CrossRef] [PubMed]

8. Velusamy, V.; Arshak, K.; Korostynska, O.; Oliwa, K.; Adley, C. An overview of foodborne pathogen detection: In the perspective of biosensors. *Biotechnol. Adv.* **2010**, *28*, 232–254. [CrossRef]

9. Callejon, R.M.; Rodriguez-Naranjo, M.I.; Ubeda, C.; Hornedo-Ortega, R.; Garcia-Parrilla, M.C.; Troncoso, A.M. Reported Foodborne Outbreaks Due to Fresh Produce in the United States and European Union: Trends and Causes. *Foodborne Pathog. Dis.* **2015**, *12*, 32–38. [CrossRef]

10. Song, S.X.; Wang, X.Y.; Xu, K.; Xia, G.M.; Yang, X.B. Visualized Detection of Vibrio parahaemolyticus in Food Samples Using Dual-Functional Aptamers and Cut-Assisted Rolling Circle Amplification. *J. Agric. Food Chem.* **2019**, *67*, 1244–1253. [CrossRef]

11. Chen, J.; Park, B. Label-free screening of foodborne Salmonella using surface plasmon resonance imaging. *Anal. Bioanal. Chem.* **2018**, *410*, 5455–5464. [CrossRef] [PubMed]

12. Zhu, L.J.; He, J.; Cao, X.H.; Huang, K.L.; Luo, Y.B.; Xu, W.T. Development of a double-antibody sandwich ELISA for rapid detection of Bacillus Cereus in food. *Sci. Rep.* **2016**, *6*, 10. [CrossRef] [PubMed]

13. Zhou, B.Q.; Liang, T.B.; Zhan, Z.X.; Liu, R.; Li, F.; Xu, H.Y. Rapid and simultaneous quantification of viable Escherichia coli 0157:H7 and Salmonella spp. in milk through multiplex real-time PCR. *J. Dairy Sci.* **2017**, *100*, 8804–8813. [CrossRef] [PubMed]

14. Mazumdar, S.D.; Barlen, B.; Kampfer, P.; Keusgen, M. Surface plasmon resonance (SPR) as a rapid tool for serotyping of Salmonella. *Biosens. Bioelectron.* **2010**, *25*, 967–971. [CrossRef]

15. Liu, Y.S.; Zhao, C.; Song, X.L.; Xu, K.; Wang, J.; Li, J. Colorimetric immunoassay for rapid detection of Vibrio parahaemolyticus. *Microchim. Acta* **2017**, *184*, 4785–4792. [CrossRef]

16. Kempe, H.; Pujolras, A.P.; Kempe, M. Molecularly Imprinted Polymer Nanocarriers for Sustained Release of Erythromycin. *Pharm. Res.* **2015**, *32*, 375–388. [CrossRef]

17. Zhao, X.; Wang, J.; Wang, J.; Wang, S. Development of water-compatible molecularly imprinted solid-phase extraction coupled with high performance liquid chromatography-tandem mass spectrometry for the detection of six sulfonamides in animal-derived foods. *J. Chromatogr. A* **2018**, *1574*, 9–17. [CrossRef]

18. Han, S.; Su, L.Q.; Zhai, M.H.; Ma, L.; Liu, S.W.; Teng, Y. A molecularly imprinted composite based on graphene oxide for targeted drug delivery to tumor cells. *J. Mater. Sci.* **2019**, *54*, 3331–3341. [CrossRef]

19. Wang, R.Y.; Pan, J.P.; Qin, M.; Guo, T.Y. Molecularly imprinted nanocapsule mimicking phosphotriesterase for the catalytic hydrolysis of organophosphorus pesticides. *Eur. Polym. J.* **2019**, *110*, 1–8. [CrossRef]

20. Chen, L.X.; Xu, S.F.; Li, J.H. Recent advances in molecular imprinting technology: Current status, challenges and highlighted applications. *Chem. Soc. Rev.* **2011**, *40*, 2922–2942. [CrossRef]

21. Shen, X.; Ye, L. Interfacial Molecular Imprinting in Nanoparticle-Stabilized Emulsions. *Macromolecules* **2011**, *44*, 5631–5637. [CrossRef] [PubMed]

22. Shen, X.; Zhou, T.; Ye, L. Molecular imprinting of protein in Pickering emulsion. *Chem. Commun.* **2012**, *48*, 8198–8200. [CrossRef] [PubMed]

23. Shen, X.T.; Bonde, J.S.; Kamra, T.; Bulow, L.; Leo, J.C.; Linke, D.; Ye, L. Bacterial Imprinting at Pickering Emulsion Interfaces. *Angew. Chem. Int. Ed.* **2014**, *53*, 10687–10690. [CrossRef]

24. Pan, J.M.; Qu, Q.; Cao, J.; Yan, D.; Liu, J.X.; Dai, X.H.; Yan, Y.S. Molecularly imprinted polymer foams with well-defined open-cell structure derived from Pickering HIPEs and their enhanced recognition of lambda-cyhalothrin. *Chem. Eng. J.* **2014**, *253*, 138–147. [CrossRef]

25. Gan, M.Y.; Pan, J.M.; Zhang, Y.L.; Dai, X.H.; Yin, Y.J.; Qu, Q.; Yan, Y.S. Molecularly imprinted polymers derived from lignin-based Pickering emulsions and their selectively adsorption of lambda-cyhalothrin. *Chem. Eng. J.* **2014**, *257*, 317–327. [CrossRef]

26. Zhu, W.J.; Ma, W.; Li, C.X.; Pan, J.M.; Dai, X.H. Well-designed multihollow magnetic imprinted microspheres based on cellulose nanocrystals (CNCs) stabilized Pickering double emulsion polymerization for selective adsorption of bifenthrin. *Chem. Eng. J.* **2015**, *276*, 249–260. [CrossRef]

27. Liang, W.X.; Hu, H.W.; Guo, P.R.; Ma, Y.F.; Li, P.Y.; Zheng, W.R.; Zhang, M. Combining Pickering Emulsion Polymerization with Molecular Imprinting to Prepare Polymer Microspheres for Selective Solid-Phase Extraction of Malachite Green. *Polymers* **2017**, *9*, 17. [CrossRef] [PubMed]

28. Sun, H.; Li, Y.; Yang, J.J.; Sun, X.L.; Huang, C.N.; Zhang, X.D.; Chen, J.P. Preparation of dummy-imprinted polymers by Pickering emulsion polymerization for the selective determination of seven bisphenols from sediment samples. *J. Sep. Sci.* **2016**, *39*, 2188–2195. [CrossRef]

29. Amjadi, M.; Jalili, R. Molecularly imprinted mesoporous silica embedded with carbon dots and semiconductor quantum dots as a ratiometric fluorescent sensor for diniconazole. *Biosens. Bioelectron.* **2017**, *96*, 121–126. [CrossRef] [PubMed]

30. Wang, X.J.; Ding, H.; Yu, X.R.; Shi, X.Z.; Sun, A.L.; Li, D.X.; Zhao, J. Characterization and application of molecularly imprinted polymer-coated quantum dots for sensitive fluorescent determination of diethylstilbestrol in water samples. *Talanta* **2019**, *197*, 98–104. [CrossRef]

31. Feng, J.W.; Tao, Y.; Shen, X.L.; Jin, H.; Zhou, T.T.; Zhou, Y.S.; Hu, L.Q.; Luo, D.; Mei, S.R.; Lee, Y.I. Highly sensitive and selective fluorescent sensor for tetrabromobisphenol-A in electronic waste samples using molecularly imprinted polymer coated quantum dots. *Microchem. J.* **2019**, *144*, 93–101. [CrossRef]

32. Huang, S.Y.; Guo, M.L.; Tan, J.A.; Geng, Y.Y.; Wu, J.Y.; Tang, Y.W.; Su, C.C.; Lin, C.C.; Liang, Y. Novel Fluorescence Sensor Based on All-Inorganic Perovskite Quantum Dots Coated with Molecularly Imprinted Polymers for Highly Selective and Sensitive Detection of Omethoate. *ACS Appl. Mater. Interfaces* **2018**, *10*, 39056–39063. [CrossRef] [PubMed]

33. Liu, H.B.; Du, X.J.; Zang, Y.X.; Li, P.; Wang, S. SERS-Based Lateral Flow Strip Biosensor for Simultaneous Detection of Listeria monocytogenes and Salmonella enterica Serotype Enteritidis. *J. Agric. Food Chem.* **2017**, *65*, 10290–10299. [CrossRef] [PubMed]

34. Wu, S.L.; Dou, J.; Zhang, J.; Zhang, S.F. A simple and economical one-pot method to synthesize high-quality water soluble CdTe QDs. *J. Mater. Chem.* **2012**, *22*, 14573–14578. [CrossRef]

35. Wongkongkatep, P.; Manopwisedjaroen, K.; Tiposoth, P.; Archakunakorn, S.; Pongtharangkul, T.; Suphantharika, M.; Honda, K.; Hamachi, I.; Wongkongkatep, J. Bacteria Interface Pickering Emulsions Stabilized by Self-assembled Bacteria-Chitosan Network. *Langmuir* **2012**, *28*, 5729–5736. [CrossRef] [PubMed]

36. Pan, J.M.; Yin, Y.J.; Gan, M.Y.; Meng, M.J.; Dai, X.H.; Wu, R.R.; Shi, W.D.; Yan, Y.S. Fabrication and evaluation of molecularly imprinted multi-hollow microspheres adsorbents with tunable inner pore structures derived from templating Pickering double emulsions. *Chem. Eng. J.* **2015**, *266*, 299–308. [CrossRef]

37. Liu, Y.; Hu, X.; Meng, M.J.; Liu, Z.C.; Ni, L.; Meng, X.G.; Qiu, J. RAFT-mediated microemulsion polymerization to synthesize a novel high-performance graphene oxide-based cadmium imprinted polymer. *Chem. Eng. J.* **2016**, *302*, 609–618. [CrossRef]

38. Liang, X.Y.; Liu, F.; Wan, Y.Q.; Yin, X.Y.; Liu, W.T. Facile synthesis of molecularly imprinted polymers for selective extraction of tyrosine metabolites in human urine. *J. Chromatogr. A* **2019**, *1587*, 34–41. [CrossRef]

39. Martinez, V.M.; Arbeloa, F.L.; Prieto, J.B.; Arbeloa, I.L. Characterization of rhodamine 6G aggregates intercalated in solid thin films of laponite clay. 2-Fluorescence spectroscopy. *J. Phys. Chem. B* **2005**, *109*, 7443–7450. [CrossRef]
40. Kim, Y.S.; Lee, Y.; Kim, Y.; Kim, D.; Choi, H.S.; Park, J.C.; Nam, Y.S.; Jeon, D.Y. Synthesis of efficient near-infrared-emitting CuInS2/ZnS quantum dots by inhibiting cation-exchange for bio application. *RSC Adv.* **2017**, *7*, 10675–10682. [CrossRef]

Article

Highly Selective Fluorescence Sensing and Imaging of ATP Using a Boronic Acid Groups-Bearing Polythiophene Derivate

Lihua Liu [1], Linlin Zhao [1,2,*], Dandan Cheng [1], Xinyi Yao [1] and Yan Lu [1,3,*]

[1] School of Materials Science & Engineering, Tianjin University of Technology, Tianjin 300384, China
[2] Key Laboratory of Display Materials & Photoelectric Devices, Ministry of Education, Tianjin University of Technology, Tianjin 300384, China
[3] Tianjin Key Laboratory for Photoelectric Materials and Devices, Tianjin University of Technology, Tianjin 300384, China
* Correspondence: luyan@tjut.edu.cn (L.Z.); luxingzhao@hotmail.com (Y.L.)

Received: 8 June 2019; Accepted: 1 July 2019; Published: 3 July 2019

Abstract: A boronic acid groups-bearing polythiophene derivate (**L**) was designed and synthesized for highly sensitive fluorescence detection of ATP based on a multisite-binding coupled with analyte-induced aggregation strategy. **L** has a polythiophene backbone as fluorophores and two functional side groups, i.e., quaternary ammonium group and boronic acid group, as multibinding sites for ATP. When various structural analogues such as ADP, AMP, and various inorganic phosphates were added into the aqueous solution of **L**, only ATP caused a remarkable fluorescence quenching of about 60-fold accompanied by obvious color changes of solution from yellow to purple. The detection limit is estimated to be 2 nM based on 3σ/slope. With the advantage of good water solubility, low toxicity, and highly selective response to ATP, **L** was successfully utilized as a probe to real-time assay activity of adenylate kinase (ADK) and map fluorescent imaging of ATP in living cells.

Keywords: ATP detection; fluorescent probe; polythiophene; multisite binding; analyte-induced aggregation

1. Introduction

Nucleotide adenosine triphosphate (ATP) is one of the critical biological anions in all living organisms. It serves as a universal energy source in various cellular events and plays significant roles in many biological processes including biosynthesis, DNA replication, cell signaling, and so on [1–5]. The intracellular ATP levels have been demonstrated to be closely relevant to pathogenesis of many diseases including Parkinson's disease, ischemia, and hypoglycemia, etc. [5–8]. Therefore, sensitive and selective detection of ATP is highly desired for biochemical research and clinical diagnosis.

In recent years, fluorescent probes are considered as powerful tools for the visible detection and identification of biological substances owing to the good selectivity, high sensitivity, real-time analysis capability, and high temporal and spatial resolution [9–14]. Thanks to the strong light-harvesting and signal-amplification properties of conjugated polymers (CPs), water-soluble CPs-based probes have been demonstrated to be efficient probes for detection of diverse bio-related species such as nucleotides, DNA, proteins, protease, etc. [15–17]. Among various CPs, polythiophene (PT) derivatives have received extensive attention because of their sensitive chain conformation transition responsive to external stimuli [7,8,18–21], providing unique advantages over other CPs on keeping the balance between simplicity (colorimetric mode) and sensitivity (fluorescence mode). In previous reports including ours, several PTs derivatives have been designed for the fluorescent detection of ATP with moderate selectivity and sensitivity [19,22–25]. These reported PT-based ATP probes contain generally cationic groups on their side chains, for example, quaternary ammonium [18,19,25,26], imidazolium [22], or phosphonium [23], to offer essential water solubility and recognition site for ATP. Thus, in most cases, electrostatic interaction is the primary driving force for the binding of polymer probes with ATP, resulting in these assays being sensitive to ionic strength. This will prevent practical application of these probes in more complex environments like cells and even living body. On the other hand, the selectivity of the reported ATP probe was not satisfying since the electrostatic attraction was nonspecific. To address these challenges, herein, we propose a multisite-binding coupled with analyte-induced aggregation strategy to develop PT-based probes for highly selective ATP detection and imaging.

It is well known that boronic acids can bind vicinal diols with high affinities via reversible boronate formation in weak-base aqueous solution, which has been widely used to develop synthetic receptors for saccharides detection [6,26–29]. In the present contribution, both boronic acid groups and quaternary ammonium groups were introduced to the side chain of polythiophene as binding sites of ribose and phosphate of ATP, respectively, to produce a new ATP probe (**L**) with strong and selective affinity for ATP (Scheme 1a). **L** and ATP come together easily by the electrostatic attraction, at the same time, the neighboring boronic acid group on the side chains of **L** further anchors ATP via formation of pentacyclic borate to restrict the rotation of **L** backbone and thus, promote polymer aggregation through the interchain π–π stacking interaction (Scheme 1b). This process will lead to the significant fluorescence quenching of **L**. Based on the sensing strategy, **L** can distinguish ATP from its analogues ADP and AMP as well as various inorganic phosphates, etc. Taking advantage of high selectivity of **L** to ATP, the activity assay of ATP-related enzyme in buffer solution and the fluorescence imaging of ATP in living cells were successfully achieved.

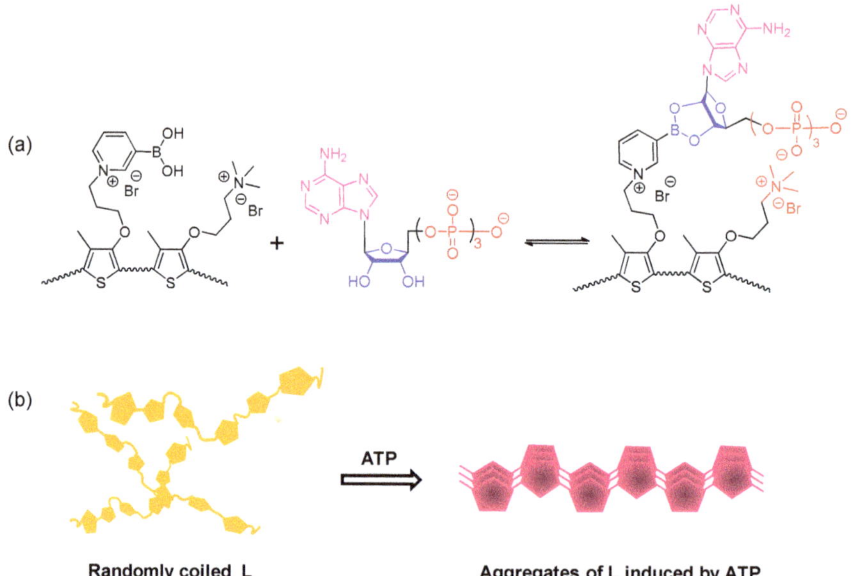

Scheme 1. (**a**) The possible binding pattern of **L** with ATP; (**b**) Conformation transition and the formation of aggregates of **L** induced by ATP.

2. Experimental

2.1. Materials

Adenosine 5′-riphosphate (ATP), adenosine 5′-diphosphate (ADP), adenosine 5′-nophosphate (AMP), and adenylate kinase (ADK) were provided by Aladdin (Shanghai, China).

3-(4,5-Dimethylthiazol-2-yl)-2,5-diphenyl-tetrazolium bromide (MTT) was purchased from Sigma-Aldrich (St. Louis, MO, USA). 2-Amino-2-hydroxymethyl-1,3-propanediol hydrochloride (*tris*-HCl) was bought from Alfa Aesar (Tianjin, China). The ultrapure water in this study was obtained by a Millipore filtration system (Millipore, Darmstadt, Germany). All other chemicals were used as received without further purification. The polymer concentration was calculated on the basis of the repeat unit.

2.2. Instruments

^{1}H NMR spectra were performed on a Bruker Avance III-400 NMR spectrometer (400 MHz, Bruker, Fällanden, Switzerland) using tetramethylsilane (TMS) as an internal standard. Gel permeation chromatography (GPC) trace was obtained with a Waters 1515 liquid chromatography system (Waters, MA, USA) equipped with a Waters 2414 refractive index detector (Waters, MA, USA) and Phenogel GPC columns (Waters, MA, USA) using pullulan as the standard and H_2O as the eluent at a flow rate of 1.0 mL min^{-1} at 35 °C. FTIR spectra were recorded with a ALPHA II FTIR spectrometer (Bruker, Rheinstetten, Germany). Circular dichroism (CD) spectra were collected on a Jasco J-715 WI spectropolarimeter (Jasco, Tokyo, Japan) operating at room temperature. UV-Vis absorption spectra were measured on a Shimadzu UV-2550 spectrometer (Shimadzu, Tokyo, Japan). Fluorescent spectra were recorded with a Hitachi F-4600 fluorescence spectrophotometer (Hitachi, Chiyoda ku, Japan) equipped with a xenon lamp excitation source. Fluorescence quantum yields were achieved by comparison with fluorescien in water as standard. The pH values were determined with a Mettler-Toledo Delta 320 pH meter (Mettler-Toledo, Greifensee, Switzerland).

2.3. Synthesis

2.3.1. Synthesis of 3-(3-Bromopropoxy)-4-methylthiophene (1)

The monomer **1** was synthesized according the procedure described in the previous work reported by our group [18]. Yield 81%; ^{1}H NMR (400 MHz, CDCl$_3$) δ 6.72 (s, 1H), 6.08 (s, 1H), 3.96 (m, 2H), 3.48 (m, 2H), 2.23 (m, 2H), 1.99 (s, 3H).

2.3.2. Synthesis of 3-(4-Methyl-3′-thienyloxy)propyltrimethylammonium Bromide (2)

Trimethylamine (10 mL) was added into a solution of **1** (2.68 g, 11.4 mmol) in THF (100 mL). The mixture was stirred 48 h at room temperature, and then evaporated to dryness. The crude product was washed with THF to give **2** (2.62 g, 78% yield). ^{1}H NMR (400 MHz, d_6-DMSO) δ 7.01 (s, 1H), 6.48 (s, 1H), 4.15 (m, 2H), 3.51 (m, 2H), 3.14 (s, 9H), 2.30 (m, 2H), 2.06 (s, 3H).

2.3.3. Synthesis of Copolymer (3)

Copolymer **3** was synthesized according the procedure as following: anhydrous FeCl$_3$ (649 mg, 4 mmol) was added to the freshly dry CHCl$_3$ (20 mL), and the mixture was stirred for 30 min at room temperature under N$_2$ atmosphere. Then, the solution of **1** (235 mg, 1 mmol) and **2** (294 mg, 1 mmol) dissolved in dry CHCl$_3$ (10 mL) was added dropwise to the above as-prepared FeCl$_3$ solution. The mixture was stirred for 48 h at 35 °C. The reaction mixture was then concentrated to about 2 mL. The residue was precipitated by addition of MeOH (200 mL). The precipitate was collected by filtration, and the resulting crude product was extracted by Soxhlet extraction with MeOH for 24 h to remove possibly residual FeCl$_3$. The residual solid was filtrated and dried under reduced pressure to give copolymer **3** (331.6 mg, 63%). GPC (H$_2$O, pullulan standard): M_n: 6.3 kDa, PDI: 1.819.

2.3.4. Synthesis of Copolymer **L**

Copolymer **3** (200 mg) was dissolved in a mixed solvent of DMF (30 mL) and THF (20 mL). 3-Pyridineboronic acid (300 mg, 2.44 mmol) was then added into the above solution. The mixture was stirred at 70 °C under a N$_2$ atmosphere for 48 h. The mixed solution was concentrated to about 2 mL, and the residue was then dropwise added into THF (60 mL). The resultant precipitate was collected by filtration, washed with THF (50 mL × 3), and dried under vacuum at room temperature to obtain polymer **L** (211 mg, 86% yield) as a dark-red solid.

2.4. Sample Preparation for Spectroscopic Analysis

All spectroscopic experiments were carried out at 25 °C. The stock solution of the probe (**L** = 2.0 × 10^{-3} M) in water was diluted to 50 μM for UV-Vis absorption spectrum and 10 μM for photoluminescence spectrum.

2.5. Cell Culture and Cytotoxicity Assays

HeLa cells were cultured in DMEM supplemented with 10% FBS and 1% penicillin–streptomycin at 37 °C in a humidified 5% CO_2–95% air atmosphere.

The cytotoxicity of **L** was evaluated by typical MTT assay. HeLa cells (1 × 10^4 cells/well) were seeded onto 96-well plates in 100 μL DMEM and allowed to attach for 24 h. After cell attachment, the medium was removed and the cells were washed with PBS three times, then 100 μL of fresh medium containing **L** with different concentrations ranging from 0 to 50 μM was added and incubated for 24 h. After 24 h incubation, the cell viability was evaluated by MTT assay.

2.6. Fluorescence Imaging

HeLa cells (1 × 10^5 cells/well) were seeded onto petri dish in 2 mL DMEM and allowed to attach for 24 h. After cell attachment, the medium was replaced with 2 mL of fresh medium containing **L** (10 μM, 2 mL) for 2 h, then the medium was removed and the cells were washed with PBS three times. Fluorescence images were recorded with a confocal laser scanning microscope (CLSM) (Olympus FV1000-IX81, Tokyo, Japan). The excitation wavelength used was 405 nm, and all images were analyzed with Olympus FV1000-ASW.

3. Results and Discussion

3.1. Synthesis of L

The synthetic route for the probe **L** is outlined in Scheme 2. The traditional chemical oxidative polymerization with $FeCl_3$ in chloroform in the presence of equivalent **1** and **2** gave the random copolymer product **3** with a moderate number-average molecular weight of 6.3 kDa, determined by GPC using pullulan as the standard and H_2O as the eluent, respectively. Subsequently, 3-pyridineboronic acid groups were linked to the side chain of **3** by nucleophilic substitution reaction to produce the target **L**. The strong absorption peaks at 1,360 and 1,485 cm^{-1} in FTIR of **L** were assigned to the characteristic symmetric and unsymmetric stretching vibration of the B–O bond, respectively (Figure S1, Supplementary Materials), suggesting that the boronic acid groups were successfully introduced into the side chain of **L**. The ratio of the three repeated units in **L** was determined to be 0.17:0.83:1 for Br, boronic acid and quaternary ammonium contained ones, respectively, by inductively coupled plasma (ICP) analysis. In other words, the molar content of boronic acid-containing moieties in the total polymer was about 41.5%.

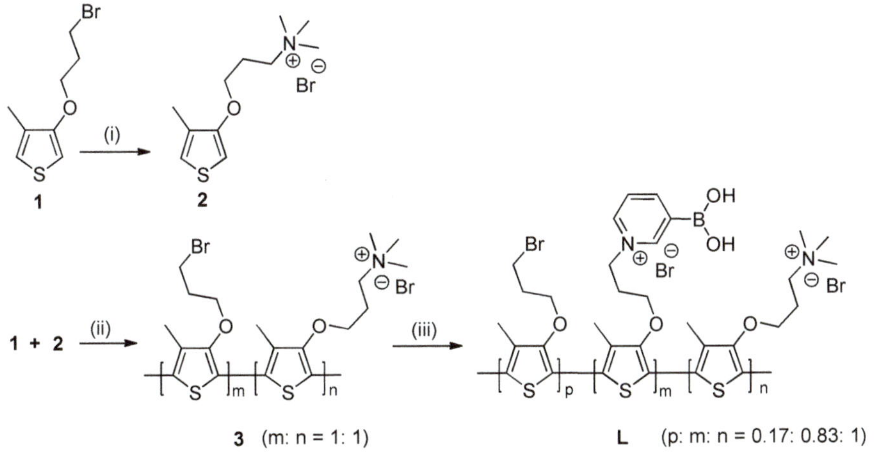

Scheme 2. Synthesis of the probe L. (i) $N(CH_3)_3$, THF, rt, 48 h; (ii) $FeCl_3$, $CHCl_3$, 35 °C, 48 h; (iii) 3-Pyridineboronic acid, DMF/THF (3:2 *v/v*), N_2, 70 °C, 48 h.

3.2. Screen of pH of Assay System

Since pH values of the aqueous solution influences strongly the formation of ester between boronic acids and vicinal diols [6,26], the pH conditions for ATP sensing using **L** as a probe were first optimized. The pyridinium hydroxyboronate moiety in our probe **L** has a very low pKa value of about 4.0 [30]. Thus, the weakly alkaline conditions with pH values ranging from 7.0–10.0 were carefully assessed. As shown in Figure S2a (Supplementary Materials), the changes on absorption spectra of **L** (50 μM) in water were almost ignored within the total tested pH ranges. Upon addition of 1 equiv. of ATP to **L** aqueous solution (50 μM) with different pH values, the absorption spectra of solutions at pH = 7.4, 8.0, and 9.0 gave the significant responses (Figure S2b, Supplementary Materials). Based on these results, the aqueous solution of physiological pH 7.4 was chosen as assay system for the evaluation of sensing properties of **L**.

3.3. Selectivity

Selectivity of the probes is one of the key parameters for evaluating their performances. We first checked the response of **L** to a wide range of the biologically relevant anions including Cl^-, PO_4^{3-}, HPO_4^{2-}, $H_2PO_4^-$, $P_2O_7^{2-}$, as well as ATP and its analogues ADP and AMP by absorption and emission spectra. As shown in Figure 1, after addition of 1 equiv. of these anions to aqueous solution of **L**, most of the solutions remained yellow with slight changes of λ_{max}, except for that containing ADP, which gave orange solutions with the shifts of the absorption peaks from 450 to 468 nm. Compared with those anions, the most remarkable influence was observed upon adding ATP, which gave a purple solution along with the significant red-shifts of λ_{max}. The significant color change indicated that **L** can serve as a "naked-eye" indicator for ATP in aqueous solution. On the other hand, the addition of 3 equiv. of ATP resulted in almost complete quenching (about 60-fold) in photoluminescence intensity of **L** (Figure 2). However, the addition of other related anions had no obvious effect on the fluorescence emission of **L**, except that adding ADP produced a slight decrease in emission intensity. These results indicated that the probe **L** had high selectivity to ATP over other tested anions in aqueous solution.

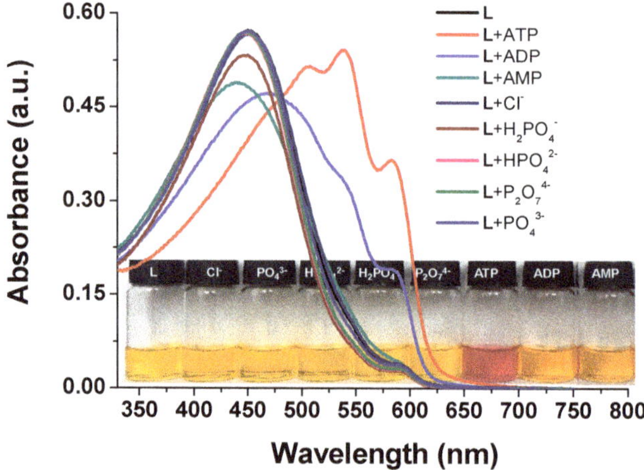

Figure 1. Absorption spectra of **L** (50 µM) upon addition of various bio-related sodium salts of anions including ATP, ADP, AMP, Cl⁻, PO_4^{3-}, HPO_4^{2-}, $H_2PO_4^-$ and $P_2O_7^{2-}$ (50 µM) in water at pH 7.4.

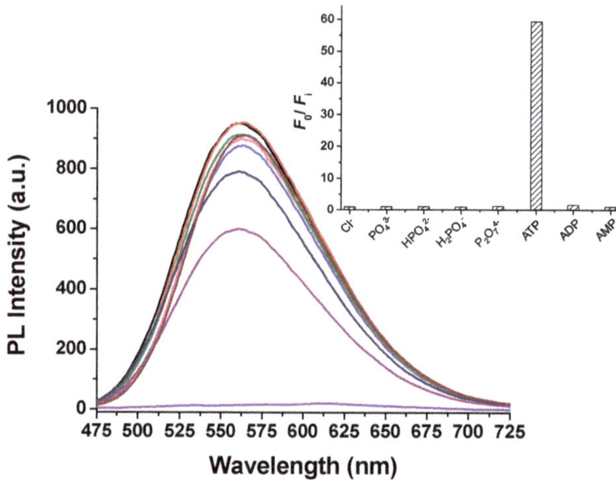

Figure 2. Fluorescent emission changes of **L** (10 µM) upon addition of various bio-related sodium salts of anions including ATP, ADP, AMP, Cl⁻, PO_4^{3-}, HPO_4^{2-}, $H_2PO_4^-$ and $P_2O_7^{2-}$ (30 µM) in water at pH 7.4.

3.4. UV-Vis Spectral Response of Probe **L** to ATP

The interaction of **L** with ATP in *tris*-HCl buffer solution (10 mM, pH = 7.4) at room temperature was carefully investigated by UV-Vis absorption spectroscopy. As displayed in Figure 3a, the *tris*-HCl buffer solution (10 mM, pH 7.4) of **L** (50 µM) exhibited a maximum absorption band at 450 nm with $\varepsilon = 1.2 \times 10^4$ M⁻¹ cm⁻¹, which is attributed to π–π^* transition of polythiophene backbone with a random-coil conformation [19,31,32]. As the concentrations of ATP increased, the λ_{max} was gradually red-shifted from 450 to 540 and 588 nm, accompanied with a distinct solution color change from yellow to purple. The significant shift and the appearance of two vibronic bands are characteristic of the aggregation of the polythiophene backbone [19]. Job's plot (Figure S3, Supplementary Materials) indicated a near 1:1 stoichiometry of complex formed between **L** and ATP. The binding constant of **L** with ATP was estimated to be about 4.02×10^5 M⁻¹ with R^2 of 0.995 based on the double-reciprocal plot of $1/(A - A_0)$ versus $1/(ATP)$, as shown in Figure 3b [33–35]. Further, the control homopolymer **PTB,** which just contains boronic acid groups on its side chains, exhibited the binding constant of 2.68×10^4 M^{-1} (Figure S4, Supplementary Materials). In summary, comparing with other PT-based ATP probes which only contained quaternary ammonium groups [18,24] or boronic acid groups on their side chains, the relatively larger binding constant of **L** with ATP implied stronger affinity between the both, probably due to synergistic interactions.

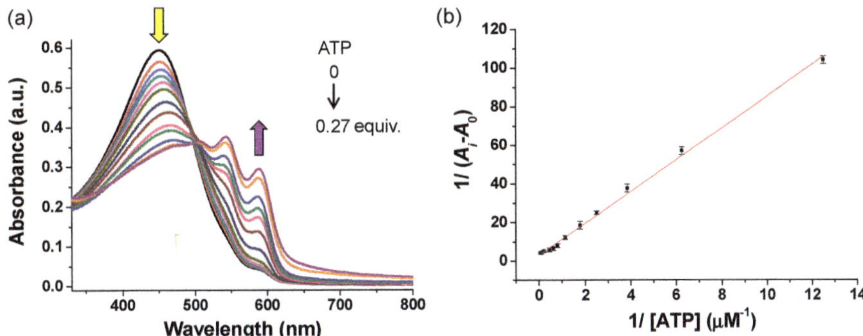

Figure 3. (a) UV-Vis absorption spectra of **L** (50 µM) with the addition of ATP with concentrations ranging from 0 to 13.6 µM in *tris*-HCl buffer solution (10 mM, pH = 7.4); (b) Absorbance at 540 nm of **L** as a function of ATP concentration. A_0 is the initial absorbance of **L** at 540 nm, and A_i is the recorded absorbance of **L** in the presence of ATP with different concentrations. Error bars represent the standard deviations of three trials.

Further, we confirmed the formation of **L** supermolecular aggregates induced by ATP by circular dichroism (CD) measurements. As shown in Figure 4, both **L** and ATP are optically inactive, and no CD pattern in the π–π^* transition region can be detected. These results indicate that free **L** adopts an achiral random-coiled conformation in water. However, upon binding with ATP, an intense split-type induced CD (ICD) in the π–π^* transition region appeared, and as the ATP concentration increased, the ICD intensity increased gradually, suggesting the formation of chiral π-stacked aggregates of **L** in the presence of ATP [36].

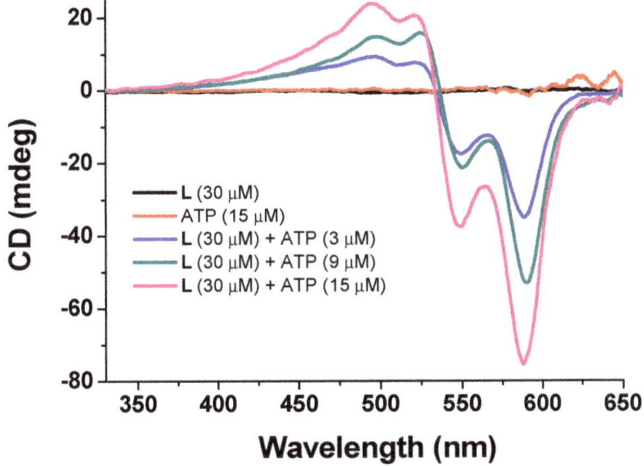

Figure 4. Circular dichroism (CD) spectra of ATP (15 µM) and **L** (30 µM) in the absence and the presence of various amounts of ATP in water (pH 7.4).

3.5. Fluorescence Spectral Response of Probe *L* to ATP

Figure 5a displays the changes of fluorescent spectra of polymer **L** (10 μM) in the presence of different amounts of ATP in *tris*-HCl buffer solution (10 mM, pH 7.4). The fluorescence spectra of **L** exhibited a strong fluorescence at 570 nm upon excitation at 450 nm ($\varphi = 0.304$). The titration of ATP into **L** buffer solution caused that the fluorescent intensities of the emission band centered at 570 nm decreased gradually with the increasing concentrations of ATP. The significant quenching of the emission of **L** could be explained by the formation of π-stacked aggregates of **L** induced by ATP, as demonstrated by UV-Vis and CD studied, quenching light emission via other nonradiative pathways [24,25]. The quenching constant K_{SV} was determined according to the well-known Stern-Volmer equation [37]. As seen from the Stern-Volmer plot (Figure 5b), there was a linear relationship between $(F_0 - F_i)/F_0$, and the concentrations of ATP ranging from 0 to 0.8 μM and K_{SV} were estimated to 1.08×10^6 M^{-1} ($R^2 = 0.997$). The high quenching efficiency indicated the strong affinity between **L** and ATP. In addition, for our system, only 0.16 equiv. of ATP almost completely quenched the fluorescence of **L**, but in case of the PT-based ATP probe bearing only quaternary ammonium groups, about 1 equiv. of ATP was required to produce such a degree of quenching under similar detection conditions [19]. Thus, it is reasonable to believe that boronic acid groups were involved in binding of **L** with ATP, which promoted the conformational transformation and subsequent aggregation of **L**.

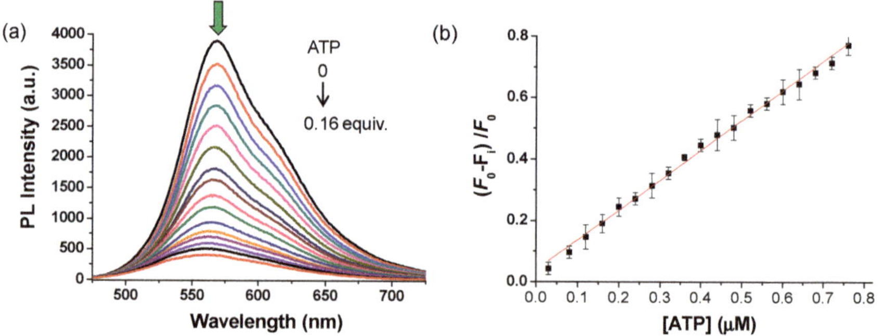

Figure 5. (a) Fluorescence spectra of **L** (10 μM) with the addition of various concentrations of ATP in *tris*-HCl buffer solution (10 mM, pH 7.4); (b) Plot of $(F_0 - F_i)/F_0$ at 570 nm vs. concentrations of ATP. F_0 is the initial emission intensity of **L** (10 μM), and F_i is the recorded emission intensities of **L** in the presence of ATP with different concentrations. Excitation wavelength: 450 nm. Error bars represent the standard deviations of three trials.

From the changes in ATP concentrations-dependent fluorescence intensities (Figure S5, Supplementary Materials), the detection limit was estimated to be 2.07×10^{-9} M based on the 3σ/slope (where σ is the standard deviation of the background and slope is the sensitivity) [24,38]. Such a low detection limit should be very competitive with most of previously reported fluorescence or colorimetric probes for ATP [18–25] (Table S1, Supplementary Materials). The high sensitivity of **L** for ATP should be closely related to the strong synergetic interactions between **L** and ATP.

In order to further confirm that cyclic borate can be formed between 3-pyridineboric acid of **L** and ribose of ATP in weakly basic aqueous solution, ^{1}H NMR spectra of 3-pyridineboric acid, ATP, and their mixture at pH 8.0 in D$_2$O were collected, as shown in Figure S6. From Figure S6 (Supplementary Materials), it is obviously seen that both the strong peak at δ 7.42 in ^{1}H NMR of 3-pyridineboronic acid (Figure S6A, Supplementary Materials) assigned as the 5-position H$_a$ and the peak at δ 6.08 in ^{1}H NMR of ATP (Figure S6B, Supplementary Materials) corresponding to H$_b$ of ribose in ATP moved towards higher field in ^{1}H NMR of 3-pyridineboronic acid/ATP mixture, along with more complex splits (Figure S6C, Supplementary Materials). It could be explained as follows: the neutral form of the boronic acid group serves as an electron-withdrawing group, but after complex with vicinal diols of ribose in ATP, it is transformed into the anionic form, which will act as an electron donor group [28]. These results indicated the formation of covalent bonds between 3-pyridineboronic acid and ATP in a weakly alkaline environment.

3.6. Assay for ADK

Since **L** can selectively recognize ATP from phosphate-containing anions, particularly ADP and AMP, **L** was applied to set up a real-time fluorescence assay for ATP-relevant enzyme activity. For instance, adenylate kinase (ADK) is an important enzyme which is involved in maintaining cellular energy homeostasis. It can catalyze the reversible reaction: ATP + AMP $\rightleftharpoons$ 2ADP. As shown in Figure 6a, **L** (50 µM) exhibited an absorption peak at about 450 nm in *tris*-HCl buffer solution (10 mM, pH 7.4) containing 0.2 mM $MgCl_2$. When the ATP (20 µM) and AMP (20 µM) were added into the above solution, the absorption band of the solution red-shifted to 535 and 578 nm along with an obvious color change (Inset, Figure 6a). After incubating with ADK ((ADK) = 100 mU/mL) for 15 minutes at 37 °C, the conversion reaction from ATP and AMP to ADP happened, resulting in that the λ_{max} returned to 450 nm and the solution color became yellow (Figure 6a). It was interesting to note that the resultant absorption spectra was almost identical with the control spectra of **L** with 40 µM of ADP, indicating the ATP and AMP can be converted to ADP thoroughly under the catalysis of ADK. These results demonstrated preliminarily that **L** can be used as a probe to monitor ADK activity in buffer solution.

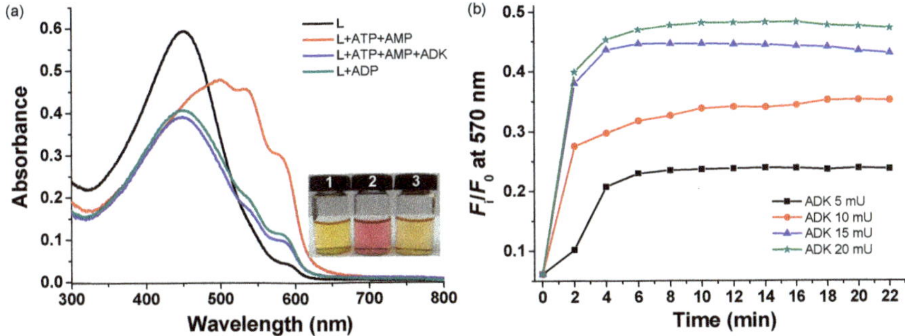

Figure 6. (a) Absorption spectra of **L** before and after incubating with ADK in buffer solution. Inset: The photographs taken from **L** (1) and L/ATP/AMP before (2) and after (3) incubating with ADK. (**L**) = 50 µM, (ATP) = (AMP) = 20 µM, (ADP) = 40 µM, (ADK) = 100 mU/mL. Incubating time was 15 min; (**b**) Time-trace plots of the conversion from ATP and AMP to ADP catalyzed by various concentration of ADK (5–20 mU/mL) monitored by the emission ratio F_i/F_0. (**L**) = 10 µM, (ATP) = (AMP) = 2 µM. Excitation wavelength: 450 nm. All measurements were performed in *tris*-HCl buffer solution (10 mM, pH 7.4) containing 0.2 mM Mg^{2+} ions.

Figure 6b shows the dependence of the fluorescence intensity ratio (F_i/F_0) at 570 nm of **L** on the concentration of ADK and incubating time. In this study, the fluorescent spectra were measured at 2 min intervals. From Figure 6b, it is seen that the reaction rate was accelerated with the increase in ADK concentration and after about 10 min, all reaction reached equilibrium. These results indicated that **L** has great potential for facile real-time monitoring of enzyme activity to elucidate their functions in biological systems.

3.7. Cell Imaging

Before applying to cell imaging, the biocompatibility of **L** was investigated by MTT assay. As shown in Figure S7 (Supplementary Materials), as the concentration of **L** increased, the average cell viability was greater than 80%, suggesting low cytotoxicity of **L** to living cells in the range of tested concentrations.

The applicability of in situ sensing ATP using **L** as a probe in living cell was further investigated by a confocal laser scanning microscope (CLSM). After incubation of HeLa cells with **L** (10 μM) for 2 h at 37 °C, the strong green fluorescence of **L** was observed in the cells (Figure S8, Supplementary Materials), demonstrating that **L** can be uptaken by HeLa cells with good biocompatibility. The photostability measurement (Figure S8, Supplementary Materials) showed almost negligible changes in the fluorescence intensities of **L** in HeLa cells after continuously irradiating at 405 nm for 100 s.

The sensing ability of **L** in HeLa cells was subsequently assessed by extracellular addition of ATP (4 μM). Figure 7 shows the representative CLSM images of **L**-loaded HeLa cells that were cocultured with ATP for various times. It was observed that the fluorescence intensity of **L** was decreased gradually with the prolonged incubation time. After about 30 min, the fluorescence of **L** was almost completely quenched. As a control, the fluorescence intensity of **L** itself in HeLa cells did not change significantly within the same test time. These results suggested that **L** had the potential to detect the fluctuation of ATP concentration in living cells.

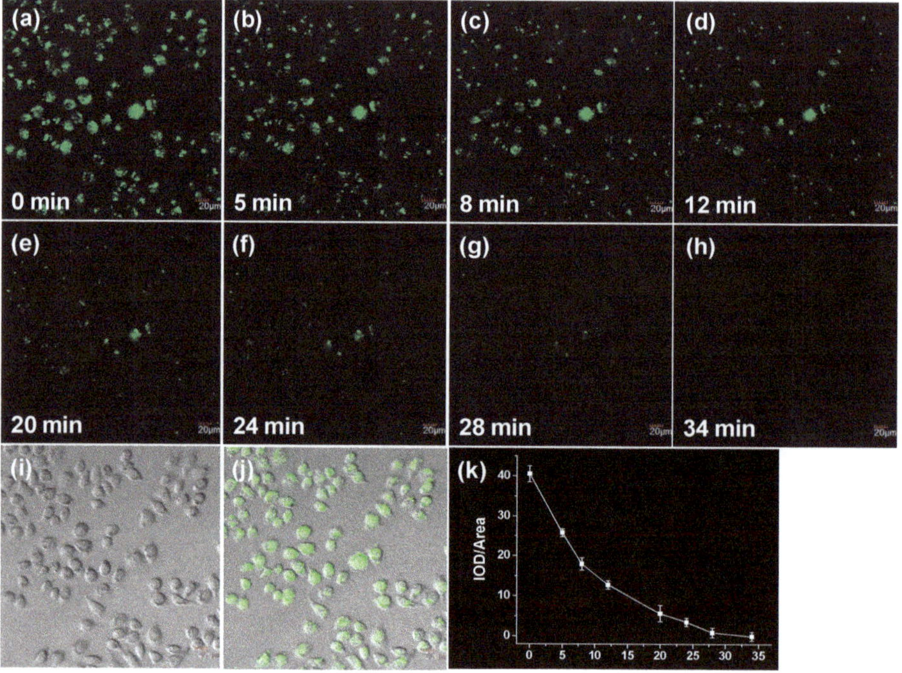

Figure 7. CLSM images of **L**-loaded HeLa cells incubated with ATP (4 μM) for various times from 0 to 34 min (**a–h**). (**i**): Brightfield; (**j**): Overlay of (**a**) and (**i**); (**k**): Dot representing the integrated optical density (IOD/area) of the probe **L** from the fluorescence images at each time point. Scale bars: 20 μm. λ_{ex} = 405 nm, λ_{em} = 564–633 nm. Error bars represent the standard deviations of three independent experiments.

4. Conclusions

A new water-soluble polythiophene-based fluorescent probe **L** with boronic acid groups on its side chains was designed and synthesized for the detection of biologically important ATP. **L** exhibited effective and selective recognition ability for ATP over other structurally similar nucleotides such as ADP and AMP in aqueous solution at physiological pH 7.4. A remarkable fluorescent quenching took place upon addition of ATP into the buffer solution of **L** due to the formation of chiral aggregates induced by ATP via multi-sites binding between **L** and ATP. The change in fluorescent intensities for ATP is much larger than that for ADP and AMP, which is large enough to discriminate ATP from ADP and AMP. Based on the highly selective recognition of **L** for ATP, **L** was successfully utilized as a fluorescent probe for continuous and facile assays of ADK activity. In addition, **L** has been demonstrated to be a useful probe for monitoring the ATP levels in living cells by fluorescence imaging. With the advantages of good water solubility, high selectivity to ATP, and low cytotoxicity, further work to inspect the practical applicability in the real-time detection of enzyme activity in living cells is being carried out in our group.

Supplementary Materials: The following are available online at http://www.mdpi.com/2073-4360/11/7/1139/s1. Scheme S1. Synthesis of the control polymer **PTB**; Figure S1. FTIR spectra of **L**; Figure S2. Effect of pH on absorption spectra of **L** and **L**/ATP complex; Figure S3. Job's plot; Figure S4. Determination of the binding constant between **PTB** and ATP; Figure S5. Determination of detection limit of **L**; Figure S6. Study on interaction of 3-pyridineboronic acid with ATP by ^{1}H NMR. Figure S7. Cytotoxicity of **L**; Figure S8. Photostability of **L**; Table S1. Comparison of cation or boronic acid-contained polythiophene-based probes for the detection of ATP.

Author Contributions: Y.L. designed the study; D.C. synthetized the probe **L**; L.L., L.Z. and X.Y. performed the measurements; all authors discussed the results and Y.L. wrote the manuscript. All authors read and approved the final version of the manuscript.

Funding: We are grateful for the National Natural Science Foundation of China (51373122, 51703163), the Natural Science Foundation of Tianjin (Grant No.: 18JCZDJC34600, 18JCYBJC86700) and the Program for Prominent Young College Teachers of Tianjin Educational Committee.

Conflicts of Interest: The authors declare no conflict of interest.

References

1. Abraham, E.H.; Okunieff, P.; Scala, S.; Vos, P.; Oosterveld, M.J.S.; Chen, A.Y.; Shrivastav, B.; Guidotti, G. Cystic fibrosis transmembrane conductance regulator and adenosine triphosphate. *Science* **1997**, *275*, 1324–1326. [CrossRef] [PubMed]

2. Newman, E.A.; Zahs, K.R. Calcium waves in retinal glial cells. *Science* **1997**, *275*, 844–847. [CrossRef] [PubMed]

3. Szewczyk, A.; Pikuła, S.A. Adenosine 5'-triphosphate: An intracellular metabolic messenger. *BBA Bioenerg.* **1998**, *1365*, 333–353. [CrossRef]

4. Eguchi, Y.; Shimizu, S.; Tsujimoto, Y. Intracellular ATP levels determine cell death fate by apoptosis or necrosis. *Cancer Res.* **1997**, *57*, 1835–1840. [PubMed]

5. Chen, J.; Liu, Y.; Ji, X.; He, Z. Target-protecting dumbbell molecular probe against exonucleases digestion for sensitive detection of ATP and streptavidin. *Biosens. Bioelectron.* **2016**, *83*, 221–228. [CrossRef] [PubMed]

6. Liu, Y.; Deng, C.; Tang, L.; Qin, A.; Hu, R.; Sun, J.Z.; Tang, B.Z. Specific detection of D-glucose by a tetraphenylethene-based fluorescent sensor. *J. Am. Chem. Soc.* **2011**, *133*, 660–663. [CrossRef] [PubMed]

7. Bush, K.T.; Keller, S.H.; Nigam, S.K. Genesis and reversal of the ischemic phenotype in epithelial cells. *J. Clin. Investig.* **2000**, *106*, 621–626. [CrossRef]

8. Dong, J.; Zhao, M. In-vivo fluorescence imaging of adenosine 5'-triphosphate. *TRAC Trend. Anal. Chem.* **2016**, *80*, 190–203. [CrossRef]

9. Quang, D.T.; Kim, J.S. Fluoro- and chromogenic chemodosimeters for heavy metal ion detection in solution and biospecimens. *Chem. Rev.* **2010**, *110*, 6280–6301. [CrossRef]

10. Zhou, Y.; Xu, Z.; Yoon, J. Fluorescent and colorimetric chemosensors for detection of nucleotides, FAD and NADH: Highlighted research during 2004–2010. *Chem. Soc. Rev.* **2011**, *40*, 2222–2235. [CrossRef]

11. Yuan, L.; Lin, W.; Zheng, K.; He, L.; Huang, W. Far-red to near infrared analyte-responsive fluorescent probes based on organic fluorophore platforms for fluorescence imaging. *Chem. Soc. Rev.* **2013**, *42*, 622–661. [CrossRef] [PubMed]

12. Liu, Z.; He, W.; Guo, Z. Metal coordination in photoluminescent sensing. *Chem. Soc. Rev.* **2013**, *42*, 1568–1600. [CrossRef] [PubMed]

13. Li, X.; Gao, X.; Shi, W.; Ma, H. Design strategies for water-soluble small molecular chromogenic and fluorogenic probes. *Chem. Rev.* **2014**, *114*, 590–659. [CrossRef] [PubMed]

14. Zhang, X.; Yin, J.; Yoon, J. Recent advances in development of chiral fluorescent and colorimetric sensors. *Chem. Rev.* **2014**, *114*, 4918–4959. [CrossRef] [PubMed]

15. Wu, W.; Bazan, G.C.; Liu, B. Conjugated-polymer-amplified sensing, imaging, and therapy. *Chem* **2017**, *2*, 760–790. [CrossRef]

16. Zhu, C.; Liu, L.; Yang, Q.; Lv, F.; Wang, S. Water-soluble conjugated polymers for imaging, diagnosis, and therapy. *Chem. Rev.* **2012**, *112*, 4687–4735. [CrossRef] [PubMed]

17. Lv, F.; Qiu, T.; Liu, L.; Ying, J.; Wang, S. Recent advances in conjugated polymer materials for disease diagnosis. *Small* **2016**, *12*, 696–705. [CrossRef] [PubMed]

18. Cheng, D.; Li, Y.; Wang, J.; Sun, Y.; Jin, L.; Li, C.; Lu, Y. Fluorescence and colorimetric detection of ATP based on a strategy of self-promoting aggregation of a water-soluble polythiophene derivative. *Chem. Commun.* **2015**, *51*, 8544–8546. [CrossRef]

19. Li, C.; Numata, M.; Takeuchi, M.; Shinkai, S. A sensitive colorimetric and fluorescent probe based on a polythiophene derivative for the detection of ATP. *Angew. Chem. Int. Ed.* **2005**, *44*, 6371–6374. [CrossRef]

20. Lin, C.; Cai, Z.; Wang, Y.; Zhu, Z.; Yang, C.J.; Chen, X. Label-free fluorescence strategy for sensitive detection of adenosine triphosphate using a loop DNA probe with low background noise. *Anal. Chem.* **2014**, *86*, 6758–6762. [CrossRef]

21. Lu, C.H.; Li, J.; Lin, M.H.; Wang, Y.W.; Yang, H.H.; Chen, X.; Chen, G.N. Amplified aptamer-based assay through catalytic recycling of the analyte. *Angew. Chem. Int. Ed.* **2010**, *49*, 8454–8457. [CrossRef] [PubMed]

22. Huang, B.H.; Geng, Z.R.; Ma, X.Y.; Zhang, C.; Zhang, Z.Y.; Wang, Z.L. Lysosomal ATP imaging in living cells by a water-soluble cationic polythiophene derivative. *Biosens. Bioelectron.* **2016**, *83*, 213–220. [CrossRef] [PubMed]

23. Huang, B.; Geng, Z.; Yan, S.; Li, Z.; Cai, J.; Wang, Z. Water-soluble conjugated polymer as a fluorescent probe for monitoring adenosine triphosphate level fluctuation in cell membranes during cell apoptosis and in vivo. *Anal. Chem.* **2017**, *89*, 8816–8821. [CrossRef] [PubMed]

24. An, N.; Zhang, Q.; Wang, J.; Liu, C.; Shi, L.; Liu, L.; Deng, L.; Lu, Y. A new FRET-based ratiometric probe for fluorescence and colorimetric analyses of adenosine 5′-triphosphate. *Polym. Chem.* **2017**, *8*, 1138–1145. [CrossRef]

25. Liu, C.; Zhang, Q.; An, N.; Wang, J.; Zhao, L.; Lu, Y. A new water-soluble polythiophene derivative as a probe for real-time monitoring adenosine 5′-triphosphatase activity in lysosome of living cells. *Talanta* **2018**, *182*, 396–404. [CrossRef] [PubMed]

26. Xue, C.; Cai, F.; Liu, H. Ultrasensitive fluorescent responses of water-soluble, zwitterionic, boronic acid-bearing, regioregular head-to-tail polythiophene to biological species. *Chem.-Eur. J.* **2008**, *14*, 1648–1653. [CrossRef]

27. Lee, S.; Lee, K.M.; Lee, M.; Yoon, J. Polydiacetylenes bearing boronic acid groups as colorimetric and fluorescence sensors for cationic surfactants. *ACS Appl. Mater. Inter.* **2013**, *5*, 4521–4526. [CrossRef]

28. DiCesare, N.; Lakowicz, J.R. Spectral properties of fluorophores combining the boronic acid group with electron donor or withdrawing groups. Implication in the development of fluorescence probes for saccharides. *J. Phys. Chem. A* **2001**, *105*, 6834–6840. [CrossRef]

29. Gao, X.; Zhang, Y.; Wang, B. New boronic acid fluorescent reporter compounds. 2. A naphthalene-based on-off sensor functional at physiological pH. *Org. Lett.* **2003**, *5*, 4615–4618. [CrossRef]

30. Mohler, L.K.; Czarnik, A.W. Ribonucleoside membrane transport by a new class of synthetic carrier. *J. Am. Chem. Soc.* **1993**, *115*, 2998–2999. [CrossRef]

31. Li, C.; Shi, G. Polythiophene-based optical sensors for small molecules. *ACS Appl. Mater. Inter.* **2013**, *5*, 4503–4510. [CrossRef] [PubMed]

32. Spangler, C.; Lang, T.; Schäferling, M. Spectral response of a methylimidazolium-functionalized polythiophene to phosphates. *Dyes Pigm.* **2012**, *95*, 194–200. [CrossRef]

33. Stephanos, J.J. Drug-protein interactions: Two-site binding of heterocyclic ligands to a monomeric hemoglobin. *J. Inorg. Biochem.* **1996**, *62*, 155–169. [CrossRef]

34. Froehlich, E.; Mandeville, J.S.; Weinert, C.M.; Kreplak, L.; Tajmir-Riahi, H.A. Bundling and aggregation of DNA by cationic dendrimers. *Biomacromolecules* **2011**, *12*, 511–517. [CrossRef] [PubMed]

35. Han, F.; Lu, Y.; Zhang, Q.; Sun, J.; Zeng, X.; Li, C. Homogeneous and sensitive DNA detection based on polyelectrolyte complexes of cationic conjugated poly(pyridinium salt)s and DNA. *J. Mater. Chem.* **2012**, *22*, 4106–4112. [CrossRef]

36. Li, C.; Numata, M.; Takeuchi, M.; Shinkai, S. Unexpected chiroptical inversion observed for supramolecular complexes formed between an achiral polythiophene and ATP. *Chem.-Asian J.* **2006**, *1*, 95–101. [CrossRef]
37. Lakowicz, J.R. Introduction to Fluorescence. In *Principles of Fluorescence Spectroscopy*, 3rd ed.; Lakowicz, J.R., Ed.; Springer: Boston, MA, USA, 2006; pp. 1–26.
38. Caballero, A.; Martínez, R.; Lloveras, V.; Ratera, I.; Vidal-Gancedo, J.; Wurst, K.; Tárraga, A.; Molina, P.; Veciana, J. Highly selective chromogenic and redox or fluorescent sensors of Hg^{2+} in aqueous environment based on 1,4-disubstituted azines. *J. Am. Chem. Soc.* **2005**, *127*, 15666–15667. [CrossRef]

Article

Oxygen-Resistant Electrochemiluminescence System with Polyhedral Oligomeric Silsesquioxane

Ryota Nakamura, Hayato Narikiyo, Masayuki Gon, Kazuo Tanaka * and Yoshiki Chujo

Department of Polymer Chemistry, Graduate School of Engineering, Kyoto University, Katsura, Nishikyo-ku, Kyoto 615-8510, Japan
* Correspondence: tanaka@poly.synchem.kyoto-u.ac.jp

Received: 19 June 2019; Accepted: 7 July 2019; Published: 10 July 2019

Abstract: We report the oxygen-resistant electrochemiluminescence (ECL) system from the polyhedral oligomeric silsesquioxane (POSS)-modified tris(2,2′-bipyridyl)ruthenium(II) complex (Ru-POSS). In electrochemical measurements, including cyclic voltammetry (CV), it is shown that electric current and ECL intensity increase in the mixture system containing Ru-POSS and tripropylamine (TPrA) on the indium tin oxide (ITO) working electrode. The lower onset potential (E_{onset}) in CV is observed with Ru-POSS compared to tris(2,2′-bipyridyl)ruthenium(II) complex (Ru(bpy)$_3{}^{2+}$). From the series of mechanistic studies, it was shown that adsorption of Ru-POSS onto the ITO electrode enhances TPrA oxidation and subsequently the efficiency of ECL with lower voltage. Moreover, oxygen quenching of ECL was suppressed, and it is proposed that the enhancement to the production of the TPrA radical could contribute to improving oxygen resistance. Finally, the ECL-based detection for water pollutant is demonstrated without the degassing treatment. The commodity system with Ru(bpy)$_3{}^{2+}$ is not applicable in the absence of degassing with the sample solutions due to critical signal suppression, meanwhile the present system based on Ru-POSS was feasible for estimating the amount of the target even under aerobic conditions by fitting the ECL intensity to the standard curve. One of critical disadvantages of ECL can be solved by the hybrid formation with POSS.

Keywords: electrochemiluminescence; POSS; ruthenium(II) complex; oxygen resistance

---

1. Introduction

Electrochemiluminescence (ECL), also called as electrogenerated chemiluminescence, is light emission from the excited species generated through multiple redox reactions on an electrode [1,2]. ECL has many advantages as a platform of sensing materials. Owing to unnecessary light excitation, background noise in the luminescence detection can be drastically reduced by removing scattering of incident light. Compared to chemiluminescence which is produced through chemical reactions, it is relatively easy to achieve temporal and spatial control for emission generation. Therefore, the ECL techniques are recognized as a powerful analytical tool and are used in various areas, such as immunoassay, genosensors, and for monitoring small molecules [1–6]. Although a variety of substances can potentially work as a source of ECL, ruthenium(II) (Ru(II)) complexes, especially tris(2,2′-bipyridyl)ruthenium(II) (Ru(bpy)$_3{}^{2+}$), are conventionally used because of many superior properties, such as high chemical stability, recyclability, redox properties, and long lifetime in the excited state [2–4]. Meanwhile, oxygen is well known to be a crucial quencher to the photo-excited state of Ru(II) complexes [7–9], followed by the reduction of the ECL intensity [10–12]. Thereby, pretreatments for deoxygenation with the analytical samples are essential before conventional ECL detection. Hence, improvement of oxygen resistance of the ECL system is strongly needed not only for improving accuracy and sensitivity in detection but also for extending applicability to future sensing technologies such as real-time monitoring with the native samples.

Polyhedral oligomeric silsesquioxane (POSS) attracted attention as an "element-block" which is the minimum unit containing heteroatoms for constructing functional materials [13,14]. We report unique and useful properties of POSS-containing hydrophilic materials [15–19]. In aqueous media, it was shown that the water-soluble POSS-containing materials can efficiently capture various types of guest molecules in the distinct hydrophobic spaces created around the POSS units and are applicable for sensors [20–24]. It is proposed that the cubic core is favorable for adsorbing various guest molecules with hydrophobic interaction. In particular, it is shown that the captured molecules in the POSS-core dendrimer are protected from photobleaching. It should be noted that stimuli responsiveness is compatibly observed from the captured guest molecules, although these molecules are isolated by POSS. Because of the steric structure of POSS, accessibility of reactive oxygen species which are promised to induce degradation can be lowered. Meanwhile, energy and electron transferring is acceptable through the POSS scaffold. Owing to these characters, environment-responsive photon upconversion including multiple energy transferring steps [23] and photo-triggered redox reactions [24] are accomplished in the POSS-containing materials.

Inspired by the enhancement of environmental resistance of POSS without losses of stimuli responsiveness, we presume that oxygen resistance in ECL systems might be reinforced. On the basis of this assumption, we synthesize the modified POSS tethered to the tris(2,2′-bipyridyl)ruthenium(II) complex (Ru-POSS) and evaluate the influence of dissolved oxygen on the ECL properties in the aqueous mixture system containing tripropylamine (TPrA, coreactant). Initially it is shown that ECL intensity and electric current can be enlarged by the connection to POSS. Moreover, from the comparison studies, it is clearly indicated that ECL quenching by oxygen is suppressed in the Ru-POSS/TPrA system. We elucidate how this phenomenon could originate from the improvement of TPrA oxidation by Ru-POSS adsorption onto the ITO electrode. Finally, we also accomplish the ECL-based detection for the water pollutant without the degassing pretreatment. The oxygen-resistance enhancement and target detection which tend to show the trade-off relationship are compatible by employing the POSS element-block.

2. Results and Discussion

2.1. Characteristics of Ru-POSS

The chemical structures and preparation of the Ru(II)-containing materials involving Ru-POSS and the model compound, Ru-Model are shown in Scheme 1. As is often the case with luminescent dyes, concentration quenching is occasionally induced through non-specific intermolecular interaction in the accumulation of the chromophores. To avoid reduction of sensitivity, we designed Ru-POSS to have the single Ru(II) complex. To obtain the target material, we used octa-substituted ammonium POSS (Amino-POSS) as a scaffold and connected it to the bipyridine ligand for complexation with the Ru(II) ion [24]. After complexation with the commodity reaction condition, the introduction ratio of the Ru(II) complex to the POSS unit in Ru-POSS was estimated from the integration ratios of the signal peaks from the side chains in the ^{1}H NMR spectrum. Purification with Ru-POSS was only precipitation and Ru-POSS was obtained as mixtures containing variable numbers of the Ru complex. The structures of synthesized compounds were confirmed by ^{1}H, ^{13}C and ^{29}Si NMR spectroscopies and high-resolution mass spectrometry (HRMS). The detailed procedures are shown in the Supporting Information. The structures of the compounds used, such as Ru(bpy)$_3^{2+}$, Ru(bpy)$_2$(dmbpy)$^{2+}$ (dmbpy = dimethylbipyridine) and tripropyl amine (TPrA), are listed in Figure S1.

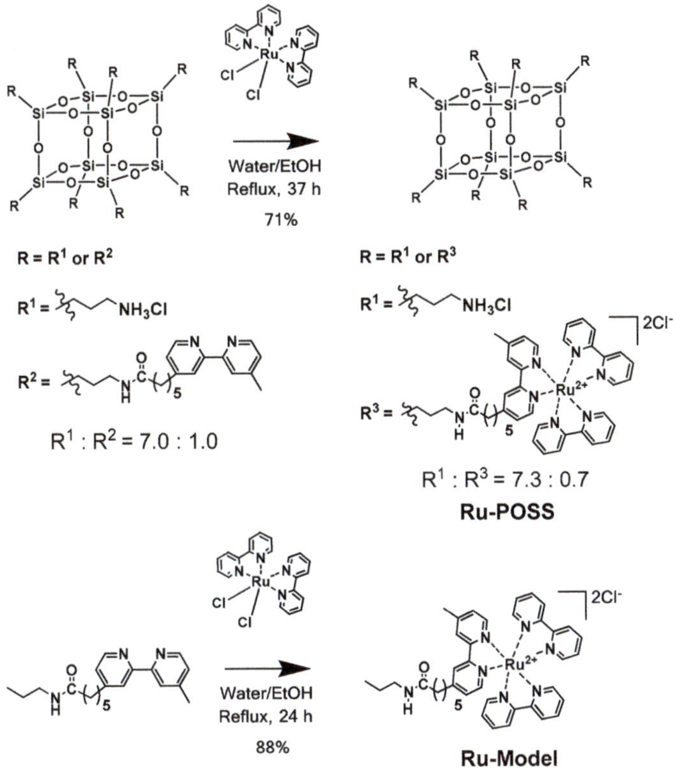

Scheme 1. Syntheses of the Ru-POSS and the Ru-Model.

2.2. ECL Properties of Ru(II)-Containing Materials

Optical measurements were performed to evaluate the influence on the electronic structure of the Ru complex moiety by the connection with POSS (Figure S2). ECL and photoluminescence (PL) spectra of the materials are shown in Figure 1 and Table 1. All compounds containing Ru(II) exhibited the emission bands at similar positions, indicating that ECL originated from the metal-to-ligand charge transfer transition in the triplet excited state (^{3}MLCT) in the Ru(II) complex moiety [2–4]. By subtracting TPrA or Ru(bpy)$_3{}^{2+}$, critical reduction of electric currents was obtained (Figure S3), supporting the conclusion that ECL is produced from the redox cycles including the Ru(II) complex and TPrA through the several proposed mechanisms (Scheme S1). The increase in the Ru(bpy)$_3{}^{2+}$ concentration was minimally influenced on the enhancement of ECL intensity (Figure S4). Concentration quenching could occur under condensed conditions. Slight bathochromic shifts were observed in UV−vis absorption and ECL spectra of the solutions containing Ru-POSS and Ru-Model compared to the ECL spectrum of Ru(bpy)$_3{}^{2+}$. It is likely that a substituent effect is responsible for these negligible changes [25]. From these results it was shown that electronic structure at the Ru(II) complex moiety is hardly influenced by connecting to the POSS unit.

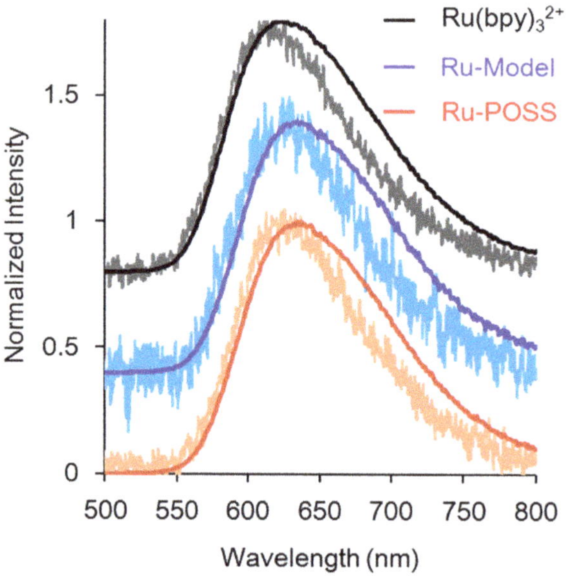

Figure 1. ECL (jagged line) and PL spectra (smooth line) of the Ru(II)-containing materials. The spectra of Ru(bpy)$_3$$^{2+}$ and Ru-Model are shifted in the *y*-axis direction for clarity.

Table 1. Luminescent properties of the Ru(II)-containing materials [a].

Compounds	λ_{abs_LC} (nm)	λ_{abs_MLCT} (nm)	λ_{em_PL} (nm) [b]	Φ_{PL} (%) [c]	τ (μs) [d]	λ_{em_ECL} (nm)	Intensity ($\times 10^4$ a.u.) [e]
Ru-POSS	286	456	634	5.6	0.44	607	13.5
Ru-Model	286	456	632	5.8	0.43	631	9.5
Ru(bpy)$_3$$^{2+}$	286	454	623	6.3	0.50	632	8.8

[a] Based on the concentration of the Ru(II) complex unit (1.0×10^{-5} M). [b] Excited at λ_{abs_MLCT}. [c] Calculated as an absolute value. [d] Measured with the excitation light at 375 nm. [e] Determined at the peak value in CV experiments.

The cyclic voltammograms and corresponding ECL profiles in the TPrA co-reactant system are shown in Figure 2. The POSS unit in Ru-POSS induced significant changes in electric and optical properties. There are two advantageous issues resulting from the connection of the Ru(II) complex to POSS. First, it was found via Ru-POSS that ECL appeared in the lower voltage region. HOMO energy levels were estimated from the onset values (E_{onset}) in the oxidative wave of the CV curves (Figure S5). Obviously, the E_{onset} of Ru-POSS decreased compared to the Ru-Model, as well as Ru(bpy)$_3$$^{2+}$. Although it was reported that the substituent effect can decrease E_{onset} [26], the degree of change was several times larger by the connection to POSS than those in the previous work. In our ECL system, another pathway for decreasing E_{onset} could exist. Additionally, even in the absence of TPrA, lower E_{onset} values were obtained from the Ru(II)-containing materials except for Ru(bpy)$_3$$^{2+}$ (Figure S6), proposing that the lowering value of E_{onset} might be induced by changes in molecular morphology on the electrode. The discussion will be made in a later section. Second, a larger anodic current and ECL intensity were obtained from Ru-POSS than those from Ru(bpy)$_3$$^{2+}$ and Ru-Model. To understand the roles of POSS in these changes, further experiments were executed.

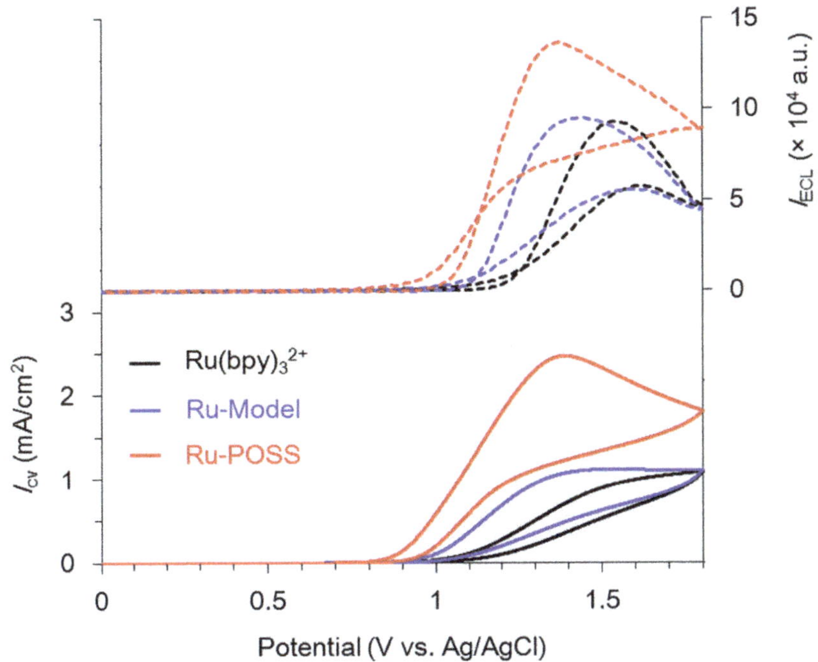

Figure 2. Cyclic voltammograms (solid lines) and corresponding ECL profiles (dashed lines) of 0.1 mM Ru(II) complexes and 100 mM TPrA in 0.20 M PBS buffer (pH 8.8) with an ITO electrode at a scan rate of 100 mV/s (*n* = 3). The concentration of Ru-POSS was based on the Ru(II) complex unit.

In the bulk electrolysis (BE) at 1.2 V vs. Ag/AgCl, it was also observed that a larger current was able to be generated from Ru-POSS than those of Ru(bpy)$_3^{2+}$ and Ru-Model (Figure S7). In particular, the initial non-faradaic current was one order magnitude larger. To explain these behaviors, including lower E_{onset}, we initially investigated the possibility that Ru-POSS could adsorb at the electrode surfaces and accelerate the redox reactions. To evaluate the validity of this speculation, we examined BE with Ru(bpy)$_3^{2+}$ in the presence of Amino-POSS under the same condition as above (Figure S8). It was clearly indicated that the connection to POSS was essential for obtaining a larger ECL. Next, the electrodes modified by immobilization of Ru(II) complexes were prepared, and their properties were examined. The ITO electrode was immersed in the ECL reaction solution containing the series of Ru(II) complexes, such as Ru-POSS, Ru(bpy)$_3^{2+}$ and Ru-Model, rinsed with deionized water and dried in vacuo. Then, the BE and ECL properties were evaluated in the buffer solution containing only TPrA without adding Ru(II)-containing molecular species in the solution (Figures S9–S11). The reduction in current intensity and disappearance of ECL were observed in the ITO electrodes treated with Ru(bpy)$_3^{2+}$ and Ru-Model, while a large current and ECL were still detected in the Ru-POSS-immersed ITO electrode. This data clearly indicated that the adsorption of Ru-POSS occurred on the ITO electrode. It has been reported that ammonium groups can form tight interaction to the ITO electrode from the studies regarding self-assembling monolayers [27–29], poly-ʟ-lysine [30–32], and cytochrome C [33]. It is likely that the multiple ammonium groups in Ru-POSS play a critical role in adsorption to the electrode. Moreover, from the adsorption experiments under variable pH conditions, enhancement of ECL was observed by using the modified ITO electrode immersed with Ru-POSS solution in an acidic condition (pH 5.2), supporting the conclusion that the ammonium-ITO interaction is responsible for anchoring Ru-POSS onto the electrode surface (Figure S12). The assembly of functional groups into the compact space as well as the high symmetry of POSS might be favorable for tight interaction.

Regarding the lowering E_{onset}, it was reported that modification of ITO electrodes with CNT [34], poly(amidoamine) (PAMAM) dendrimer [35], Au nanoparticles [36], and thin film consisting of vertically aligned silica mesochannels [37] lowered the onset potentials and enhanced ECL intensities by facilitating the redox cycles at the electrode surfaces. Similar to these modifications, it is proposed that Ru-POSS could promote the redox cycles for the generation of ECL. Bard et al. reported about the influence of molecular morphology on intensity. They claimed that some degree of distance would be needed for obtaining emission from the Ru(II) complex adsorbed to electrodes, otherwise direct quench by the electrode would occur [38]. From our experiments, although POSS could induce adsorption of Ru(II) species on the electrode, larger ECL was obtained. It is implied that steric hindrance of the silica cube might prevent quenching by the electrode.

2.3. Resistance to Oxygen Quenching

Oxygen resistance of this ECL system was evaluated (Figure 3 and Table 2). To examine the influence of oxygen on the luminescent and electric properties, the dissolved oxygen (DO) levels in the samples were tuned by argon (DO: less than 1 mg/mL) and oxygen bubbling (ca. 20 mg/mL) before monitoring. As readily expected, the aerated sample containing Ru-complex showed critical annihilation of ECL prepared by oxygen bubbling, whereas significant influence were hardly observed in the electric currents (Figure S13). It is well known that DO readily facilitates non-radiation decay of the triplet-excited state, resulting in emission annihilation. In contrast, it was shown that the intensity level from the hypoxic sample containing Ru-POSS was maintained at the higher DO level. These data clearly indicated that the connection of Ru-complex to POSS was responsible for improving oxygen resistance in the ECL system.

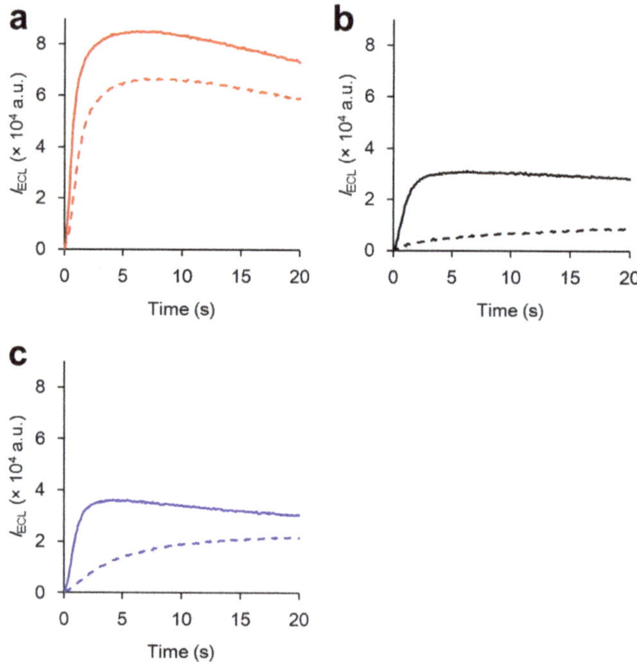

Figure 3. Time courses of ECL with (**a**) Ru-POSS, (**b**) Ru(bpy)$_3^{2+}$ and (**c**) Ru-Model in BE with 0.1 mM Ru(II) complexes and 100 mM TPrA in 0.20 M PBS buffer (pH 8.8) with Ar (solid line) and O$_2$ bubbling (dashed line) with an ITO electrode at 1.2 V vs. Ag/AgCl. The concentration of Ru-POSS was based on a Ru(II) complex unit.

Table 2. Residual rates of maximum ECL intensities and onset increase rates in BE in the aerated solutions compared to the hypoxic ones.

Compounds	$I_{ECL}{}^{Max}{}_{\_O2}/I_{ECL}{}^{Max}{}_{\_Ar}$ (%)	$m_{ECL}{}^{Max}{}_{\_O2}/m_{ECL}{}^{Max}{}_{\_Ar}$ (%)
Ru-POSS	78	51
$Ru(bpy)_3{}^{2+}$	38	9
Ru-Model	61	13

m: Slope of onset intensity curve.

To obtain further insight regarding the detailed mechanism and the role of POSS in enhanced oxygen resistance, we performed measurements under various conditions. It was shown that the connection of the Ru(II) complex was necessary for expressing oxygen resistance in the former section (Figure S8 and Table S1). From the PL measurement with the Ru-POSS sample under the aerated condition, it was found that emission annihilation was induced (Figure S14), meaning that oxygen access for the Ru(II) complex moiety could be maintained even in the presence of POSS. Therefore, it was proposed that the rate of TPrA oxidation could be significantly enhanced in the Ru-POSS system. The pH effect on ECL of Ru-POSS was initially investigated (Figure S15 and Table S2). The ECL intensities of all Ru(II)-containing materials drastically decreased under acidic conditions. It should be emphasized that even in the Ru-POSS system, which had high resistance against DO, a lower intensity was shown at pH 5.2 than that at pH 8.8. According to previous literature, deprotonation in TPrA followed by oxidation of TPrA [39–41] could be suppressed under acidic conditions. Therefore, the redox cycle for generating ECL should be inhibited. It is already known that there are several pathways in the ECL mechanism of $Ru(bpy)_3{}^{2+}$/TPrA system (Scheme S1) [12,40–44]. In these mechanisms, the TPrA oxidation should be the critical step before presenting luminescence. Furthermore, it was reported that the excess amount of TPrA radicals (TPrA$^\bullet$) generated by the TPrA oxidation can contribute to reducing of the amount of DO near the ECL reaction layer and subsequently can suppress ECL quenching [12]. In summary, the enhancement of the TPrA oxidation by Ru-POSS might be responsible for improving oxygen resistance in the ECL system as well as ECL intensity.

2.4. Analysis of Antibiotics without Deoxygenation

Because ECL is usually obtained from the triplet-excited state, most conventional ECL systems always suffer from emission quenching by DO in the pristine sample. Therefore, degassing for lowering DO levels in the sample was previously needed in practical usage. To demonstrate the advantage of oxygen-resistant ECL based on Ru-POSS, we finally performed the detection without degassing as a pre-treatment. By evaluating the decrease in ECL intensity, the target concentration was determined [45–49]. By adding oxytetracycline (OTC), which is an anti-bacterial agent for domestic animals that often induces water pollution near livestock barns, to the samples, the changes in ECL intensity were monitored. The OTC aqueous solution was added to the buffer containing 0.1 mM Ru(II) complex and 0.1 M TPrA with stirring and then ECL measurements in BE (1.2 V vs. Ag/AgCl) of the mixtures were taken (Figure 4). Apparently, ECL intensity from the $Ru(bpy)_3{}^{2+}$/TPrA system was too small to discriminate the target, meanwhile critical decrease in emission intensity was detected from the Ru-POSS sample, meaning that the existence of OTC was able to be detected even in the aerobic sample. In particular, the linear relationship between ΔI_{ECL} and the concentration in the range from 1.3×10^{-5} to 2.0×10^{-4} M ($R^2 = 0.993$) was obtained when using Ru-POSS. The limit of detection was 9.8×10^{-6} M (S/N = 3). It was suggested that ECL quenching may be caused by oxidized OTC at the electrode, which is a benzoquinone derivative [45]. Benzoquinone groups are well known to have the ability to quench ECL [48,49]. The decay of the excited state of the Ru(II) complex and TPrA intermediate-radical quenching were induced by benzoquinone and the latter pathway was predominant when TPrA direct oxidation occurred [49]. Regarding the detection limit, it has been reported that the OTC concentrations in wastewater are in the range between 10^{-6} and 10^{-9} M order [47]. The detection limit of our system was not so high compared with the other ECL system [45], however, our method could be directly applicable for detecting serious contaminated conditions.

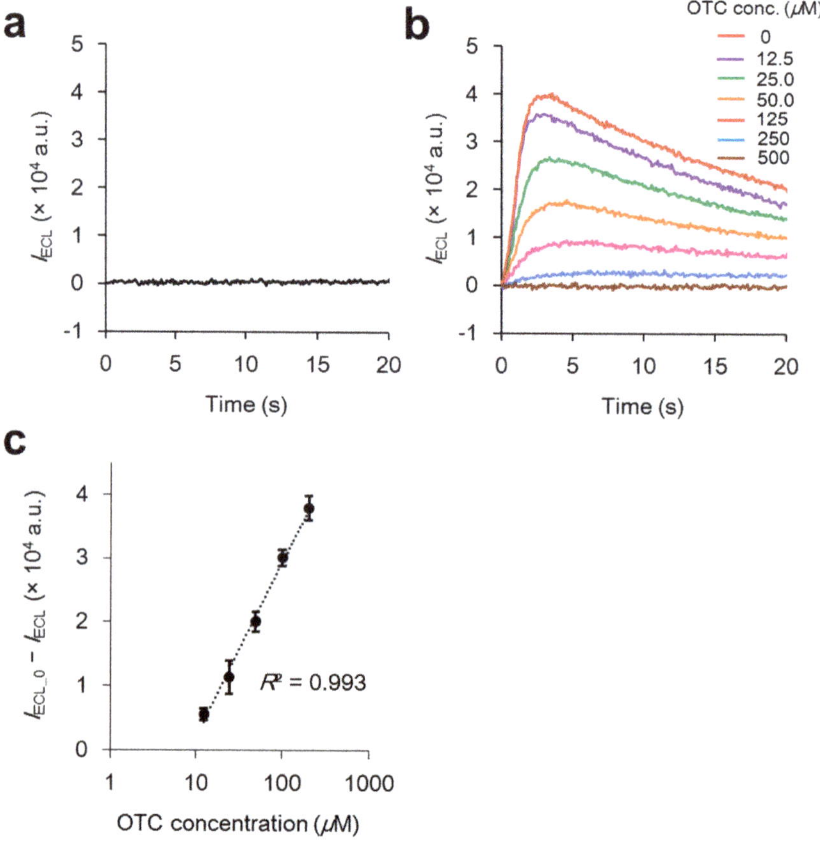

Figure 4. Time courses of ECL with (**a**) $Ru(bpy)_3^{2+}$ and (**b**) Ru-POSS in BE of 0.1 mM Ru(II) complexes and 100 mM TPrA and various concentrations of OTC in 0.20 M PBS buffer (pH 8.8) stirred for 20 min with an ITO electrode at 1.2 V vs. Ag/AgCl. (**c**) Standard curve prepared from the data from (**b**) (I_{ECL} and I_{ECL_0} represent the maximum ECL intensity in the time course of BE with and without OTC, respectively).

3. Conclusions

We demonstrate the Ru(II) complex-modified POSS and its application for oxygen-resistant ECL system. By employing POSS, various beneficial effects were obtained. Electronic currents and ECL intensities were enhanced in the electro- and optical measurements, and E_{onset} in CV was lowered compared to $Ru(bpy)_3^{2+}$. Moreover, ECL quenching by oxygen was suppressed in the Ru-POSS/TPrA system. From our mechanistic studies, adsorption of Ru-POSS to the ITO electrode, followed by efficient TPrA oxidation on the ITO electrode were suggested. Finally, oxygen quenching was compensated by generation of the TPrA radical. Our findings and materials could be a scaffold for developing future sensing technologies based on ECL, such as real-time monitoring for small molecular pollutants which it is difficult to detect with conventional environment sensors. Furthermore, compared to the conventional system, the oxygen-resistance can be improved by connecting to POSS at this stage. As mentioned in the introduction, POSS has significant properties for capturing hydrophobic molecules under aerobic conditions. In combination with the unique capturing property of the POSS unit, improvement of sensitivity and selectivity for the target detection might be proposed.

Supplementary Materials: The following are available online at http://www.mdpi.com/2073-4360/11/7/1170/s1, Figure S1: Structures of $Ru(bpy)_3{}^{2+}$, $Ru(bpy)_2(dmbpy)^{2+}$ and TPrA, Figure S2: (a) UV−vis absorption and (b) PL spectra of 1.0×10^{-5} M solutions containing $Ru(bpy)_3{}^{2+}$, Ru-Model and Ru-POSS in H_2O. The excitation light at λ_{abs_MLCT} was used for PL measurements. Inset figure in (a) is the inset around the MLCT band. The concentration of Ru-POSS was based on the Ru(II) complex unit, Figure S3: Cyclic voltammograms of 0.10 mM $Ru(bpy)_3{}^{2+}$ and/or 0.1 M TPrA in 0.20 M PBS buffer (pH 8.8) at ITO electrode at a scan rate of 100 mV/s (n = 3), Figure S4: Time courses of ECL in BE of 0.50 mM $Ru(bpy)_3{}^{2+}$ and 100 mM TPrA in 0.20 M PBS buffer (pH 8.8) with deoxygenation (solid line) or aeration (dashed line) with an ITO electrode at 1.2 V vs. Ag/AgCl, Figure S5: The determination of onset potential (E_{onset}). The onset potentials were determined from the intersection of the tangents between the baseline and the signal current, Figure S6: Cyclic voltammograms of 0.10 mM Ru(II)-containing materials in 0.20 M PBS buffer (pH 8.8) with an ITO electrode at a scan rate of 100 mV/s (n = 3). The concentration of Ru-POSS was based on the Ru(II) complex unit, Figure S7: Initial time courses of current in BE of 0.10 mM Ru(II) complexes and 100 mM TPrA in 0.20 M PBS buffer (pH 8.8) at ITO electrode at 1.2 V vs. Ag/AgCl (n = 3). The concentration of Ru-POSS was based on the Ru(II) complex unit, Figure S8: Time courses of ECL with 0.1 mM $Ru(bpy)_3{}^{2+}$, 0.1 mM Amino-POSS and 100 mM TPrA in 0.20 M PBS buffer (pH 8.8) with Ar (solid line) and O_2 bubbling (dashed line) with an ITO electrode at 1.2 V vs. Ag/AgCl, Figure S9: Time courses of ECL in BE of 100 mM TPrA in 0.20 M PBS buffer (pH 8.8) with the modified ITO electrodes at 1.2 V vs. Ag/AgCl, Figure S10: Cyclic voltammograms (solid lines) and corresponding ECL curves (dashed lines) at a scan rate of 100 mV/s with 0.1 mM Ru-POSS and 100 mM TPrA in 0.20 M PBS buffer (pH 8.8) with an ITO electrode (Red line) and 100 mM TPrA in 0.20 M PBS buffer (pH 8.8) with the modified ITO electrode (Blue line). The concentration of Ru-POSS was based on the Ru(II) complex unit, Figure S11: Time courses of currents with (a) Ru-POSS (b) $Ru(bpy)_3{}^{2+}$ (c) Ru-Model in BE of 100 mM TPrA in 0.20 M PBS buffer (pH 8.8) with (solid line) or without 0.10 mM Ru(II) complexes (dashed line) at ITO electrode at 1.2 V vs. Ag/AgCl (n = 3). The concentration of Ru-POSS was based on the Ru(II) complex unit, Figure S12: Time courses of ECL in BE of 100 mM TPrA in 0.20 M PBS buffer (pH 8.8) at 1.2 V vs. Ag/AgCl with the modified ITO electrodes immersed into the solution containing 0.1 mM Ru-POSS (based on the Ru(II) complex unit) with 100 mM TPrA in 0.20 M PBS buffer with variable pH values (pH 5.2: orange line and pH 8.8: blue line), Figure S13: Time courses of electric currents with (a) Ru-POSS, (b) $Ru(bpy)_3{}^{2+}$, (c) Ru-Model and (d) $Ru(bpy)_3{}^{2+}$ + Amino-POSS in BE of 0.10 mM Ru(II) complexes and 100 mM TPrA in 0.20 M PBS buffer (pH 8.8) with Ar (solid line) and O_2 bubbling (dashed line) with and ITO electrode at 1.2 V vs. Ag/AgCl. The concentration of Ru-POSS was based on a Ru(II) complex unit. The concentration of Amino-POSS in (d) was adjusted to the same as the POSS unit in (a), Figure S14: Residual rates of emission intensities of the photo-excited Ru(II) complexes in aerated solutions consist of 0.01 mM Ru(II) complexes and 0.1 M TPrA in 0.2 M PBS solution (pH 8.8). Excitation wavelengths were λ_{abs_MLCT} Residual rates were calculated from the equation, I_{PL_Air}/I_{PL_Ar}, where I_{PL_Air} is the intensity at $\lambda_{em}{}^{max}$ in aerated solutions and I_{PL_Ar} is in the hypoxic solutions, Figure S15: Cyclic voltammograms (solid lines) and corresponding ECL curves (dotted lines) of 0.1 mM Ru(II) complexes and 100 mM TPrA in 0.20 M PBS buffer (pH 5.2) with an ITO electrode at a scan rate of 100 mV/s (n = 3). The concentration of Ru-POSS was based on a Ru(II) complex unit, Table S1: Residual rates of maximum ECL intensities and onset increase rates in BE in the aerated solution compared to the hypoxic one, Table S2: E_{onset} in various conditions. Scheme S1: The summary of the ECL mechanisms in the $Ru(bpy)_3{}^{2+}$/TPrA system ($TPrA^{\bullet+} = Pr_3N^{\bullet+}$, $TPrA^{\bullet} = Pr_2NC^{\bullet}HCH_2CH_3$, $P_1 = Pr_2N^+CHCH_2CH_3$).

Author Contributions: Conceptualization, K.T.; data curation, R.N., H.N. and M.G.; funding acquisition, K.T. and Y.C.; investigation, R.N. and H.N.; project administration, K.T.; supervision, K.T.; writing—original draft, R.N., M.G. and K.T.; writing—review and editing, Y.C.

Funding: This work was partially supported by the Kurita Water and Environment Foundation (for K.T.), a Grant-in-Aid for Scientific Research (B) (JP17H03067) and (A) (JP 17H01220) and for Scientific Research on Innovative Areas "New Polymeric Materials Based on Element-Blocks (No.2401)" (JP24102013).

Conflicts of Interest: The authors declare no conflicts of interest.

References

1. Richter, M.M. Electrochemiluminescence (ECL). *Chem. Rev.* **2008**, *104*, 3003–3036. [CrossRef]

2. Miao, W. Electrogenerated Chemiluminescence and Its Biorelated Applications. *Chem. Rev.* **2008**, *108*, 2506–2553. [CrossRef]

3. Fähnrich, K.A.; Pravda, M.; Guilbault, G.G. Recent applications of electrogenerated chemiluminescence in chemical analysis. *Talanta* **2001**, *54*, 531–559. [CrossRef]

4. Gorman, B.A.; Francis, P.S.; Barnett, N.W. Tris(2,2′-bipyridyl)ruthenium(II) chemiluminescence. *Analyst* **2006**, *131*, 616–639. [CrossRef]

5. Qi, H.; Peng, Y.; Gao, Q.; Zhang, C. Applications of nanomaterials in electrogenerated chemiluminescence biosensors. *Sensors* **2009**, *9*, 674–695. [CrossRef]

6. Li, L.; Chen, Y.; Zhu, J.J. Recent Advances in Electrochemiluminescence Analysis. *Anal. Chem.* **2017**, *89*, 358–371. [CrossRef]

7. Demas, J.N.; Diemente, D.; Harris, E.W. Oxygen quenching of charge-transfer excited states of ruthenium(II) complexes. Evidence for singlet oxygen production. *J. Am. Chem. Soc.* **1973**, *95*, 6864–6865. [CrossRef]

8. Lin, C.T.; Sutin, N. An investigation of the micellar phase of sodium dodecyl sulfate in aqueous sodium chloride solutions using quasielastic light scattering spectroscopy. *J. Phys. Chem.* **1976**, *80*, 97–105. [CrossRef]

9. O'Keeffe, G.; McDonagh, C.M.; O'Kelly, B.; Vos, J.G.; Keyes, E.T.; McGilp, J.F.; MacCraith, B.D. ibre optic oxygen sensor based on fluorescence quenching of evanescent-wave excited ruthenium complexes in sol–gel derived porous coatings. *Analyst* **1993**, *118*, 385–388.

10. Fiaccabrino, G.C.; Koudelka-Hep, M.; Hsueh, Y.T.; Collins, S.D.; Smith, R.L. Electrochemiluminescence of Tris(2,2′-bipyridine)ruthenium in Water at Carbon Microelectrodes. *Anal. Chem.* **1998**, *70*, 4157–4161. [CrossRef]

11. Tomita, I.N.; Bulhões, L.O.S. Electrogenerated chemiluminescence determination of cefadroxil antibiotic. *Anal. Chim. Acta* **2001**, *442*, 201–206. [CrossRef]

12. Zheng, H.; Zu, Y. Emission of Tris(2,2′-bipyridine)ruthenium(II) by Coreactant Electrogenerated Chemiluminescence: From O_2-Insensitive to Highly O_2-Sensitive. *J. Phys. Chem. B* **2005**, *109*, 12049–12053. [CrossRef]

13. Kuo, S.W.; Chang, F.C. POSS related polymer nanocomposites. *Prog. Polym. Sci.* **2011**, *36*, 1649–1696. [CrossRef]

14. Chujo, Y.; Tanaka, K. New Polymeric Materials Based on Element-Blocks. *Bull. Chem. Soc. Jpn.* **2015**, *88*, 633–643. [CrossRef]

15. Gon, M.; Tanaka, K.; Chujo, Y. Recent Progress in the Development of Advanced Element-Block Materials. *Polym. J.* **2018**, *50*, 109–126. [CrossRef]

16. Gon, M.; Kato, K.; Tanaka, K.; Chujo, Y. Elastic and mechanofluorochromic hybrid films with POSS-capped polyurethane and polyfluorene. *Mater. Chem. Front.* **2019**, *3*, 1174–1180. [CrossRef]

17. Gon, M.; Sato, K.; Kato, K.; Tanaka, K.; Chujo, Y. Preparation of Bright-Emissive Hybrid Materials Based on Light-Harvesting POSS Having Radially Integrated Luminophores and Commodity π-Conjugated Polymers. *Mater. Chem. Front.* **2019**, *3*, 314–320. [CrossRef]

18. Ueda, K.; Tanaka, K.; Chujo, Y. Optical, Electrical and Thermal Properties of Organic–Inorganic Hybrids with Conjugated Polymers Based on POSS Having Heterogeneous Substituents. *Polymers* **2019**, *11*, 44. [CrossRef]

19. Ueda, K.; Tanaka, K.; Chujo, Y. Fluoroalkyl POSS with Dual Functional Groups as a Molecular Filler for Lowering Refractive Indices and Improving Thermomechanical Properties of PMMA. *Polymers* **2018**, *10*, 1332. [CrossRef]

20. Narikiyo, H.; Gon, M.; Tanaka, K.; Chujo, Y. Control of Intramolecular Excimer Emission in Luminophore-Integrated Ionic POSSs Possessing Flexible Side-Chains. *Mater. Chem. Front.* **2018**, *2*, 1449–1455. [CrossRef]

21. Narikiyo, H.; Kakuta, T.; Matsuyama, H.; Gon, M.; Tanaka, K.; Chujo, Y. Development of the Optical Sensor for Discriminating Isomers of Fatty Acids Based on Emissive Network Polymers Composed of Polyhedral Oligomeric Silsesquioxane. *Bioorg. Med. Chem.* **2017**, *25*, 3431–3436. [CrossRef]

22. Kakuta, T.; Jeon, J.-H.; Tanaka, K.; Chujo, Y. Development of Highly-Sensitive Detection System in ^{19}F NMR for Bioactive Compounds Based on the Assembly of Paramagnetic Complexes with Fluorinated Cubic Silsesquioxanes. *Bioorg. Med. Chem.* **2017**, *25*, 1389–1393. [CrossRef]

23. Tanaka, K.; Okada, H.; Ohashi, W.; Jeon, J.-H.; Inafuku, K.; Chujo, Y. Hypoxic Conditions-Selective Upconversion via Triplet-Triplet Annihilation Based on POSS-Core Dendrimer Complexes. *Bioorg. Med. Chem.* **2013**, *21*, 2678–2681. [CrossRef]

24. Jeon, J.-H.; Tanaka, K.; Chujo, Y. Light-Driven Artificial Enzymes for Selective Oxidation of Guanosine Triphosphate Using Water-Soluble POSS Network Polymers. *Org. Biomol. Chem.* **2014**, *12*, 6500–6506. [CrossRef]

25. McClanahan, S.F.; Dallinger, R.F.; Holler, F.J.; Kincaid, J.R. Mixed-ligand poly(pyridine) complexes of ruthenium(II). Resonance Raman spectroscopic evidence for selective population of ligand-localized ^{3}MLCT excited states. *J. Am. Chem. Soc.* **1985**, *107*, 4853–4860. [CrossRef]

26. Juris, A.; Balzani, V.; Belser, P.; VON Zelewsky, V.A. Characterization of the Excited State Properties of Some New Photosensitizers of the Ruthenium (Polypyridine) Family. *Helv. Chim. Acta* **1981**, *64*, 2175–2182. [CrossRef]

27. Oh, S.Y.; Yun, Y.J.; Kim, D.Y.; Han, S.H. Formation of a Self-Assembled Monolayer of Diaminododecane and a Heteropolyacid Monolayer on the ITO Surface. *Langmuir* **1999**, *15*, 4690–4692. [CrossRef]

28. Kim, K.; Song, H.; Park, J.T.; Kwak, J. C_{60} Self-Assembled Monolayer Using Diamine as a Prelayer. *Chem. Lett.* **2000**, *29*, 958–959. [CrossRef]

29. Bermudez, V.M.; Berry, A.D.; Kim, H.; Piqué, A. Functionalization of Indium Tin Oxide. *Langmuir* **2006**, *22*, 11113–11125. [CrossRef]

30. Bearinger, J.P.; Vörös, J.; Hubbell, J.A.; Textor, M. Electrochemical optical waveguide lightmode spectroscopy (EC-OWLS): A pilot study using evanescent-field optical sensing under voltage control to monitor polycationic polymer adsorption onto indium tin oxide (ITO)-coated waveguide chips. *Biotechnol. Bioeng.* **2003**, *82*, 465–473. [CrossRef]

31. Yang, H.J.; Sugiura, Y.; Ishizaki, I.; Sanada, N.; Ikegami, K.; Zaima, N.; Shrivas, K.; Setou, M. Imaging of lipids in cultured mammalian neurons by matrix assisted laser/desorption ionization and secondary ion mass spectrometry. *Surf. Interface Anal.* **2010**, *42*, 1606–1611. [CrossRef]

32. Choi, Y.; Yagati, A.K.; Cho, S. Electrochemical Characterization of Poly-L-Lysine Coating on Indium Tin Oxide Electrode for Enhancing Cell Adhesion. *J. Nanosci. Nanotechnol.* **2015**, *15*, 7881–7885. [CrossRef]

33. Koller, K.B.; Hawkridge, F.M. Temperature and electrolyte effects on the electron-transfer reactions of cytochrome c. *J. Am. Chem. Soc.* **1985**, *107*, 7412–7417. [CrossRef]

34. Valenti, G.; Zangheri, M.; Sansaloni, S.E.; Mirasoli, M.; Penicaud, A.; Roda, A.; Paolucci, F. Transparent Carbon Nanotube Network for Efficient Electrochemiluminescence Devices. *Chem. Eur. J.* **2015**, *21*, 12640–12645. [CrossRef]

35. Kim, Y.; Kim, J. Modification of Indium Tin Oxide with Dendrimer-Encapsulated Nanoparticles To Provide Enhanced Stable Electrochemiluminescence of $Ru(bpy)_3^{2+}$/Tripropylamine While Preserving Optical Transparency of Indium Tin Oxide for Sensitive Electrochemiluminescence-Based Analyses. *Anal. Chem.* **2014**, *86*, 1654–1660.

36. Pan, S.; Liu, J.; Hill, C.M. Observation of Local Redox Events at Individual Au Nanoparticles Using Electrogenerated Chemiluminescence Microscopy. *J. Phys. Chem. C* **2015**, *119*, 27095–27103. [CrossRef]

37. Su, B.; Xu, L.; Yang, Q.; Zhou, Z.; Guo, W. Two orders-of-magnitude enhancement in the electrochemiluminescence of $Ru(bpy)_3^{2+}$ by vertically ordered silica mesochannels. *Anal. Chim. Acta* **2015**, *886*, 48–55.

38. Miller, C.J.; McCord, P.; Bard, A.J. Study of Langmuir monolayers of ruthenium complexes and their aggregation by electrogenerated chemiluminescence. *Langmuir* **1991**, *7*, 2781–2787. [CrossRef]

39. Leland, J.K.; Powell, M.J. Electrogenerated Chemiluminescence: An Oxidative-Reduction Type ECL Reaction Sequence Using Tripropyl Amine. *J. Electrochem. Soc.* **1990**, *137*, 3127–3131. [CrossRef]

40. Miao, W.; Choi, J.P.; Bard, A.J. Electrogenerated Chemiluminescence 69: The Tris(2,2′-bipyridine)ruthenium(II), $(Ru(bpy)_3^{2+})$/Tri-*n*-propylamine (TPrA) System Revisited−A New Route Involving TPrA$^{\bullet+}$ Cation Radicals. *J. Am. Chem. Soc.* **2002**, *124*, 14478–14485. [CrossRef]

41. Zu, Y.; Li, F. Characterization of the low-oxidation-potential electrogenerated chemiluminescence of tris(2,2′-bipyridine)ruthenium(II) with tri-*n*-propylamine as coreactant. *Anal. Chim. Acta* **2005**, *550*, 47–52. [CrossRef]

42. Wightman, R.M.; Forry, S.P.; Maus, R.; Badocco, D.; Pastore, P. Rate-Determining Step in the Electrogenerated Chemiluminescence from Tertiary Amines with Tris(2,2′-bipyridyl)ruthenium(II). *J. Phys. Chem. B* **2004**, *108*, 19119–19125. [CrossRef]

43. Kanoufi, F.; Zu, Y.; Bard, A.J. Homogeneous Oxidation of Trialkylamines by Metal Complexes and Its Impact on Electrogenerated Chemiluminescence in the Trialkylamine/$Ru(bpy)_3^{2+}$ System. *J. Phys. Chem. B* **2001**, *105*, 210–216. [CrossRef]

44. Gross, E.M.; Pastore, P.; Wightman, R.M. High-Frequency Electrochemiluminescent Investigation of the Reaction Pathway between Tris(2,2′-bipyridyl)ruthenium(II) and Tripropylamine Using Carbon Fiber Microelectrodes. *J. Phys. Chem. B* **2001**, *105*, 8732–8738. [CrossRef]

45. Pang, Y.Q.; Cui, H.; Zheng, H.S.; Wan, G.H.; Liu, L.J.; Yu, X.F. Flow injection analysis of tetracyclines using inhibited Ru(bpy)$_3{}^{2+}$/tripropylamine electrochemiluminescence system. *Luminescence* **2005**, *20*, 8–15. [CrossRef]
46. Zhu, Y.; Li, B.; Liu, J.; Chen, K. Determination of p-aminobenzenesulfonic acid based on the electrochemiluminescence quenching of tris(2,2′-bipyridine)-ruthenium (II). *Luminescence* **2013**, *28*, 363–367. [CrossRef]
47. Tagiri-Endo, M.; Suzuki, S.; Nakamura, T.; Hatakeyama, T.; Kawamukai, K. Rapid determination of five antibiotic residues in swine wastewater by online solid-phase extraction-high performance liquid chromatography-tandem mass spectrometry. *Anal. Bioanal. Chem.* **2009**, *393*, 1367–1375. [CrossRef]
48. McCall, J.; Alexander, C.; Richter, M.M. Quenching of Electrogenerated Chemiluminescence by Phenols, Hydroquinones, Catechols, and Benzoquinones. *Anal. Chem.* **1999**, *71*, 2523–2527. [CrossRef]
49. Zheng, H.; Zu, Y. Highly Efficient Quenching of Coreactant Electrogenerated Chemiluminescence by Phenolic Compounds. *J. Phys. Chem. B* **2005**, *109*, 16047–16051. [CrossRef]

Article

pH Behavior of Polymer Complexes between Poly(carboxylic acids) and Poly(acrylamide derivatives) Using a Fluorescence Label Technique

Yuriko Matsumura [1,*] **and Kaoru Iwai** [2,*]

[1] Postgraduate School of Healthcare, Tokyo Healthcare University, 4-1-17 Higashi-Gotanda, Shinagawa-ku, Tokyo 141-8648, Japan

[2] Professor Emeritus at Nara Women's University, Kitauoya-Nishimachi, Nara 630-8506, Japan

* Correspondence: y-matsumura@thcu.ac.jp (Y.M.); iwai@cc.nara-wu.ac.jp (K.I.); Tel.: +81-3-5421-7685 (Y.M.)

Received: 19 June 2019; Accepted: 11 July 2019; Published: 17 July 2019

Abstract: In order to clarify the local environment during interpolymer complex formation between poly(carboxylic acids) and poly(acrylamide derivatives) with different N-substitutions, a fluorescence label technique was used. 3-(2-propenyl)-9-(4-N,N-dimethylaminophenyl) phenanthrene (VDP) was used as an intramolecular fluorescence probe. All polymers were synthesized by free radical polymerization. Interpolymer complexation was monitored by charge transfer emission from the VDP unit. Both of the poly(carboxylic acids) formed interpolymer complexes with poly(N,N-dimethylacrylamide) (polyDMAM). The micro-environments around the VDP unit in the acidic pH region for the poly(methacrylic acid) (polyMAAc) and polyDMAM mixed systems were more hydrophobic than those of the poly(acrylic acid) (polyAAc) and polyDMAM mixed systems, as the α-methyl group of the MAAc unit contributed to hydrophobicity around the polymer chain during hydrogen bond formation. This suggests that, when the poly(carboxylic acids) and poly(acrylamide derivatives) were mixed, with a subsequent decrease in the solution pH, a hydrogen bond was partially formed, following which the hydrophobicity of the micro-environment around the polymer chains was changed, resulting in the formation of interpolymer complexes. Moreover, the electron-donating ability of the carbonyl group in the poly(acrylamide derivatives) had an effect on complexation with poly(carboxylic acids).

Keywords: poly(acrylic acid); poly(methacrylic acid); poly(acrylamide derivatives); complexation; fluorescence label technique

1. Introduction

Poly(acrylic acid) (polyAAc) has been shown to form interpolymer complexes with poly(ethylene glycol) and poly(acrylamide) [1]. Interpolymer complexes have been recently intensively investigated in the field of developing new bio-compatible materials [2]. These interpolymer complexes are formed by hydrogen bonding between polymers containing hydrogen bond-donating groups and polymers containing hydrogen bond-accepting groups. Interpolymer complexes of polymer solutions or polymer blends have been widely investigated in the pharmaceutical field, especially in the study of drug delivery techniques [3]. Recently, polymer–polymer complexes between poly(carboxylic acids) and thermo-responsive polyacrylamides were investigated in an organic solvent [4]. However, the processes of interpolymer complex formation, including the local environment around the polymer chains, have not been studied much.

Fluorescence methods are powerful tools for investigating the conformational changes of polymers and micelle formation. This method has been widely used in the study of polymer–polymer interactions and micelle formation [5]. When a fluorescent probe is added into the interpolymer complex formation system, the environment around the fluorescent probe, located somewhere in the system, can be determined. However, information on the environment of the polymer chains cannot be obtained. In our previous paper, we reported the thermo-responsive behavior and micro-environments of poly(*N*-isopropylacrylamide) microgel particles labeled with the polarity-sensitive fluorescent molecule 3-(2-propenyl)-9-(4-*N*,*N*-dimethylaminophenyl)phenanthrene (VDP) (chemical structure shown in Figure 1) dispersed in water [6]. The VDP units inside the microgel particles became hydrophobic in conjunction with the phase transition of the microgel particles. Thus, information around the polymer chain where the VDP unit exists can be obtained by using VDP as an intrapolymer fluorescent probe.

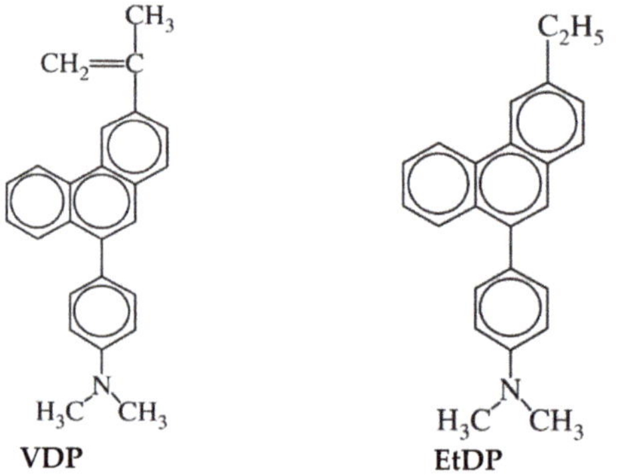

Figure 1. Chemical structures of 3-(2-propenyl)-9-(4-*N*,*N*-dimethylaminophenyl)phenanthrene (VDP) and 3-ethyl-9-(4-*N*,*N*-dimethylaminophenyl) phenanthrene (EtDP).

In this paper, the fluorescent labeling method was used in order to clarify the local environment during interpolymer complex formation. Poly(acrylic acid) (polyAAc) and poly(methacrylic acid) (polyMAAc) were the poly(carboxylic acids) used, and three kinds of poly(acrylamide derivatives) with different *N*-substituents, poly(*N*,*N*-dimethylacrylamide) (polyDMAM), poly(*N*-ethyl-*N*-methylacrylamide) (polyEMAM), and poly(*N*,*N*-diethylacrylamide) (polyDEAM), were employed.

2. Materials and Methods

2.1. Materials

Acrylic acid (AAc) and methacrylic acid (MAAc) were purchased from Wako Pure Chemicals (Osaka, Japan) and purified by vacuum distillation. *N*,*N*-Dimethylacrylamide (DMAM) was purchased from Tokyo Kasei Kogyo (Tokyo, Japan) and purified by vacuum distillation. α,α'-azobisisobutyronitrile (AIBN) was purchased from Wako Pure Chemicals (Osaka, Japan) and recrystallized twice from methanol. *N*-ethyl-*N*-methylacrylamide (EMAM) was prepared by the acrylation of *N*-ethyl-*N*-methylamine with acryloyl chloride. *N*,*N*-diethylacrylamide (DEAM) was prepared by the acrylation of *N*,*N*-diethylamine with acryloyl chloride. All other reagents were of guaranteed reagent grade and used without further purification. The fluorescent probe monomer VDP and its monomer unit model compound 3-ethyl-9-(4-*N*,*N*-dimethylaminophenyl) phenanthrene (EtDP) (chemical structure shown in Figure 1) were used as previously prepared [6].

2.2. Synthesis of VDP-Labeled Polymers

All polymers were synthesized by radical polymerization, using AIBN as an initiator, as reported previously [7]. PolyAAc and poly(VDP-*co*-AAc) were synthesized as follows: AAc (1 mol/L), fluorescent monomer VDP (0 or 1 mmol/L), and AIBN (5 mmol/L) were dissolved in methanol. The solution was degassed by the freeze-pump-thaw method, heated to 60 °C for 6 h, and then cooled to room temperature. The reaction mixture was poured into a 20 times amount of ethyl acetate. The obtained polymer was purified by re-precipitation, using methanol as a solvent and ethyl acetate as a precipitant. The obtained polymers were dried under vacuum for 24 h. The VDP unit contents in the polymers were determined from the absorbance of methanol solutions (8–10 mg/10 mL), as compared to EtDP (ε = 16,100 mol^{-1}·L·cm^{-1} at 314 nm) [6] as a model compound. The absorption spectra were measured on a HITACHI U-3200 spectrophotometer. PolyMAAc, poly(VDP-*co*-MAAc), polyDMAM, poly(VDP-*co*-DMAM), poly(VDP-*co*-EMAM), and poly(VDP-*co*-DEAM) were prepared by a similar procedure, as shown in Table 1.

Table 1. Recipes for polymerizations [1].

Polymer	Solvent	Precipitant	Reaction Time (h)	mol% of VDP Monomer		Yield (%)
				In Feed	In Polymer	
PolyAAc	methanol	ethyl acetate	6	0	0	72.7
Poly(VDP-*co*-AAc)	methanol	ethyl acetate	6	0.1	0.12	74.1
PolyMAAc	methanol	diethyl ether	6	0	0	68.0
Poly(VDP-*co*-MAAc)	methanol	diethyl ether	6	0.1	0.15	61.0
PolyDMAM	benzene	*n*-hexane	1	0	0	52.9
Poly(VDP-*co*-DMAM)	benzene	*n*-hexane	1	0.1	0.17	50.5
Poly(VDP-*co*-EMAM) [2]	benzene	*n*-hexane	1	0.1	0.14	47.4
Poly(VDP-*co*-DEAM)	benzene	*n*-hexane	1	0.1	0.15	57.6

[1] AIBN = 5 mmol/L, $\sum$ monomer = 1 mol/L at 60 °C. [2] $\sum$ monomer = 0.5 mol/L.

2.3. Fluorescence Measurements

The stock polymer solutions were separately prepared at a concentration of 0.01 w/v% using distilled water. The polymer solutions were mixed with equal amounts of polymer solution and distilled water and bubbled with nitrogen gas at 25 °C for 30 min. The pH of the solution was changed by adding 0.02 mol/L of NaOH or 0.02 mol/L of HCl solution under a nitrogen atmosphere and measured using a Horiba D-13 pH meter. The fluorescence spectra of the polymer solutions were measured at 25 °C under a nitrogen atmosphere using a Hitachi F-2500 fluorescence spectrophotometer. The excitation wavelength was set at 320 nm to excite the VDP unit in the polymers.

2.4. Fraction of Dissociated Carboxyl Group (α) of Poly(Carboxylic Acids)

A pH titration was performed by adding 0.02 mol/L NaOH solution to 50 mL of polymer solution (0.005 w/v%). The fraction of dissociated carboxylic group of the poly(carboxylic acids) (α) was calculated using the equation:

$$\alpha = (C_a + C_{H+} - C_{OH-})/C_p \tag{1}$$

where C_a, C_{H+}, C_{OH-}, and C_p denote the concentrations of sodium ions (mol/L), protons (mol/L), hydroxide ions (mol/L), and the initial polymer (unit mol/L), respectively.

3. Results and Discussion

3.1. pH Responsive Behavior of VDP-Labeled Polymers

Typical fluorescence spectra of poly(VDP-*co*-AAc) and poly(VDP-*co*-DMAM) solutions at various pH are shown in Figure 2. All of the VDP-labeled polymer solutions exhibited emission at 340–400 nm from protonation of the nitrogen atom of the VDP units, and a broad intramolecular charge transfer (ICT) emission (400–600 nm) from the VDP units. The wavelength at the maximum intensity (λ_{max}) for ICT emission at pH 4.3 of the poly(VDP-*co*-AAc) solution shifted to a shorter wavelength, compared with that at pH 10.2. On the other hand, the λ_{max} values of the poly(VDP-*co*-DMAM) solution were not changed with varying pH. This blue-shift of the λ_{max} value reflects an increase of hydrophobicity around the VDP units, as reported previously [6]. When the pH of the poly(carboxylic acids) solutions was changed from acidic to alkaline, the carboxylic groups dissociate and electrostatic repulsion occurred between the carboxylate anions. The conformation of the polymer chains was changed from a hypercoiled structure to a water-swollen state with a decrease in the solution pH [8,9]. As the λ_{max} value for poly(VDP-*co*-DMAM) was not affected by the pH of the polymer solution, the VDP units in the polymers could sense and report this formation of the polymer chains in response to the pH.

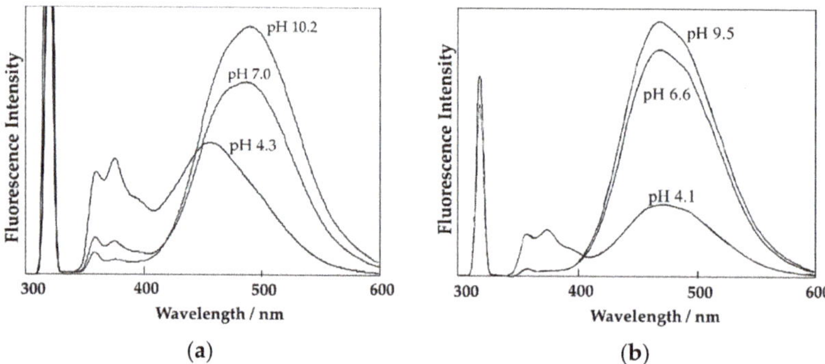

Figure 2. Fluorescence spectra of the VDP-labeled polymer solutions: (**a**) poly(VDP-*co*-AAc) and (**b**) poly(VDP-*co*-DMAM).

As discussed in detail, the fraction of dissociated carboxyl group (α) and the λ_{max} values for poly(VDP-*co*-AAc) and poly(VDP-*co*-MAAc) are plotted against pH in Figure 3. The dissociation constants for poly(VDP-*co*-AAc) and poly(VDP-*co*-MAAc) were 7.3 and 8.1, respectively, meaning that poly(VDP-*co*-MAAc) was a weaker acid than poly(VDP-*co*-AAc). Poly(VDP-*co*-AAc) showed a one-step dissociation, whereas poly(VDP-*co*-MAAc) showed a two-step dissociation. It has been reported that polyMAAc presents small-scale rearrangements in structure around the acidic region, rather than a large-scale expansion, which is then followed by a macroscopic change in dimension at the neutralization point [8]. The λ_{max} value gradually blue-shifted with decreasing pH for both polymer solutions. The λ_{max} values were almost constant at 490 nm, with above 50% dissociation of the carboxyl group for both poly(carboxylic acids), and was blue-shifted with a decrease in the dissociation degree, meaning that the hydrophobicity around the VDP unit increased. In the acidic region, the λ_{max} values for poly(VDP-*co*-MAAc) (420 nm at pH 3.3) were smaller than those for poly(VDP-*co*-AAc) (453 nm at pH 3.3). The λ_{max} values were linearly correlated with the solvent polarity (ε), and the following equation was obtained by least-squares analysis (where the r value is the correlation coefficient), as reported previously [6]:

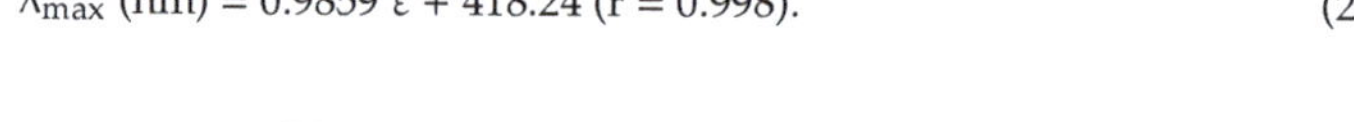

$$\lambda_{max} \text{ (nm)} = 0.9859\,\varepsilon + 418.24 \ (r = 0.998). \tag{2}$$

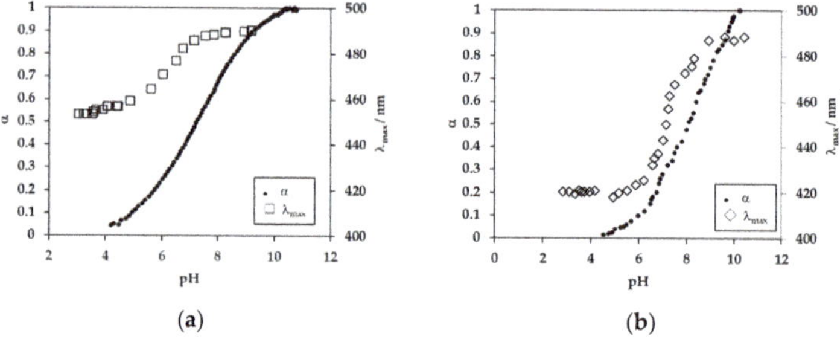

Figure 3. The fraction of dissociated carboxylic group (α) and fluorescence λ_{max} values for 0.005 w/v% poly(VDP-*co*-carboxylic acids) solutions: (**a**) poly(VDP-*co*-AAc) and (**b**) poly(VDP-*co*-MAAc).

The micro-environmental polarity near the VDP units in the polymers was estimated from the observed λ_{max} value using Equation (2). The estimated ε value for poly(VDP-*co*-MAAc) at pH 3.3 was <8, which was smaller than that for poly(VDP-*co*-AAc) (estimated ε = 35). The estimated ε values were 73 at pH 9 for both polymers. The MAAc unit has an α-methyl group, which affects the polymer conformation where the hydrogen bond is formed between the carboxylate groups. This means that the hydrophobicity around the VDP units in poly(VDP-*co*-MAAc) is large, compared to that in poly(VDP-*co*-AAc). On the other hand, in the alkaline pH region, where the electric repulsion causes the hypercoiled form, the VDP units sense a similar hydrophilic environment.

3.2. Complexation Between PolyDMAM With Poly(Carboxylic Acids)

Polymers containing hydrogen-donating groups were mixed with polymers containing hydrogen-accepting groups, resulting in the formation of interpolymer complexes, as shown in Figure 4. When poly(VDP-*co*-AAc) was mixed with polyDMAM, the scattering intensity increased in the acidic region (under pH 4), as shown in Figure 5a. The increase in scattering intensity means that the turbidity of the solution increased. This increase in scattering intensity was also observed for the polyAAc and poly(VDP-*co*-DMAM) mixed system. These phenomena were also observed for the polyMAAc and polyDMAM mixed systems, as shown in Figure 5b. Poly(carboxylic acids) can form complexes with polyDMAM by hydrogen bonding between the H atom of the carboxylic acid in the AAc unit and the O atom of the carbonyl group in the DMAM unit [10,11]. This hydrogen bond formation results in interpolymer complex formation and an increased turbidity of the mixed solution.

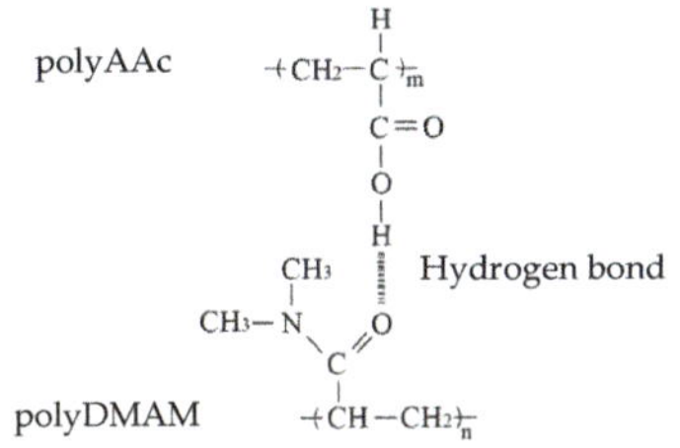

Figure **4.** The polymer complexation of poly(acrylic acid) (polyAAc) with poly(*N*,*N*-dimethylacrylamide) (polyDMAM) by hydrogen bonding.

When poly(VDP-*co*-DMAM) was mixed with polyAAc or polyMAAc, the fluorescent λ_{max} value blue-shifted with a decrease in the solution pH, at around pH 6, as shown in Figure 5c,d. As interpolymer complexation caused an increase in turbidity, this blue-shift means that the hydrophobicity around the VDP unit increased, as discussed above, and that interpolymer interaction occurred even above pH values where the interpolymer complexations were occurred. Thus, these findings suggest that, with decreasing solution pH, interpolymer hydrogen bonds were partially formed, following which the interpolymer complex formed. The fluorescent λ_{max} profile for the poly(VDP-*co*-DMAM) and PAAc mixed system was different, compared to that of the polyDMAM and poly(VDP-*co*-AAc) mixed system. The fluorescent λ_{max} values for poly(VDP-*co*-AAc) were smaller than those of poly(VDP-*co*-DMAM) in the polymer mixed systems, in the range of pH 4–6. This blue-shift was also observed under pH 4, where the interpolymer complex had formed. On the other hand, the fluorescent λ_{max} value for poly(VDP-*co*-MAAc) was of a shorter wavelength than that for poly(VDP-*co*-DMAM), in the range of pH 6.5–7, and was the same under pH 6.5. The fluorescent λ_{max} profiles differed with the poly(carboxylic acids) used. The micro-environment around the VDP unit in polyMAAc was more hydrophobic than polyAAc in the acidic pH region, as the α-methyl group in the MAAc unit affected the hydrophobicity around the polymer chain during hydrogen bond formation.

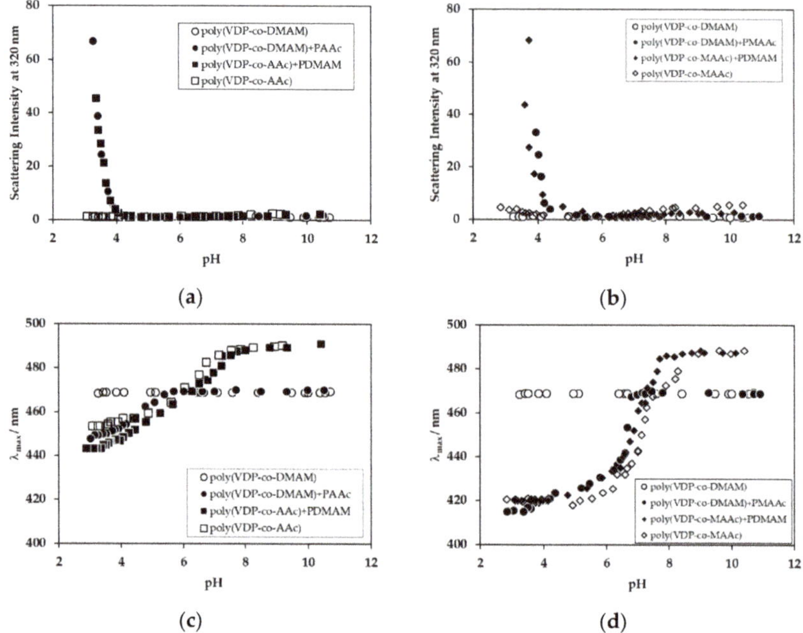

Figure 5. Scattering intensity at 320 nm (**a,b**) and fluorescence λ_{max} values (**c,d**) as a function of pH: PolyDMAM and polyAAc mixed systems (**a,c**) and polyDMAM and poly(methacrylic acid) (polyMAAc) mixed systems (**b,d**).

3.3. Complexation Between PolyAAc and VDP-Labeled Poly(Acrylamide Derivatives)

Drastic changes in the micro-environments around the polymer chains were observed during complex formation when polyDMAM was labeled with the VDP unit. In the case of complex formation between polyAAc and poly(acrylamide derivatives), the VDP unit labeled the poly(acrylamide derivatives). When polyAAc was mixed with poly(VDP-*co*-acrylamide derivatives), the scattering intensities at 320 nm were increased for all polymer mixed solutions with decreased pH, where the threshold pHs were 4.2, 4.8, and 4.9 for poly(VDP-*co*-DMAM), poly(VDP-*co*-DEAM), and poly(VDP-*co*-EMAM), respectively (Figure 6a). Complex formation was observed for all poly(acrylamide derivatives) and polyAAc mixed systems. On the other hand, the fluorescence λ_{max} values were shifted to short wavelengths at a pH of 6.3, 5.7, and 5.3 for poly(VDP-*co*-DEAM), poly(VDP-*co*-EMAM), and poly(VDP-*co*-DMAM), respectively (Figure 6b). These pH values were higher than those of the interpolymer complex formations. This suggests that, when the poly(carboxylic acids) and poly(acrylamide derivatives) were mixed and then the pH was decreased, a hydrogen bond was partially formed and, when the micro-environment around the polymer chains became hydrophobic, the formation of the interpolymer complexes resulted. It has been reported that complexation between polyMAAc and polyDEAM occurs due to the formation of hydrogen bonds, and a ladder model was proposed at low pH [12]. In the case of our study, this proposed ladder model was also applicable to the structures of the polyAAc and poly(acrylamide derivative) complexes. The N atom is an electron donor, which will increase the electron density on the carbonyl O. Thus, the carbonyl O in poly(acrylamide derivatives) may increase hydrogen bond probability. The electron-donating ability of the N side-chain has been explained in terms of the oxidation potentials of amines with different N-substituents. The oxidation potentials for trimethylamide, dimethyl-ethylamine, and triethylamine were 0.76, 0.74, and 0.69 eV (vs. SCE in Britton–Robinson Buffer solution at pH 11.9), respectively [13]. The electron-donating ability of the ethyl group is larger than that of the methyl group. The electron-donating ability of the carbonyl group in poly(acrylamide derivatives) is affected by the electron-donating ability of the amine group. As the polymers with higher electron-donating ability had a strong proton acceptor, poly(VDP-*co*-DEAM) formed an interpolymer complex with polyAAc at the highest pH region of the three poly(acrylamide derivatives). The micro-environment around the VDP unit in the acidic pH region was affected not only by the proton-accepting ability, but also by the hydrophobicity of the poly(acrylamide derivatives) used.

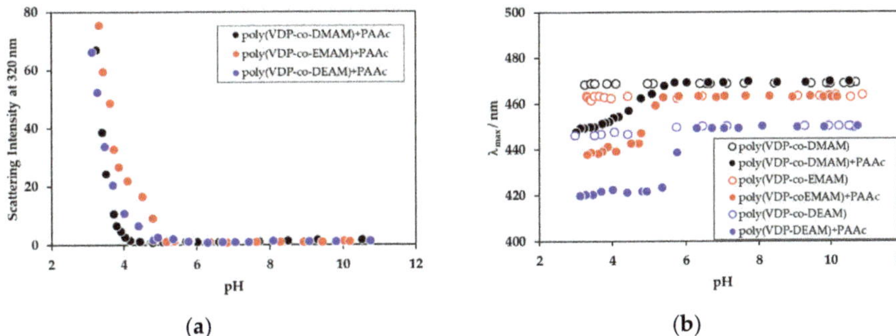

Figure 6. Scattering intensity at 320 nm and fluorescence λ_{max} values of the mixed solutions of poly(VDP-*co*-acrylamide derivatives) and polyAAc, as a function of pH.

4. Conclusions

In this paper, the interpolymer complexes between two kinds of poly(carboxylic acids) and three kinds of poly(acrylamide derivatives) with different *N*-substituent groups were investigated. Both of the poly(carboxylic acids) formed interpolymer complexes with polyDMAM. The micro-environments around the VDP unit in the acidic pH region for the polyMAAc and polyDMAM mixed systems were more hydrophobic than those in the polyAAc and polyDMAM mixed systems, as the α-methyl group of the MAAc unit affected the hydrophobicity around the polymer chain during hydrogen bond formation. This suggests that when the poly(carboxylic acids) and poly(acrylamide derivatives) were mixed and then, the solution pH was decreased, a hydrogen bond was partially formed, following which the micro-environment around the polymer chains changed in hydrophobicity, resulting in the formation of interpolymer complexes. The electron-donating ability of the carbonyl group in the poly(acrylamide derivatives) affected the complexation with polyAAc.

Author Contributions: Conceptualization, Y.M. and K.I.; Y.M. carried out experiments and analyzed the data; writing—original draft preparation, review and editing Y.M.; K.I. supervised the work.

Funding: This research received no external funding.

Conflicts of Interest: The authors declare no conflict of interest.

References

1. Tsuchida, E.; Abe, K. Interactions between macromolecules in solution and intermacromolecular complexes. *Adv. Polym. Sci.* **1982**, *45*, 1–99.

2. Khutoryanskiy, V.V. Hydrogen-bonded interpolymer complexes as materials for pharmaceutical applications. *Int. J. Pharm.* **2007**, *334*, 15–26. [CrossRef] [PubMed]

3. Lankalapalli, S.; Kolapalli, V.R.M. Polyelectrolyte Complexes: A Review of their Applicability in Drug Delivery Technology. *Indian J. Pharm. Sci.* **2009**, *71*, 481–487. [CrossRef] [PubMed]

4. Ruiz-Rubio, L.; Laza, J.M.; Pérez, L.; Rioja, N.; Bilbao, E. Polymer–polymer complexes of poly(*N*-isopropylacrylamide) and poly(*N,N*-diethylacrylamide) with poly(carboxylic acids): A comparative study. *Colloid Polym. Sci.* **2014**, *292*, 423–430. [CrossRef]

5. Winnik, F.M.; Regismond, S.T.A. Fluorescence methods in the study of the interactions of surfactants with polymers. *Colloids Surf. A Physicochem. Eng. Asp.* **1996**, *118*, 1–39. [CrossRef]

6. Matsumura, Y.; Iwai, K. Synthesis and thermo-responsive behavior of fluorescent labeled microgel particles based on poly(*N*-isopropylacrylamide) and its related polymers. *Polymer* **2005**, *46*, 10027–10034. [CrossRef]

7. Uchiyama, S.; Matsumura, Y.; De Silva, A.P.; Iwai, K. Fluorescent molecular thermometers based on polymers showing temperature-induced phase transitions and labeled with polarity-responsive Benzofurazans. *Anal. Chem.* **2003**, *75*, 5926–5935. [CrossRef] [PubMed]

8. Ruiz-Pérez, L.; Pryke, A.; Sommer, M.; Battaglia, G.; Soutar, I.; Swanson, L.; Geoghegan, M. Conformation of Poly(methacrylic acid) Chains in Dilute Aqueous Solution. *Macromolecules* **2008**, *4*, 2203–2211. [CrossRef]

9. Laguecir, A.; Ulrich, S.; Labille, J.; Fatin-Rouge, N.; Stoll, S.; Buffle, J. Size and pH effect on electrical and conformational behavior of poly(acrylic acid): Simulation and experiment. *Eur. Polym. J.* **2006**, *42*, 1135–1144. [CrossRef]

10. Wang, Y.; Morawetz, H. Fluorescence study of the complexation of poly(acrylic acid) with poly(*N*,*N*-dimethylacrylamide-*co*-acrylamide). *Macromolecules* **1989**, *22*, 164–167. [CrossRef]

11. Aoki, T.; Kawashima, M.; Katono, H.; Sanui, K.; Ogata, N.; Okano, T.; Sakurai, Y. Temperature-Responsive Interpenetrating Polymer Networks Constructed with Poly(acrylic acid) and Poly(*N*,*N*-dimethylacrylamide). *Macromolecules* **1994**, *27*, 947–952. [CrossRef]

12. Liu, S.; Fang, Y.; Gao, G.; Liu, M.; Hu, D. Fluorescence probe studies on the complexation between poly(methacrylic acid) and poly(*N*,*N*-diethylacrylamide). *Spectrochim. Acta A Mol. Biomol. Spectrosc.* **2005**, *61*, 887–892. [CrossRef] [PubMed]

13. Weinberg, N.L. Technique of electroorganic synthesis Part II. In *Techniques of Chemistry*; John Wiley & Sons: Hoboken, NJ, USA, 1975; p. 668.

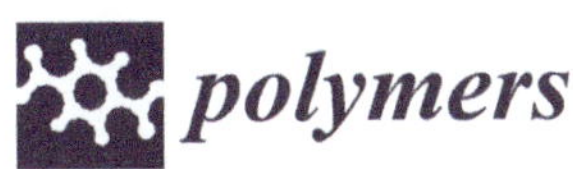 

Article

Two-Step Energy Transfer Dynamics in Conjugated Polymer and Dye-Labeled Aptamer-Based Potassium Ion Detection Assay

Inhong Kim [1,†], Ji-Eun Jung [2,†], Woojin Lee [3], Seongho Park [3], Heedae Kim [4], Young-Dahl Jho [1], Han Young Woo [2] and Kwangseuk Kyhm [3,*]

[1] School of Electrical and Computer Science, Gwangju Institute of Science and Technology (GIST), Gwangju 61005, Korea
[2] Department of Chemistry, Korea University, Seoul 02841, Korea
[3] Department of Optics & Mechatronics Engineering, Pusan National University, Busan 46241, Korea
[4] School of Physics, Northeast Normal University, Changchun 130024, China
* Correspondence: kskyhm@pusan.ac.kr; Tel.: +82-051-510-2728
† These authors were equally contributed.

Received: 17 June 2019; Accepted: 18 July 2019; Published: 19 July 2019

Abstract: We recently implemented highly sensitive detection systems for photo-sensitizing potassium ions (K^+) based on two-step Förster resonance energy transfer (FRET). As a successive study for quantitative understanding of energy transfer processes in terms of the exciton population, we investigated the fluorescence decay dynamics in conjugated polymers and an aptamer-based 6-carboxyfluorescein (6-FAM)/6-carboxytetramethylrhodamine (TAMRA) complex. In the presence of K^+ ions, the Guanine-rich aptamer enabled efficient two-step resonance energy transfer from conjugated polymers to dyed pairs of 6-FAM and TAMRA through the G-quadruplex phase. Although the fluorescence decay time of TAMRA barely changed, the fluorescence intensity was significantly increased. We also found that 6-FAM showed a decreased exciton population due the compensation of energy transfer to TAMRA by FRET from conjugated polymers, but a fluorescence quenching also occurred concomitantly. Consequently, the fluorescence intensity of TAMRA showed a 4-fold enhancement, where the initial transfer efficiency (~300%) rapidly saturated within ~0.5 ns and the plateau of transfer efficiency (~230%) remained afterward.

Keywords: FRET; time-resolved photoluminescence; two-step FRET; potassium ion detection

---

1. Introduction

In the context of various optical sensing applications, Förster resonance energy transfer (FRET) has been extensively investigated over decades due to its superior capacity to detect unknown particles and their conformational change at the molecular scale as well as their energy harvesting nature via amplification in the selected spectral windows [1–6]. FRET is a distance-dependent phenomenon, which is built on the basis of non-radiative energy transfer from energy donors to energy acceptors within close proximity (~10 nm) via long-range dipole–dipole interactions [7]. Thus, the design of FRET-based optical sensing assays always needs a platform for a proper intermolecular distance and a specificity for target molecules. Most of the myriad of recent FRET configurations are relevant to this intrinsic sensitivity to nanoscale change between two dipoles and the selection of proper materials, including fluorophores and recognition elements [8].

Conjugated polymers (CPs) have been utilized as an optical platform for many bio- or chemical applications due to their useful optical and electronic properties characterized by delocalized π-electrons [5,9–18]. In particular, cationic CPs with terminal quaternary ammonium groups were recently used in optical DNA sequence detection through electrostatic complexation, which provides a noble route to molecular distance control [1,2,5].

Aptamers are nucleic acid molecules that bind to specific targets, forming a secondary-folded structure [19,20]. Recently, they have attracted attention as an alternative conventional recognition component such as antibodies and various biosensor applications [21]. The advantages of aptamers compared with conventional recognition elements lies in their cost-effective production, easy modification, and low immunogenicity [22]. In particular, some specific single-stranded aptamers with guanine (G)-rich base sequences have a high affinity and high specificity for alkali metal ion. They can construct a secondary-folded structure, a so-called G-quadruplex, in the presence of specific alkali metal ions through hydrogen bonding [23,24].

As a one of the main cations in intracellular fluids in living bodies, potassium ion plays an important role in physiological activities as well as biological processes, for example, in maintenance of muscular strength, extracellular osmolality, enzyme activation, and apoptosis [25–27]. Because many diseases like diabetes, anorexia, bulimia, and heart disease are also closely related to abnormal potassium ion concentration, monitoring of potassium levels is crucial for clinical diagnosis [28]. Various studies for the detection of K^+ ions have been reported; however, selectivity against other intra/extra-cellular cations (Na^+) and detection sensitivity still need to be improved.

Recently, we demonstrated a noble potassium ion detection assay consisting of water-soluble CPs and dye-labeled aptamers based on FRET [1]. In this FRET system, dye-labeled aptamers play two roles simultaneously, as not only a scaffold for FRET signaling but also a receptor for metal ions. The presence of K^+ ions within a solution results in the conformational change of complex molecules consisting of positively charged CPs and negatively charged aptamers. This phenomenon was observed through a dramatic fluorescence enhancement. Nevertheless, the dynamics of sequential energy transfers are completely unknown.

When FRET is occurring, donor fluorophores absorb the energy under the irradiation of incident light, then transfer the excited energy to nearby acceptor materials. In the presence of proper acceptors, efficient energy transfer leads to significantly quenched donor fluorescence intensity, providing the amplified acceptor fluorescence. This intensity variation is often measured by time-integrated fluorescence measurement. However, the fluorescence intensity can easily vary due to the changes in intensity fluctuations of excitation light, photobleaching, and light scattering [29]. In particular, the presence of metallic particles can alter the surrounding conditions, which may influence the optical properties of molecules. They may also act as collisional quenchers of fluorescence [30]. Moreover, we have to separately distinguish complexation-induced quenching from FRET-based fluorescence signals to increase our understanding of the molecular dynamics. In general, the correlation between FRET efficiency and changes in donor lifetime can be supported by the equation below:

$$\phi = 1 - \frac{\tau_{DA}}{\tau_D} \tag{1}$$

where τ_{DA} and τ_D are the fluorescence lifetimes of the FRET donor in the presence and absence of the FRET acceptor, respectively [31]. Since FRET efficiency is inversely proportional to the fluorescence lifetime of the donor fluorophore, the higher the FRET efficiency means, the shorter donor lifetime, suggesting a decrease in the excited lifetime of the donor is great evidence of FRET. Observing the time-related fluorescence of the FRET system will be helpful to optimize the condition for maximized FRET efficiency as well as to understand dynamic events involved in the intermolecular energy transfer phenomenon.

In this paper, we investigated the fluorescence decay dynamics of the conjugated polymer and aptamer-based 6-FAM/TAMRA complex. Following our previous demonstration of the two-step FRET-based K^+ ion detection assay, we studied the dynamics of sequential energy transfer processes in terms of exciton population variation of FRET partners. When CPs were excited with 380 nm light, the population dynamics of CPs, 6-FAM, and TAMRA were compared in the absence and presence of K^+

ions, respectively. Regarding the intermediate energy level of 6-FAM located in between the high-level CPs and the low-level TAMRA, we also excited 6-FAM selectively using 490 nm light. This enables the study of FRET from 6-FAM to TAMRA selectively. Those results allowed us to investigate the two different FRET processes separately, whereby the detection of K^+ ions was evaluated quantitatively.

2. Materials and Methods

2.1. Materials

All chemicals were purchased from Aldrich Chemical Co. (St. Louis, MA, USA) for measurement at room temperature. As a FRET donor, the polyfluorene-based CP was synthesized via the Suzuki coupling reaction of 2,7-bis(4,4,5,5-tetramethyl-1,3,2-dioxaborolan-2-yl)-9,9-bis(6′-bromohexyl)fluorene and 2,7-dibromo-9,9-bis(3,4-bis(2-(2-methoxyethoxy)ethoxy)phenyl)fluorene using $(PPh_3)_4Pd(0)$ as a catalyst in toluene/water (2:1, volume ratio) at 85 °C for 36 h (yield: 66%), followed by successive quaternization with condensed trimethylamine at room temperature. As a FRET acceptor, 6-FAM and TAMRA dyes were used. Both dyes were labeled to a high-pressure liquid chromatography (HPLC)-purified molecular aptamer with 15 bases (5′-6-FAMGGTT GGTG TGGT TGG-6-TAMRA-3′) obtained from Sigma-Genosys (The Woodlands, TX, USA). For sensing of K^+ ions, 20 μL of the aptamer stock solution was diluted to 2 mL buffer, and the resulting solution was incubated at 60 °C for 30 min with and without K^+ ions. Detailed information about the sample fabrication including chemical synthesis and polymerization can be found elsewhere [1].

2.2. Time-Integrated Fluorescence and Quantum Yield Measurement

The fluorescence spectra of three fluorophores (CPs, 6-FAM, and TAMRA) dissolved in water solution were obtained by spectrofluorometer (Jasco, FP-6500, Hachioji, Tokyo, Japan), where a xenon lamp was used as an excitation source. The corresponding quantum yield of the CPs was estimated relative to a freshly prepared fluorescein solution in water at pH = 11.

2.3. Time-Resolved Fluorescence Measurement

The ultrafast decay dynamics of three fluorophores in the absence and presence of K^+ ions were measured using a conventional time-correlated single photon counting (TCSPC) system (SPC-130EM, Becker & Hickl GmbH, Berlin, Germany). For excitation light sources, 380 and 490 nm were used, which were obtained by second harmonic generation from the fundamental laser light (680~1080 nm) of a femtosecond oscillator (Chameleon Ultra-II, Coherent, Santa Clara, CA, USA). The pulse duration and repetition rate of femtosecond light was 70 fs and 80 MHz, respectively. For the detection of time-resolved fluorescence, a high-speed single photon detector (PMH-100, Becker & Hickl GmbH) and a photomultiplier tube (PMT) were used. The temporal resolution was around 190 ps.

3. Results and Discussion

3.1. Two-Step FRET Process in CPs and Two-Dye-Labeled Aptamer Complex

Two-step FRET systems with three fluorophores have advantages compared to one-step FRET systems, such as efficiency enhancement through relay stations and better detection sensitivity due to the lower background fluorescence of acceptors [32–34]. We carefully designed the optical potassium detection system based on the sequential energy transfer. As the energy donors and acceptors for FRET, three fluorophores (CPs, 6-FAM, and TAMRA) were selected. The three fluorophores emit different colors (blue, green, and red). Based on Bazan's scheme [2], which enables tuning of intermolecular distance at the molecular scale by Coulomb interaction between donor and acceptor fluorophores, CPs as FRET donors were positively charged due to their side chains, including cations. On the other hand, a guanine (G)-rich aptamer containing 15 bases (GGTT GGTG TGGT TGG) as a scaffold for FRET was negatively charged. Two dyes (6-FAM and TAMRA) were labeled to both ends of the aptamer.

In particular, the molecular interaction between CPs and aptamer, such as electrostatic and hydrophobic interaction, can be controlled by varying the charge ratio of CPs and aptamer. For efficient attractive interaction (i.e., all aptamers were complexed with CPs), the concentration of CPs was adjusted by changing its repeating units on the backbone.

Figure 1 shows our potassium ion detection system schematically, which consists of CPs and the two-dye-labeled aptamer complex. Since both CPs and aptamer are flexible in water solvent, they will be randomly distributed in a solution after complexation. The complexation by the attractive interaction between CPs and aptamer keeps the two dyes (i.e., 6-FAM and TAMRA) in close vicinity of the CPs. While the excitons generated optically in the CPs will be diffused over the backbone of CPs due to its delocalized electronic nature, the excitation energy of CPs can be transferred to the two adjacent dyes of 6-FAM and TAMRA by dipole–dipole coupling. The dipole–dipole interaction depends on the dipole orientation factor (κ^2) [30], and κ^2 is calculated by the consideration of the spatial distribution of donor and acceptor dipoles and the freedom of dipole motion, which are limited by surrounding environment. Nevertheless, the allowed range of the dipole motion may be narrowly defined or close to isotropic [35]. In our FRET system, we assumed that κ^2 was 2/3 by the dynamic averaging of fluorophores [36].

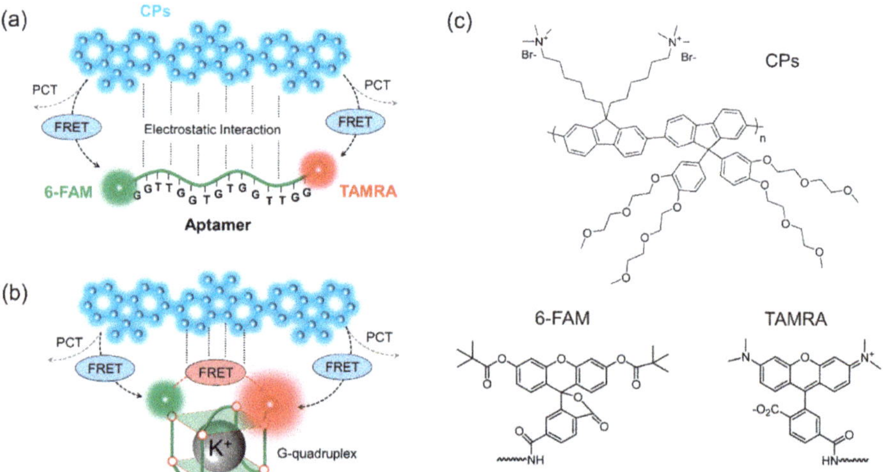

Figure 1. Schematics for potassium detection based on FRET. (**a**) In the absence of K$^+$ ions, the dominant energy transfer process occurs from CPs to 6-FAM or TAMRA, that is, one-step FRET; (**b**) In the presence of K$^+$ ions, by preferred molecular interaction between metal ions and guanine bases, the secondary structure (g-quadruplex) leads to the sequential energy transfer from CPs to 6-FAM to TAMRA, that is, two-step FRET; (**c**) The chemical structure of CPs, 6-FAM, and TAMRA.

In order to understand the energy migration mechanisms within a system, we assumed the simple case of a single donor–aptamer pair and the effective intermolecular distance for the energy transfer. The average separation between CPs and aptamer was in the 1~10 nm range. As shown in Figure 1a, the energy of CPs was transferred to both 6-FAM and TAMRA simultaneously. It is noticeable that the size of an aptamer is much longer than the effective FRET distance. Thus, only one-step energy transfer from CPs to 6-FAM or TAMRA is possible, i.e., CPs are a FRET donor while both 6-FAM and TAMRA will act as FRET acceptors. The energy transfer efficiency by FRET will be determined by the extent of the resonance coupling among each of the donor–acceptor pairs. However, FRET partners will also result in unwanted fluorescence quenching by the surrounding environment. The environmental factors, including electrostatic complex between charged molecules and physical response (hydrophobic or hydrophilic) of constituents over the solvent can affect fluorescence intensity.

As a result, some of the energy can be dissipated through the exciton deactivation process, resulting in fluorescence quenching. In the case of organic molecules, various fluorescence quenchings often occur by molecular contact between the fluorophores and quenchers. Thus, if the intermolecular distance is sufficiently short and the energy levels of acceptors is low enough compared to that of donors, excitons can be separated into individual charges or the separated electrons and holes of donor can be transferred to acceptors by photo-induced charge transfer (PCT) [37]. When quenchers are randomly distributed, the fluorescence quenching near the radiation boundary is determined by the encounter distance and the quencher concentration [30]. Furthermore, delocalized excitons migrating along the CPs' chains can also be scattered or dissociated into charge carriers by hole polarons or trap sites [38]. We have assumed that the intrinsic optical properties of constituents are unaffected or unchanged by the complexation and the addition of new molecules in order to simplify our understanding on phenomenological dynamics. Therefore, total energy transfer will be determined by the competition between energy harvesting one-step FRET and energy-wasting PCT because both processes have an intermolecular distance-dependent nature: FRET rate (k_{FRET}) is inversely proportional to the sixth power of the donor–acceptor distance ($\sim 1/R_{\mathrm{DA}}^6$), while PCT rate (k_{PCT}) has an exponential distance dependence ($k_{\mathrm{PCT}} \propto e^{-R_{\mathrm{DA}}}$) [39].

The presence of K^+ ions leads to a new molecular configuration due to the specific binding between aptamers and metal ions, as shown in Figure 1b. This molecular configuration is known as g-quadruplex, with a planar motif generated from the pairing of four guanine residues through Hoogsteen-like hydrogen bonding [40,41]. Guanine (G)-rich base sequences of aptamers can make secondary-folded structures in the presence of metal ions. In practice, our guanine-rich aptamer is designed for this specific capturing of K^+ ions, while there is no spectral shift in the intrinsic emission of 6-FAM and TAMRA [1]. In this case, one-step FRET is no more a dominant emission process. Since guanine tetrad (g-quartet) shortens the intermolecular distance among two dyes attached to both edges of an aptamer, both additional energy transfer between the two dyes and the sequential energy transfer from CPs to TAMRA via 6-FAM (two-step FRET) are possible. The 6-FAM will act as not only an intermediator for the two-step FRET, but also as a FRET acceptor in the one-step FRET. On the other hand, TAMRA always performs as a FRET acceptor due to its low bandgap energy. Consequently, TAMRA's emission will be enhanced by CPs and 6-FAM. In other words, the fluorescence enhancement in TAMRA indicates the detection of K^+ ions. The detailed molecular structures of CPs, 6-FAM, and TAMRA are given in Figure 1c.

3.2. The Spectral Overlap between Conjugated Polymer, 6-FAM, and TAMRA

Figure 2a–c shows the intrinsic optical properties of three fluorophores (i.e., CPs, 6-FAM, and TAMRA), respectively. Absorption spectra were shown in terms of the molar extinction coefficient and emission spectra were normalized by the maximum fluorescence intensity. The three fluorophores of CPs, 6-FAM, and TAMRA show a broad absorption and emission spectrum with a Stoke shift. Their mirror symmetric shape between absorption and emission spectra is attributed to the overlap of the initial and final wave functions relevant to the vibronic transition at thermal equilibrium [42], and the spectral intensity is determined by the probability amplitude of energy levels involved in the vibronic transition. From the absorption measurement, the molar extinction coefficient of CPs calculated by the Beer–Lambert law was $0.52 \times 10^4 \ \mathrm{M}^{-1} \cdot \mathrm{cm}^{-1}$ and CPs had an absorption maximum at 397.8 nm. Unlike absorption spectra, the blue color emission spectrum of CPs showed two peaks (i.e., 426.6 and 451.0 nm). The quantum efficiency of CPs was 0.58 in a water solvent. Both the spectrum of 6-FAM and TAMRA showed perfect mirror symmetry with high molar absorption coefficients ($1.25 \times 10^4 \ \mathrm{M}^{-1} \cdot \mathrm{cm}^{-1}$ for 6-FAM and $1.96 \times 10^4 \ \mathrm{M}^{-1} \cdot \mathrm{cm}^{-1}$ for TAMRA). The maximum absorption peaks of two dyes appear at 490.4 and 567.6 nm, respectively. 6-FAM shows a greenish color emission with the peak wavelength at 514.4 nm. The TAMRA showed a reddish color emission with the emission maximum at 590.2 nm.

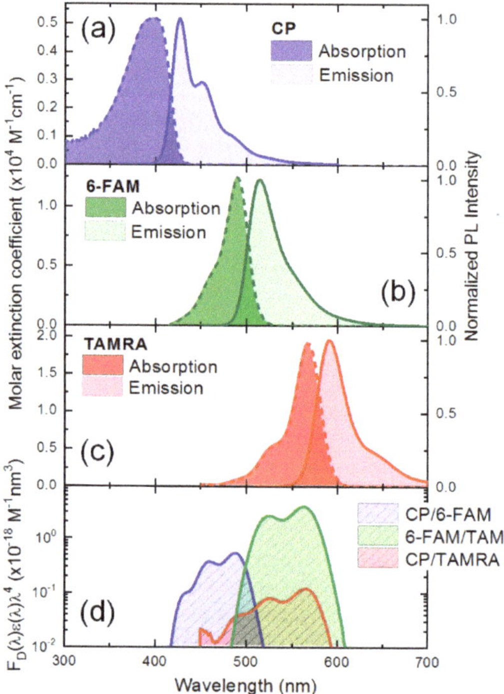

Figure 2. The optical properties of the three fluorophores (i.e., CPs, 6-FAM, and TAMRA) and the spectral overlap among them. The absorption and the emission spectra of (**a**) CPs; (**b**) 6-FAM; and (**c**) TAMRA were characterized in terms of the molar extinction coefficient and normalized by the maximum fluorescence intensity, respectively; (**d**) The spectral overlap between the three fluorophores was calculated by integrating the product of the absorption spectra and the emission spectra.

Since FRET phenomenon is analogous to the interaction of a coupled oscillation system, where two oscillators harmonically interact with each other at a separation R, the spectral overlap among FRET partners represents the extent of resonant coupling [43]. In FRET measurement, spectral overlap (J) can be simply calculated by integrating an overlap area between the normalized emission spectra of the donor $f_D(\lambda)$ and the absorption of the acceptor with a molar extinction coefficient ($\varepsilon_A(\lambda)$) [30]:

$$J = \int f_D(\lambda)\varepsilon_A(\lambda)\lambda^4 d\lambda \tag{2}$$

Figure 2d shows a spectral overlap in three fluorophores. The calculated spectral overlap between CPs and 6-FAM (2.47×10^{-26} $M^{-1}\cdot nm^4$) was higher than that between CPs and TAMRA (8.00×10^{-27} $M^{-1}\cdot nm^4$). This result implies that the excitation energy of CPs can be transferred effectively to 6-FAM compared to TAMRA when only one-step FRET process is considered in the absence of K^+ ions. It is noticeable that the spectral overlap between 6-FAM and TAMRA (5.00×10^{-26} $M^{-1}\cdot nm^4$) was higher than that involved with CPs. Furthermore, we can calculate a characteristic Förster distance (R_0) from the spectral overlap, where the FRET efficiency becomes 50% [8]. Regarding the three FRET partners, the three Förster distances were calculated for CPs/6-FAM (33.1 Å), CPs/TAMRA (27.4 Å), and 6-FAM/TAMRA (44.8 Å), respectively. These results imply that a selection of FRET partners and a system design is theoretically suitable for efficient FRET. Along with the spectral overlap, the Förster distance also provides a clue for the conformational change in molecular scale, and this can help us estimate a relative donor–acceptor distance through a theoretical FRET efficiency estimation [30,44].

However, the accuracy of the relative intermolecular distance calculation is limited due to the presence of various quenching processes because the original FRET equation assumes that the variation in energy transfer efficiency has only resulted in FRET. Although the intermolecular distance calculation for individual energy transfer processes is difficult due to the complexity of system, we can estimate total energy transfer efficiency regardless of the number of transfer steps.

3.3. The Fluorescence Decay Dynamics of the Conjugated Polymer, 6-FAM, and TAMRA

For a better understanding of the energy transfer processes among FRET partners in the absence and presence of K^+ ions, we measured the time-resolved fluorescence. First of all, we measured the fluorescence decay of fluorophores before the complexation with the two-dye-labeled aptamer (denoted by "Free") was measured individually to understand the intrinsic fluorescence decay time of fluorophores. Then, we compared it with the measured data in the absence and presence of K^+ ions. The measured fluorescence decay curve was fitted by single- or multi-exponential decay function taking into account the change in the decay curvature [30]. In the fluorescence decay curve, the multiple decay components generally implied the contribution of additional decay pathways due to the molecular interaction to the whole fluorescence decay. Thus, we distinguished fast and slow decay components from the multi-exponential decay curvature by taking into account the intrinsic fluorescence decay time of fluorophores.

To calculate the energy transfer efficiency of FRET, we used a rate equation or its time-dependent differential equation. In general, the fluorescence decay rate (k_F) is defined by the sum of all radiative (k_{rad}) and non-radiative decay (k_{nonrad}) components [30].

$$k_F = k_{rad} + k_{nonrad} = \frac{1}{\tau_{rad}} + \frac{1}{\tau_{nonrad}} \tag{3}$$

where τ_{rad} and τ_{nonrad} represent radiative and non-radiative decay time, respectively. Since the fluorescence intensity (I) is proportional to the exciton population (N), the time-dependent fluorescence decay can be characterized by a differential equation form of exponential function [39,45].

$$I(t) = \int N(t)dt \tag{4}$$

$$\frac{dN(t)}{dt} = g - \frac{N(t)}{\tau_{int}} \tag{5}$$

where g indicates carrier (exciton) generation function by excitation light. τ_{int} represents an intrinsic fluorescence decay time including all radiative and non-radiative decay components. When decay pathways of FRET and charge transfer are involved, the rate equation can be modified by an additional non-radiative decay time (τ^*_{nonrad}) [45].

$$\frac{dN(t)}{dt} = g - \frac{N(t)}{\tau_{int}} - \frac{N(t)}{\tau^*_{nonrad}} \tag{6}$$

In particular, this differential rate equation effectively provides the relation between an individual fluorescence decay rate and a time-dependent exciton population.

3.3.1. The Fluorescence Decay of Conjugated Polymers

Figure 3 shows the fluorescence decay curve of CPs under 380 nm excitation where this wavelength corresponds to the absorption maximum of CPs and minimizes the direct absorption by 6-FAM. Free CPs before the complexation with aptamer show a single exponential decay curve with 0.38 ns (without K^+) and 0.43 ns (with K^+) decay time. In the absence of K^+ ions, the fluorescence decay time of multi-exponential components in CPs were 0.35 ns (fast) and 1.76 ns (slow), respectively. In particular, the fluorescence decay curve of CPs with K^+ rarely changed compared to that of Free CPs, and at initial decay time, one-step FRET occurs. This result represents that the complexation effect hardly changed the intrinsic fluorescence decay time of CPs. On the other hand, the addition of K$^+$ ions significantly affected the fluorescence decay dynamics of CPs. Within 0.50 ns, the time-resolved fluorescence intensity of CPs showed a significant decrease, whereby two decay times of 0.19 ns (fast) and 1.46 ns (slow) were obtained. Regarding the FRET process occurring within several hundred ps time, we can infer that the energy transfer became effective due to the emergence of additional decay pathways relevant to the presence of K$^+$ ions, i.e., two-step FRET.

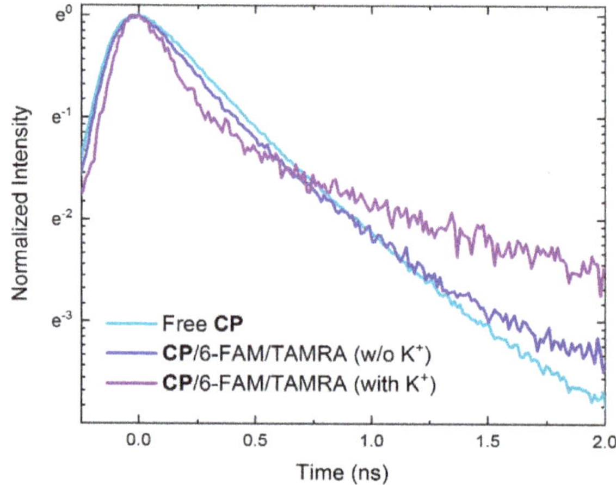

Figure 3. Fluorescence decay dynamics of CPs without the complexation with the two-dye-labeled (i.e., 6-FAM and TAMRA) aptamer (Free CP), in the absence (*w/o*) and presence (with) of K$^+$ ions after the complexation with the two-dye-labeled aptamer.

Given the fluorescence decay rate equation including an additional exciton decay, we estimated the decay rate of CPs. The total rate equation and its variation ($\Delta k(t)$) before and after molecular interaction can be described as:

$$k(t) = g - \left\{ \frac{dN(t)}{dt} \right\} N(t)^{-1} \tag{7}$$

$$\Delta k(t) = k_i(t) - k_f(t) \tag{8}$$

where subscripts denote initial (Free) and final states (complexation or after K$^+$ ion addition). In the presence of K$^+$ ions, the decay rate of CPs (3.58 ns^{-1}) involved in the two-step FRET was the fastest compared to other cases (2.29 ns^{-1} for Free CPs and 2.58 ns^{-1} for complexation) at the start moment of fluorescence decay. Afterward, the decay rate of CPs with K$^+$ ions drastically decreased while the decay rate of other cases barely changed until 1 ns. This result shows our prediction of an additional non-radiative decay pathway is plausible.

3.3.2. The Fluorescence Decay of 6-FAM

The fluorescence decay dynamics of 6-FAM or TAMRA in the complexation phase of CPs and aptamer were also measured. The fluorescence decay of individual dyes (i.e., Free 6-FAM and Free TAMRA before complexation) was undistinguishable due to their low absorption coefficient at 380 nm excitation. First, we investigated the fluorescence decay dynamics of 6-FAM. The 6-FAM performs the role of an intermediator in the two-step FRET process in the presence of CPs, but it always becomes a FRET donor in the absence of CPs. To clarify the different roles of 6-FAM associated with CPs, we measured the fluorescence decay dynamics of 6-FAM in the absence and presence of K$^+$ ions, as shown in Figure 4a. Two kinds of excitation lasers (i.e., 380 and 490 nm) were used. The 380 nm excitation excited carriers mostly in CPs, giving rise to a FRET from CPs to 6-FAM. On the other hand, 490 nm excitation generated carriers in 6-FAM selectively without any excitation in CPs. Hence, no FRET occurred from CPs to 6-FAM. Regardless of K^+ ions, the fluorescence intensity of 6-FAM at the moment of excitation rarely changed under 380 nm excitation. However, under 490 nm excitation, the presence of K^+ ions led to a significant decrease of the fluorescence intensity at 6-FAM. When FRET occurs, a decrease in the fluorescence intensity is generally observed in a donor molecule as evidence of the energy transfer. Thus, this decreased intensity at 6-FAM resulting from 490 nm excitation indicated a one-step FRET from 6-FAM to TAMRA, where 6-FAM was a donor in the FRET process. One should be reminded that no carrier was excited in CPS when 490 nm excitation was used. However, with 380 nm excitation, 6-FAM seemed to play a role as the two-step FRET intermediator. Initially, carriers were excited in CPs, then transferred to 6-FAM. Because the energy transfer from 6-FAM to TAMRA was efficient, the carriers in 6-FAM are immediately transferred to TAMRA. As a result, the population of 6-FAM remains constant due to the dynamic compensation effect. The intrinsic decay time of Free 6-FAM was 3.72 ns under 490 nm excitation. However, we observed multi-exponential decay in the absence of K^+ ions. The two decay times were 0.36 ns (fast) and 2.09 ns (slow) at 490 nm excitation, respectively. The fast decay component also indicated FRET from 6-FAM to TAMRA. On the contrary, the fluorescence decay at 6-FAM with 380 nm excitation showed a single exponential shape with a longer decay time of 0.49 ns compared to that with 490 nm excitation. The elongated decay time possibly resulted from the population increase resulting from the energy transfer from CPs to 6-FAM.

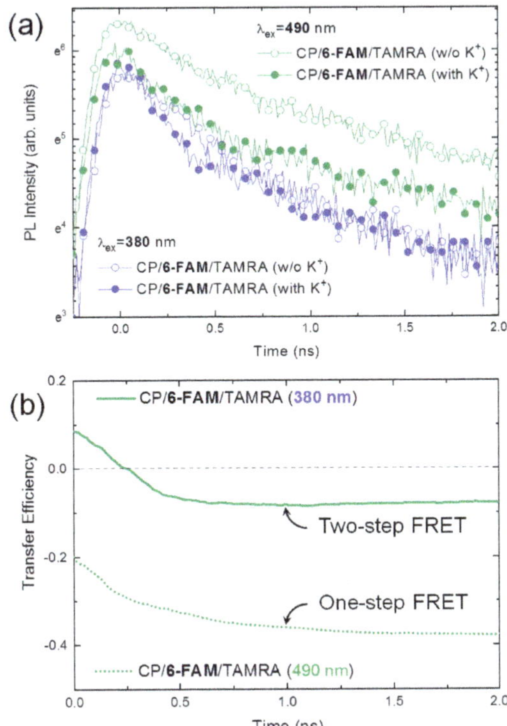

Figure 4. (a) The fluorescence decay of 6-FAM in the absence and presence of K^+ ions. Time-resolved fluorescence decay was measured by two kinds of excitation wavelength (**blue** for 380 nm and **green** for 490 nm); (b) The energy transfer efficiency of 6-FAM for one-step FRET (**dotted line** for 490 nm excitation) and two-step FRET (**solid line** for 380 nm excitation) was calculated using Equation (9), respectively.

Upon the addition of K^+ ions, the fluorescence lifetime of 6-FAM at initial decay times decreased regardless of the excitation wavelengths. In the case of 490 nm excitation, the fast decay time was 0.25 ns and the slow decay time was 2.18 ns, respectively. With 380 nm excitation, the fluorescence decay time was 0.33 ns. Those results suggest the presence of K^+ ions accelerated the fluorescence decay of 6-FAM through two-step FRET in addition to one-step processes. In particular, it was very noticeable that the fast decay component within 0.50 ns were identically observed at the different excitation wavelengths, while the decay curvatures barely changed afterward. This implies that each energy transfer pathways can be resolved in different timescales within temporal resolution of our experimental system. Nevertheless, the fluorescence intensity and decay curvature were determined by the combination of all decay pathways.

For a quantitative evaluation of the two-step FRET, we calculated the time-dependent energy transfer efficiency in the absence and the presence of K^+ ions. From the change in time-resolved fluorescence intensity before and after K^+ ion addition, the total energy transfer efficiency ($E_{\text{total}}(t)$) of 6-FAM can be theoretically calculated by [44]:

$$E_{\text{total}}(t) = \frac{k_{\text{FRET}}(t)}{k_{\text{nonrad}}(t) + k_{\text{FRET}}(t)} = \frac{I_{\text{D}}(t) - I_{\text{DA}}(t)}{I_{\text{D}}(t)} \tag{9}$$

where $I_{\text{D}}(t)$ and $I_{\text{DA}}(t)$ are the fluorescence intensity of a donor (D) in the absence and presence of an acceptor (A), respectively. Similar to the fluorescence decay of CPs, we found that the fast decrease of the energy transfer efficiency occurs within 0.5 ns regardless of the excitation wavelength as shown in Figure 4b. Afterward, the total transfer efficiency remained constant. Those results confirm the presence of a two-step energy transfer gain. In particular, the minus sign of the transfer efficiency indicating the reduction of the total energy was noticeable. Interestingly, we found that a loss of exciton population due to the energy transfer and the fluorescence quenching at 6-FAM was compensated by CPs.

3.3.3. The Fluorescence Decay of TAMRA

Finally, we measured the fluorescence decay dynamics of TAMRA in the absence and presence of K^+ ions, respectively. To discern individual energy migration by one- or two-step FRET from the total energy transfer process, the fluorescence decay of TAMRA without 6-FAM (single dye-labeled aptamer denoted by "CP/TAMRA") was also measured. Figure 5a shows the fluorescence decay of TAMRA under 380 nm laser excitation. As discussed above, the addition of K^+ ions resulted in an enhancement of the acceptor's exciton population by not only an additional energy transfer but also the fundamental interaction of the complexation with CPs. Consequently, in the presence of K^+ ions, the fluorescence intensity of TAMRA significantly increased compared to that of the fundamental one-step energy transfer. The four-folded enhancement of fluorescence intensity can be understood as a result of the energy transfer since TAMRA is always an energy acceptor in our FRET system. Interestingly, TAMRA without 6-FAM had the highest fluorescence intensity at the moment of excitation despite their weak absorption coefficient at 380 nm and the narrow spectral overlap. Its fluorescence intensity was much higher than that of others. This unusual feature may be attributed to the number of FRET participants regarding the total amount of energy CPs offered was fixed. In the case that there is no 6-FAM, the energy exchange with TAMRA is the only possible pathway. But, the preferential direction of the energy migration is decided by the spatial distance from CPs to dyes as well as the probability wavefunction distribution of the two dyes in the solution. Assuming that the total amount of energy transferred from CPs is equal to the fluorescence intensity of TAMRA without 6-FAM, the fluorescence intensity ratio at the moment of the excitation are $I_{\text{FT}}/I_{\text{T only}} \approx 0.19$ without K^+ ions and $I_{\text{FT}}/I_{\text{T only}} \approx 0.64$ with K^+ ions, where $I_{\text{T only}}$ and I_{FT} indicate the fluorescence intensities of TAMRA-labeled and two-dye-labeled aptamers, respectively. Since the emission of TAMRA is irrelevant to that of 6-FAM in the absence of K^+ ions, we can predict the probabilistic apportionment of the excitation energy. On the other hand, as expected by the Förster distance, this result may indicate that one-step FRET from CPs to TAMRA is preferred compared to 6-FAM in spite of a weak absorption and narrow spectral overlap. Since the size of the aptamer is smaller than that of polymers, CPs may surround the aptamer within the system and the average separation of TAMRA from CPs may be closer. Considering that the quenching effect was not observed in TAMRA unlike 6-FAM, this assumption may be reasonable.

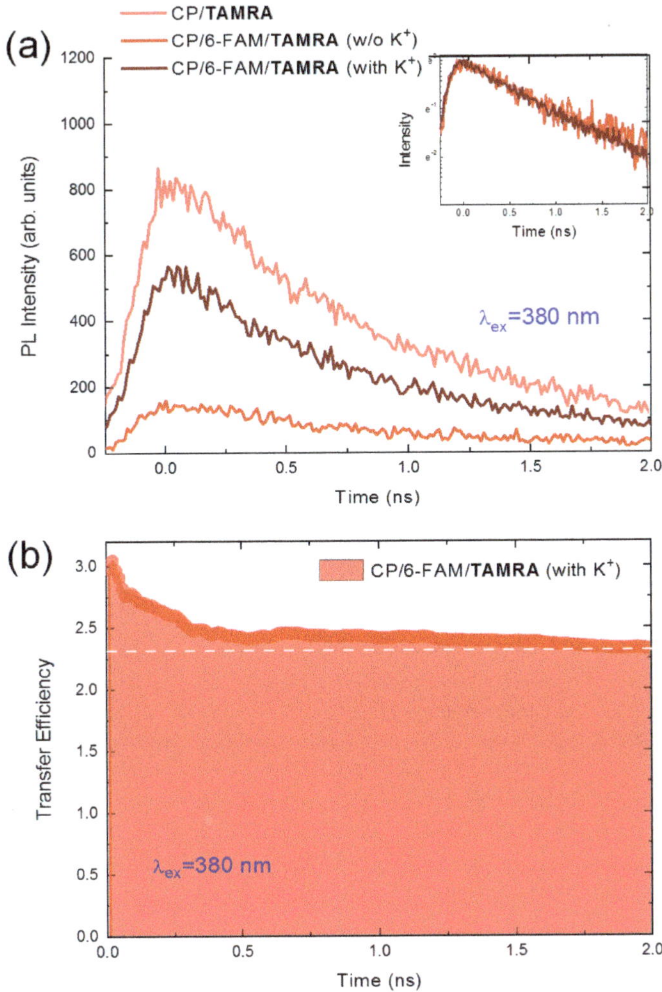

Figure 5. (a) To confirm the energy transfer from CPs to TAMRA, the fluorescence decay of TAMRA without 6-FAM was measured (**light red**, CP/TAMRA). The fluorescence decay of TAMRA in the absence (**red**) and presence (**wine**) of K$^+$ ions; (b) The total transfer efficiency TAMRA gained by two-step FRET was calculated using Equation (10).

Despite the fluorescence enhancement in the presence of K$^+$ ions, the fluorescence decay curvature (~1 ns decay time) of TAMRA barely changed due to the non-radiative transfer nature of the FRET, as shown in the inset of Figure 5a. Furthermore, we estimated the energy transfer efficiency ($\eta(t)$) of TAMRA described in Reference [44]:

$$\eta(t) = \frac{\int I_{\text{with K}^+}(t)dt - \int I_{\text{w/o K}^+}(t)dt}{\int I_{\text{w/o K}^+}(t)dt} \tag{10}$$

where $I_{\text{w/o K}^+}(t)$ and $I_{\text{with K}^+}(t)$ indicate the fluorescence intensity of TAMRA at time t in the absence and presence of K^+ ions, respectively. This equation was described in the view of the gain or loss of the exciton population during the two-step FRET process. In the absence of K^+ ions, the population of TAMRA was supported by CPs through one-step FRET. The fluorescence intensity of TAMRA in the absence of K^+ ions was 100 at 2 ns in arbitrary units. On the other hand, if K^+ ions were involved, an additional two-step FRET occurred from CPs to TAMRA via 6-FAM. The TAMRA plays an important role as an intermediator, where the energy transfer from CPs to TAMRA is mediated. The energy transfer efficiency equation enables to estimate the transferred fluorescence from TAMRA in terms of the normalized ratio to the fluorescence intensity of TAMRA in the absence of K^+ ions. When the increased fluorescence intensity of TAMRA at 2 ns was 330, the energy transfer efficiency of 230% was obtained. It is surprising that two-step FRET resulted in a remarkable population enhancement. Such a large enhancement can be attributed to the delocalized electronic nature of CPs, giving rise to the energy harvesting effect.

In summary, Table 1 shows the fitted parameters for the fluorescence decay of CPs, 6-FAM, and TAMRA. The fluorescence decay time is calculated by single- or multi-exponential fitting, as given below [30]:

$$I(t) = I_0 + a_1 \exp\left(-\frac{t}{\tau_1}\right) + a_2 \exp\left(-\frac{t}{\tau_2}\right) \tag{11}$$

where a_1 and a_2 are the weight factors of each of the decay components. τ_1 and τ_2 are decay time. In the presence of additional decay pathways due to the conformational change or molecular interaction, the fluorescence decay time decrease and its curvature can be changed from single exponential to multi-exponential shapes as well as intensity variation. Before and after complexation, this phenomenon representing occurrence of one- or two-step FRET was commonly observed in CPs and 6-FAM. In the case of TAMRA, the reduction in decay time may result in a change in the surrounding environment. In the presence of K^+ ions, the two-step FRET process can be accelerated owing to a shortened intermolecular distance and, therefore, fluorescence decay time decreases more.

Table 1. The fitted parameters for the fluorescence decay of CPs, 6-FAM, and TAMRA.

Fluorophores	λ_{ex} (nm)	Free (ns)	w/o K$^+$ (ns)		(ns)		with K$^+$ (ns)		(ns)
		τ_1	a_1	τ_1	a_2	τ_2	a_1	τ_1	a_2	τ_2
CPs	380	0.43 (with K$^+$)	0.94	0.35	0.06	1.76	0.73	0.19	0.27	1.46
6-FAM	380	-	-	0.49	-	-	-	0.33	-	-
	490	3.72	0.51	0.36	0.49	2.09	0.61	0.25	0.39	2.18
TAMRA	380	1.01 (with CPs)	-	0.87	-	-	-	0.90	-	-

4. Conclusions

We investigated the ultrafast fluorescence decay dynamics in CPs and the aptamer-based 6-FAM/TAMRA complex. In this FRET system, the conformational change by K^+ ions promoted the sequential energy transfer from CPs, through 6-FAM, to TAMRA, as well as the fundamental one-step FRET due to the complexation. To understand the energy migration mechanisms within the system, we phenomenologically analyzed and modelled in terms of the exciton population. Also, we experimentally observed the evidence of two-step FRET and the role of three fluorophores thorough the time-resolved spectroscopic technique. As we expected, the presence of the two-step FRET process due to the presence of the K^+ ion was identified by the fluorescence enhancement (4 folded) and the change in the decay curvature. In particular, we found the role of the transition of 6-FAM as an intermediator as well as the energy receiver during the energy transfer process. To distinguish the individual energy transfer step, we also calculated the energy transfer efficiency by the rate equation. Our transfer efficiency calculation, relevant only to total energy variation, minimizes the overestimation due to the theoretical FRET rate calculus. Consequently, a steady energy transfer efficiency (230%) was observed within the period of radiative decay time.

Author Contributions: Conceptualization, J.-E.J. and H.Y.W.; Data curation, I.K. and J.-E.J.; Formal analysis, I.K. and K.K.; Funding acquisition, K.K.; Investigation, I.K., J.-E.J., W.L. and S.P.; Methodology, J.-E.J. and H.Y.W.; Project administration, K.K.; Resources, J.-E.J. and H.Y.W.; Supervision, H.Y.W. and K.K.; Visualization, I.K. and K.K.; Writing—original draft, I.K.; Writing—review and editing, H.K., Y.-D.J. and K.K.

Funding: This research was supported by a two-year research grant from Pusan National University.

Conflicts of Interest: The authors declare no conflict of interest.

References

1. Nguyen, B.L.; Jeong, J.E.; Jung, I.H.; Kim, B.; Le, V.S.; Kim, I.; Kyhm, K.; Woo, H.Y. Conjugated polyelectrolyte and aptamer based potassium assay via single- and two-Step fluorescence energy transfer with a tunable dynamic detection range. *Adv. Funct. Mater.* **2014**, *24*, 1748–1757. [CrossRef]

2. Gaylord, B.S.; Heeger, A.J.; Bazan, G.C. DNA detection using water-soluble conjugated polymers and peptide nucleic acid probes. *Proc. Natl. Acad. Sci. USA* **2002**, *99*, 10954–10957. [CrossRef] [PubMed]

3. Medintz, I.L.; Clapp, A.R.; Mattousi, H.; Goldman, E.R.; Fisher, B.; Mauro, J.M. Self-assembled nanoscale biosensors based on quantum dot FRET donors. *Nat. Mater.* **2003**, *2*, 630–683. [CrossRef] [PubMed]

4. Zhang, C.-Y.; Yeh, H.-C.; Kuroki, M.T.; Wang, T.-H. Single-quantum-dot-based DNA nanosensor. *Nat. Mater.* **2005**, *4*, 826–831. [CrossRef] [PubMed]

5. Kang, M.; Nag, O.K.; Nayak, R.R.; Hwang, S.; Suh, H.; Woo, H.Y. Signal amplification by changing counterions in conjugated polyelectrolyte-based FRET DNA detection. *Macromolecules* **2009**, *42*, 2708–2714. [CrossRef]

6. Clapp, A.R.; Medintz, I.L.; Mauro, J.M.; Fisher, B.R.; Bawendi, M.G.; Mattoussi, H. Fluorescence resonance energy transfer between quantum dot donors and dye-labeled protein acceptors. *J. Am. Chem. Soc.* **2004**, *126*, 301–310. [CrossRef]

7. Wang, Y.; Gao, D.; Zhang, P.; Gong, P.; Chen, C.; Gao, G.; Cai, L. A near infrared fluorescence resonance energy transfer based aptamer biosensor for insulin detection in human plasma. *Chem. Commun.* **2014**, *50*, 811–813. [CrossRef]

8. Sapsford, K.E.; Berti, L.; Medintz, I.L. Materials for fluorescence resonance energy transfer analysis: beyond traditional donor–acceptor combinations. *Angew. Chem. Int. Ed.* **2006**, *45*, 4562–4588. [CrossRef]

9. McQuade, D.T.; Pullen, A.E.; Swager, T.M. Conjugated polymer-based chemical sensors. *Chem. Rev.* **2000**, *100*, 2537–2574. [CrossRef]

10. Liu, B.; Bazan, G.C. Homogeneous fluorescence-based DNA detection with water-soluble conjugated polymers. *Chem. Mater.* **2004**, *16*, 4467–4476. [CrossRef]

11. Wang, S.; Gaylord, B.S.; Bazan, G.C. Fluorescein provides a resonance gate for FRET from conjugated polymers to DNA intercalated dyes. *J. Am. Chem. Soc.* **2004**, *126*, 5446–5451. [CrossRef] [PubMed]

12. Jiang, G.; Susha, A.S.; Lutich, A.A.; Stefani, F.D.; Feldmann, J.; Rogach, A.L. Cascaded FRET in conjugated polymer/quantum dot/dye-labeled DNA complexes for DNA hybridization. *ACS Nano* **2009**, *12*, 4127–4131. [CrossRef] [PubMed]

13. Du, C.; Hu, Y.; Zhang, Q.; Guo, Z.; Ge, G.; Wang, S.; Zhai, C.; Zhu, M. Competition-derived FRET-switching cationic conjugated polymer-Ir(III) complex probe for thrombin detection. *Biosens. Bioelectron.* **2018**, *100*, 132–138. [CrossRef] [PubMed]

14. Liu, Z.; Zhang, L.; Shao, M.; Wu, Y.; Zeng, D.; Cai, X.; Duan, J.; Zhang, X.; Gao, X. Fine-tuning the quasi-3D geometry: Enabling efficient nonfullerene organic solar cells based on perylene diimides. *ACS Appl. Mater. Interfaces* **2018**, *10*, 762–768. [CrossRef] [PubMed]

15. Wen, S.; Wu, Y.; Wang, Y.; Li, Y.; Liu, L.; Jiang, H.; Liu, Z.; Yang, R. Pyran-bridged indacenodithiophene as a building block for constructing efficient A-D-A-type nonfullerene acceptors for polymer solar cells. *ChemSusChem* **2018**, *11*, 360–366. [CrossRef] [PubMed]

16. Liu, Z.; Wu, Y.; Zhang, Q.; Gao, X. Non-fullerene small molecule acceptors based on perylene diimides. *J. Mater. Chem. A* **2016**, *4*, 17604–17622. [CrossRef]

17. Liu, Z.; Gao, Y.; Dong, J.; Yang, M.; Liu, M.; Zhang, Y.; Wen, J.; Ma, H.; Gao, X.; Chen, W.; et al. Chlorinated wide-bandgap donor polymer enabling annealing free nonfullerene solar cells with the efficiency of 11.5%. *J. Phys. Chem. Lett.* **2018**, *9*, 6955–6962. [CrossRef]

18. Liu, Z.; Zeng, D.; Gao, X.; Li, P.; Zhang, Q.; Peng, X. Non-fullerene polymer acceptors based on perylene diimides in all-polymer solar cells. *Sol. Energ. Mat. Sol. C.* **2019**, *189*, 103–117. [CrossRef]

19. Dunn, M.D.; Jimenez, R.M.; Chaput, J.C. Analysis of aptamer discovery and technology. *Nat. Rev. Chem.* **2017**, *1*, 1–16. [CrossRef]

20. Song, S.; Wang, L.; Li, J.; Zhao, J.; Fan, C. Aptamer-based biosensors. *TrAC-Trend Anal. Chem.* **2008**, *27*, 108–117. [CrossRef]

21. Han, K.; Liang, Z.; Zhou, N. Design strategies for aptamer-based biosensors. *Sensors* **2010**, *10*, 4541–4557. [CrossRef] [PubMed]

22. Song, K.-M.; Lee, S.; Ban, C. Aptamers and their biological applications. *Sensors* **2012**, *12*, 612–631. [CrossRef] [PubMed]

23. Davis, J.T. G-quartets 40 years later: From 5′-GMP to molecular biology and supramolecular chemistry. *Angew. Chem. Int. Ed.* **2004**, *43*, 668–698. [CrossRef] [PubMed]

24. Xu, Y.; Sugiyama, H. Formation of the G-quadruplex and i-motif structures in retinoblastoma susceptibility genes (Rb). *Nucleic Acids Res.* **2006**, *34*, 949–954. [CrossRef] [PubMed]

25. Yu, S.P.; Canzoniero, L.M.; Choi, D.W. Ion homeostasis and apoptosis. *Curr. Opin. Cell Biol.* **2001**, *13*, 405–411. [CrossRef]

26. Walz, W. Role of astrocytes in the clearance of excess extracellular potassium. *Neurochem. Int.* **2000**, *36*, 291–300. [CrossRef]

27. Lippard, S.L.; Berg, J.M. *Principles of Bioinorganic Chemistry, Chapter 1*; University Science Books: Mill Valley, CA, USA, 1994.

28. Schwartz, A.B. Potassium-related cardiac arrhythmias and their treatment. *Angiology* **1978**, *29*, 194–205. [CrossRef]

29. Lakowicz, J.R.; Szmacinski, H. Fluorescence lifetime-based sensing of pH, Ca^{2+}, KS and glucose. *Sensor Actuat. B* **1993**, *11*, 133–143. [CrossRef]

30. Lakowicz, J.R. *Principles of Fluorescence Spectroscopy*, 3rd ed.; Springer: New York, NY, USA, 2006.

31. Gopich, I.V.; Szabo, A. Theory of the energy transfer efficiency and fluorescence lifetime distribution in single-molecule FRET. *Proc. Natl. Acad. Sci. USA.* **2012**, *109*, 7747–7752. [CrossRef]

32. Kawahara, S.I.; Uchimaru, T.; Murata, S. Sequential multistep energy transfer: enhancement of efficiency of long-range fluorescence resonance energy transfer. *Chem. Commun.* **1999**, 563–564. [CrossRef]

33. Song, X.; Shi, J.; Nolan, J.; Swanson, B. Detection of multivalent interactions through two-tiered energy transfer. *Anal. Biochem.* **2001**, *291*, 133–141. [CrossRef]

34. Watrob, H.M.; Pan, C.-P.; Barkley, M.D. Two-step FRET as a structural tool. *J. Am. Chem. Soc.* **2003**, *125*, 7336–7343. [CrossRef]

35. Dale, R.E.; Eisinger, J.; Blumberg, W.E. The orientational freedom of molecular probes. The orientation factor in intramolecular energy transfer. *Biophys. J.* **1979**, *26*, 161–194. [CrossRef]

36. Khrenova, M.; Topol, I.; Collins, J.; Nemukhin, A. Estimating orientation factors in the FRET theory of fluorescent proteins: The TagRFP-KFP pair and beyond. *Biophys. J.* **2015**, *108*, 126–132. [CrossRef]

37. Woo, H.Y.; Vak, D.; Korystov, D.; Mikhailovsky, A.; Bazan, G.C.; Kim, D.-Y. Cationic conjugated polyelectrolytes with molecular spacers for efficient fluorescence energy transfer to dye-labeled DNA. *Adv. Funct. Mater.* **2007**, *17*, 290–295. [CrossRef]

38. Bardeen, C. Exciton quenching and migration in single conjugated polymers. *Science* **2011**, *331*, 544–545. [CrossRef]

39. Kim, I.; Kyhm, K.; Kang, M.; Woo, H.Y. Ultrafast combined dynamics of Förster resonance energy transfer and transient quenching in cationic polyfluorene/fluorescein-labelled single-stranded DNA complex. *J. Lumin.* **2014**, *149*, 185–189. [CrossRef]

40. Cao, Q.; Li, Y.; Freisinger, E.; Qin, P.Z.; Sigel, R.K.O.; Mao, Z.-W. G-quadruplex DNA targeted metal complexes acting as potential anticancer drugs. *Inorg. Chem. Front.* **2017**, *4*, 10–32. [CrossRef]

41. Lane, A.N.; Chaires, J.B.; Gray, R.D.; Trent, J.O. Stability and kinetics of G-quadruplex structures. *Nucleic Acids Res.* **2008**, *36*, 5482–5515. [CrossRef]
42. Fox, M. *Optical Properties of Solids.*; Oxford University Press: New York, NY, USA, 2010.
43. Clegg, R.M. The history of FRET: From conception through the labors of birth. In *Reviews in Fluorescence*; Springer: New York, NY, USA, 2006; Volume 3, pp. 1–45.
44. Clegg, R.M. Fluorescence resonance energy transfer. *Curr. Opin. Biotech.* **1995**, *6*, 103–110. [CrossRef]
45. Kyhm, K.; Kim, I.; Kang, M.; Woo, H.Y. Ultrafast dynamics of Förster resonance energy transfer and photo-induced charge transfer in cationic polyfluorene/dye-labeled DNA complex. *J. Nanosci. Nanotechnol.* **2012**, *12*, 7733–7738. [CrossRef]

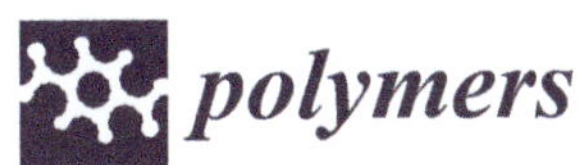 

Article

Magnetic Fluorescence Molecularly Imprinted Polymer Based on FeO$_x$/ZnS Nanocomposites for Highly Selective Sensing of Bisphenol A

Xin Zhang [1,2,*], Shu Yang [2], Weijie Chen [2], Yansong Li [1], Yuping Wei [1] and Aiqin Luo [2]

1   School of Life Science and Technology, Nanyang Normal University, Nanyang 473061, China
2   School of Life Science, Beijing Institute of Technology, No.5 Zhongguancun South Street, Beijing 100081, China
*   Correspondence: firey-sky@163.com; Tel.: +86-156-5299-9486

Received: 15 May 2019; Accepted: 9 July 2019; Published: 19 July 2019

Abstract: In this study, magnetic fluorescence molecularly imprinted polymers were fabricated and used for the selective separation and fluorescence sensing of trace bisphenol A (BPA) in environmental water samples. The carboxyl-functionalized FeO$_x$ magnetic nanoparticles were conjugated with mercaptoethylamine-capped Mn^{2+} doped ZnS quantum dots to prepare magnetic FeO$_x$ and ZnS quantum dot nanoparticles (FeO$_x$/ZnS NPs). Additionally, molecular imprinting on the FeO$_x$/ZnS NPs was employed to synthesize core-shell molecularly imprinted polymers. The resulting nanoparticles were well characterized using transmission electron microscopy, Fourier transform infrared spectra, vibrating sample magnetometer and fluorescence spectra, and the adsorption behavior was investigated. Binding experiments showed that the molecularly imprinted FeO$_x$/ZnS NPs (FeO$_x$/ZnS@MIPs) exhibited rapid fluorescent and magnetic responses, and high selectivity and sensitivity for the detection of bisphenol A (BPA). The maximum adsorption capacity of FeO$_x$/ZnS@MIPs was 50.92 mg·g^{-1} with an imprinting factor of 11.19. Under optimal conditions, the constructed fluorescence magnetic molecularly imprinted polymers presented good linearity from 0 to 80 ng mL^{-1} with a detection limit of 0.3626 ng mL^{-1} for BPA. Moreover, the proposed fluorescence magnetic polymers were successfully applied to on-site magnetic separation and real-time fluorescence analysis of target molecule in real samples.

Keywords: molecularly imprinted polymer; fluorescence magnetic polymer; fluorescence sensing; magnetic FeO$_x$/ZnS quantum dots; bisphenol A

---

1. Introduction

Multifunctional polymers with special physical and chemical properties (such as optical, electrical, thermal, chirality, and magnetic characteristics) have drawn increasing attention due to their potential application in numerous areas [1–4]. As a kind of luminescent nanomaterial, quantum dots (QDs) have high quantum yields, excellent photostability, broad excitation and narrow symmetric emission spectrum, and a size-dependent band gap, therefore, they have attracted considerable attention [5,6]. On the other hand, magnetic nanoparticles (MNPs), which are used as an important magnetic nanomaterial, have great potential applications in magnetic resonance imaging (MRI), drug delivery, catalysis, chemo/biosensors, and medicine diagnosis [7–10]. MNPs have a large surface area and high mass transference based on their size, which enables them to promote fast electron transfer [11–13]. Moreover, MNPs can be directly separated; this facilitates ultra-trace analyte enrichment without going through centrifugation and filtration steps in bioassay, and improves the detection efficiency and sensitivity.

By combining quantum dots with magnetic nanocrystals, an advanced nanocomposite polymer with excellent functionalities can be prepared, which simultaneously integrates the optical and magnetic properties [14,15]. A magnetic nanoparticle coupled with quantum dots can be directly separated and is then able to generate readable optical signals for analysis. The ability to combine the purification process with the detection procedure in one step means that magnetic quantum dot polymers have a promising future and novel applications in bio-detection, biomedicine, drug delivery, and environmental monitoring [16–18]. However, when dealing with a complex matrix, the synthesized magnetic quantum dots demonstrate nonspecific binding in the separation process, which results in high level background fluorescence response and restricts their specificity and sensitivity for analysis. To enhance the specificity and sensitivity of magnetic quantum dots, a molecular imprinted polymer (MIP) layer can be loaded on the surface of the magnetic quantum dot nanocomposite to tailor the selectivity of analytes using molecular imprinting technology [19–22].

Molecular imprinting is a well-established technique to design an artificial molecular recognition unit that involves polymerizing functional monomers in the presence of template molecules [23]. MIPs exhibit excellent selectivity and affinity with the template molecules [24]. The MIP layer coated on the surface of magnetic quantum dots will introduce the selectivity recognition sites to the nanocomposites and prevent interfering molecules from binding with the nanocomposites. MIPs have been introduced as promising recognition elements with high selectivity for detecting trace analyte. Zhang et al. developed imprinted polymer coating CdTe quantum dots for specific recognition of BHb [25]. Zhao et al. prepared ZnS QDs-based molecularly imprinted polymer composite nanospheres for fluorescent quantification of pesticides [26]. Zhao et al. used molecularly imprinted water-soluble CdTe QDs for Listeria monocytogenes detection in food samples [27]. As CdTe QDs suffer from bio-compatibility issues, ZnS QDs based MIPs with low biotoxicity are preferable [28]. Furthermore, magnetic nanoparticles involving fluorescent multifunctional nanoparticles can facilitate magnetic separation, also the fluorescent response is quicker because the magnetic nanoparticles promote faster electron transfer. So together with MIPs' selectivity, this novel multifunctional sensor is expected to demonstrate improved properties for trace target detection.

In this work, an innovative magnetic fluorescent sensor based on $FeO_x/ZnS@MIPs$ was prepared for the separation and detection of trace BPA in complex samples. Bisphenol A (2,2-bis(4-hydroxyphenyl)propane) is an estrogenic endocrine disruptor. It is a common chemical that has been extensively used in the manufacture of polycarbonate plastics and epoxy resins for the linings of food, beverage packaging, and other consumer products [29]. Increasing evidence indicates that BPA can migrate from containers into foods and beverages, and trace residue levels of BPA can be released into environmental water through diffusion or degradation, which carries a high risk of causing adverse effects on human reproductive health and the ecosystem [30,31]. Because of BPA's ubiquity in nature and its potential implications in human health and the ecological environment, various analytical methods have been developed for monitoring or detecting BPA [32–41]. However, many of those analytical techniques for BPA need sophisticated instrumentation, and the procedures involve pre-concentration, extraction, purification or derivatization [42–44]. Those methods are costly and time-consuming, and require well trained and experienced personnel to guarantee the accuracy of results. Therefore, it would be highly advantageous to develop a low-cost, selective method for the convenient and rapid separation and detection of BPA. The prepared $FeO_x/ZnS@MIP$ enables fluorescence analysis and the efficient separation of BPA without any expensive instruments or time-consuming procedures. Based on the fluorescence quenching being proportional to the concentration of the target molecule, $FeO_x/ZnS@MIP$ was successfully applied to the direct fluorescence sensing of BPA without any further pretreatment. Notably, $FeO_x/ZnS@MIP$ has the advantage that it can be easily recycled and rapidly removed from the trace contaminant in the environment. The $FeO_x/ZnS@MIPs$ could specifically bind and magnetically enrich trace BPA, which avoids interfering substances in complex matrices and enhances the detection efficiency and sensitivity. Additionally, the proposed methods for the preparation of $FeO_x/ZnS@MIPs$ exhibits promising potential to isolate and detect proteins and other target molecules in biology detection applications.

2. Materials and Methods

2.1. Materials

All of the solvents and chemicals used were of analytical grade and were used without further purification. Zinc sulfate ($ZnSO_4 \cdot 7H_2O$), sodium sulphide ($Na_2S \cdot 9H_2O$), manganese chloride ($MnCl_2 \cdot 4H_2O$), ferric chloride crystal ($FeCl_3 \cdot 6H_2O$), anhydrous sodium acetate (NaAc) and trisodium citrate dihydrate ($Na_3Cit \cdot 2H_2O$) were purchased from Beijing Chemical Works (Beijing, China). Bisphenol A, 4,4′-bisphenol (BP) and ethylene glycol (EG) were obtained from Sinopharm Chemical Reagent Co. Ltd. (Beijing, China). 1.1-Bis (4-hydroxyphenyl) cyclohexane (BPZ) was received from Tokyo Chemical Industry Co. Ltd. (Tokyo, Japan). Mercaptoethylamine (MEA), 1-ethyl-3-(3-dimethylaminopropyl) carbodiimide (EDC), *N*-hydroxysuccinimide (NHS) were purchased from Macklin Biochemical Co. Ltd. (Shanghai, China). Tetraethyl silicate (TEOS), 4-tert-butylphenol (PTBP), 3-aminopropyltriethoxylsilane (APTES) were purchased from Beijing J&K Chemical Technology Co. Ltd. (Beijing, China). All solutions were prepared with double deionized water (DDW).

2.2. Instrumentation

All the fluorescence measurements were recorded using a spectrofluorometer (Shimadzu, RF-6000, Kyoto, Japan). BPA adsorption data were recorded on a UV–Vis spectrophotometer (Shimadzu, Kyoto, Japan). The magnetization measurements of the FeO_x MNPs, MNP/QDs and MNP/QD@MIPs were carried out using a vibrating sample magnetometer (VSM, Lake Shore 7307, Columbus, OH, USA) at room temperature. The morphology of the QDs was characterized by a JEM-2100F high resolution transmission electron microscopy (TEM, JEOL, Tokyo, Japan). The shape and structures of the magnetic nanomaterials were examined using a H-800 transmission electronic microscopy (Hitachi, Tokyo, Japan). The Fourier transform infrared (FT-IR) spectra were determined on a Vertex 70v spectrometer (Bruker, Karlsruhe, Germany).

2.3. Synthesis of Amino-Modified ZnS: Mn^{2+} QDs and Carboxyl-Functionalized MNPs

The MEA-modified Mn^{2+}-doped ZnS QDs (ZnS: Mn^{2+} QDs@MEA) and the carboxyl-functionalized FeO_x magnetic nanoparticles ($FeO_x@COOH$ MNPs) were synthesized based on previously reported methods with some modification [45–47]. The preparation procedures are presented in the Supplementary Materials.

2.4. Synthesis of FeO_x/ZnS Nanoparticles

The FeO_x/ZnS NPs was synthesized via an EDC/NHS reaction process. Briefly, 20 mg of $FeO_x@COOH$ MNPs were first dispersed in 20 mL citrate buffer solution (0.02 mol L^{-1}, pH = 6.4) to prepare magnetic fluids. Then, 2 mL of magnetic fluids were added in 50 mL EDC/NHS activating agent (1:1.2 g L^{-1}). The above mixture was stirred for 2 h, followed by 30 mg of as-prepared amino-modified QDs being added and incubated at 30 °C for 20 h. The brown solutions were collected by magnetic decantation, and washed with water and ethanol to remove the residual substance, then redispersed in buffer solution for further use.

2.5. Fabrication of $FeO_x/ZnS@MIPs$

The $FeO_x/ZnS@MIPs$ for BPA were prepared through molecular imprinting on the surface of magnetic fluorescence nanoparticle. For the fabrication of $FeO_x/ZnS@MIPs$, 10 mL of a methanol solution containing 10 mg BPA and 60 μL APTES were first added to a 25 mL flask and stirred for 30 min. Then, 60 mg of the as-prepared FeO_x/ZnS nanoparticle and 100 μL TEOS were added in sequence. After continuously stirring for another 30 min, 2.5 mL of 5% $NH_3 \cdot H_2O$ (the catalyst) was added.

The solution was deoxygenated by purging with nitrogen and stirred overnight. Non-imprinted polymers (FeO$_x$/ZnS@NIPs) were prepared by the same procedure but without the addition of BPA. The resultant products were magnetically decanted and washed with a mixture of methanol and acetic acid (9:1, *v/v*) to remove the template molecules until the fluorescence intensity of FeO$_x$/ZnS@MIPs was not changed and was similar to that of the FeO$_x$/ZnS@NIPs ones.

2.6. Fluorescence Sensing of BPA

All of the fluorescence measurements were examined using the same condition. The excitation wavelength was set to be 311 nm with the fluorescence intensity recorded at 586 nm. The recording fluorescence emission spectrum was ranged from 350 to 700 nm.

The standard solution of bisphenol A was prepared first. Then, various samples in concentrations ranging from 1.0 to 80 ng mL^{-1} were made by diluting with citrate buffer solution (0.02 mol L^{-1}, pH = 6.4). Two milligrams of FeO$_x$/ZnS@MIPs or FeO$_x$/ZnS@NIPs were dispersed in 10 mL testing samples. After incubating the samples for 5 min at room temperature, the changes in the fluorescence intensity of the solutions were recorded using a spectrofluorometer. This fluorescence quenching model was in accordance with the Stern–Volmer equation [48]:

$$F_0/F = 1 + K_{SV}C_{BPA} \tag{1}$$

where F_0 is the initial fluorescence intensity in the absence of the BPA, F is the fluorescence intensity in the presence of the BPA, K_{SV} is the quenching constant, and C_{BPA} is the concentration of BPA.

2.7. Binding Selectivity

The selectivity experiments were carried out with PTBP, BP and BPZ as structural analogs of BPA to run a batch rebinding test. Briefly, PTBP, BP and BPZ samples in the concentration ranged from 1.0 to 80 ng mL^{-1} and were made by diluting with citrate buffer solution. Two milligrams of FeO$_x$/ZnS@MIPs or FeO$_x$/ZnS@NIPs were dispersed in 10 mL testing samples and incubated for 5 min. Then the changes in the fluorescence intensity of different samples were recorded, and the imprinting factor (IF) and selectivity coefficient (SC) were used to evaluate the selectivity properties of FeO$_x$/ZnS@MIPs and FeO$_x$/ZnS@NIPs toward the template BPA and structural analogs [25]:

$$IF = K_{MIP}/K_{NIP} \tag{2}$$

$$SC = IF/IF' \tag{3}$$

where K_{MIP} and K_{NIP} are the slopes of the linear equation of FeO$_x$/ZnS@MIPs and FeO$_x$/ZnS@NIPs with the target molecule, respectively, IF and IF' are the imprinting factor for template BPA and structural analogs, respectively.

2.8. Analysis of Real Samples

The prepared FeO$_x$/ZnS@MIP was directly applied to the detection of BPA in drinking water, tap water, and lake water. The drinking water was commercial pure water purchased from the market. The tap water was collected from the laboratory. The lake water samples were collected from three different lakes (located in Beijing, China). All the samples were filtered through a 0.45 μm filter and stored at 4 °C.

The detection strategy was divided into two steps: (a) the enrichment and separation of BPA and (b) fluorescence detection (Figure S1). Firstly, the FeO$_x$/ZnS@MIPs were dispersed in the water samples. The target BPA molecule was specifically bound onto the MIP layer of the FeO$_x$/ZnS@MIPs after incubating at room temperature. Secondly, the fluorescence quench of FeO$_x$/ZnS@MIPs in each sample was recorded and the concentration of analytes in the samples was calculated. There were no other pretreatment procedures employed in the sample preparation. To evaluate the developed method, a recovery test was carried out by using the samples spiked with BPA standard solution.

3. Results and Discussion

3.1. Synthesis of the FeO$_x$/ZnS@MIPs

The design of the FeO$_x$/ZnS@MIPs was mainly based on coating the molecularly imprinted polymer layer on the surface of FeO$_x$/ZnS NPs. Figure 1 illustrates the two major synthetic steps of the proposed fluorescence sensing polymer. In the first step, mercaptoethylamine was grafted onto the surface of the Mn^{2+}-doped ZnS QDs. The mercapto of MEA was tightly bound at the surface of the bare QDs through ligand competition. The introduction of amino group to the QDs not only increases the water dispersion ability of QDs, but also provides the possibility of being combined with FeO$_x$ magnetic nanocrystals. FeO$_x$ magnetic nanocrystals were successfully synthesized by a modified solvothermal method. The –COOH was grafted onto the surface of FeO$_x$ in just one synthesis step. The carboxyl-functionalized FeO$_x$ NPs were conjugated with MEA capped QDs to prepare the multifunctional nanocomposites. The FeO$_x$/ZnS NPs integrated the distinct properties of the optical characteristics of QDs, and the magnetic separation ability of MNPs through an EDC/NHS reaction process. In the second step, APTES was chosen as a functional monomer that had noncovalent interactions with bisphenol A [49]. The monomer (APTES) interacted with the template molecule (BPA) through a hydrogen bond to form a "pre-polymerization" complex (Figure 1). The resultant FeO$_x$/ZnS NPs were used as substrate, TEOS and NH$_3$·H$_2$O were used as the crosslinker and catalyst, respectively. The pre-polymerization complex was subsequent immobilized on the surface of FeO$_x$/ZnS NPs through a facile molecular imprinting process. The imprinting layer was coated on the FeO$_x$/ZnS NPs to produce a "core-shell" structure. This core-shell composite provides selectivity to the template and prevents other interfering molecules from contacting the FeO$_x$/ZnS NPs. After the removal of the template molecule BPA, the MIP layer with imprinted cavities complementary to the BPA in size, shape, and functional groups was obtained. The quantum yield (QY) of the FeO$_x$/ZnS NPs was 20.6%, as calculated by equation S1. The resultant FeO$_x$/ZnS@MIPs, as an ideal candidate material, was able to be used as a multifunctional sensor for high selectivity and sensitivity magnetic separation and the fluorescent detection of target BPA.

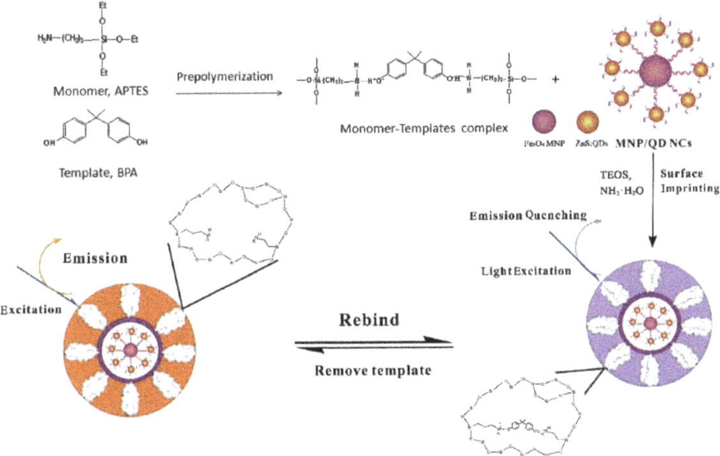

Figure 1. Schematic illustration of the process for the fabrication of the FeO$_x$/ZnS@MIP-based sensor. BPA: bisphenol A.

3.2. Characterization

The morphology of the obtained FeO_x/ZnS NPs and $FeO_x/ZnS@MIPs$ was investigated by transmission electron microscopy. As shown in Figure 2a, the particles of FeO_x/ZnS NPs were spherical in morphology, and they were about 135.2 ± 16.7 nm. The magnetic nanoparticles were tightly surrounded by ZnS: Mn^{2+} QDs@MEA through the EDC/NHS reaction, which created a rough surface on the FeO_x/ZnS NPs. After the molecular imprinting process, the size of the resulting $FeO_x/ZnS@MIPs$ increased to about 198.6 ± 13.5 nm and displayed a smooth surface (Figure 2b). As can be seen, an MIP layer had been well coated on the surface of FeO_x/ZnS NPs, and an interface could be clearly distinguished between the inner FeO_x/ZnS NPs core and the imprinting polymer shell.

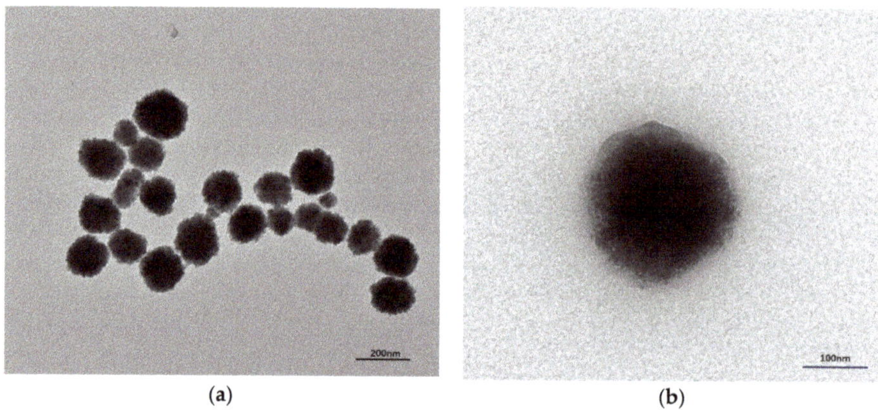

Figure 2. TEM images of (**a**) FeO_x/ZnS NPs and (**b**) $FeO_x/ZnS@MIPs$.

The structure of ZnS: Mn^{2+} QDs@MEA, $FeO_x@COOH$ and FeO_x/ZnS NPs was analyzed by FT-IR spectroscopy. As shown in Figure 3, the characteristic Fe–O bands were present in FeO_x/ZnS (632 cm^{-1}) and $FeO_x@COOH$ MNPs (594 cm^{-1}), respectively. Peaks at 1598 cm^{-1} (amino groups), 925 cm^{-1} (bending vibration) and the disappeared peaks at 2550–2670 cm^{-1} (S–H thiol group) in the curve b indicated that the MEA was modified on the surface of ZnS: Mn^{2+} QDs through covalent bonds formed between thiols and Zn^{2+} surface atoms [50]. The Peak at 1413 cm^{-1} corresponded to C–O on the FeO_x, and a peak at 1605 cm^{-1} corresponded to the vibration of water molecules adsorbed on Fe_3O_4 [13]. Compared with the ZnS: Mn^{2+} QDs@MEA and $FeO_x@COOH$ (curve b and c), FeO_x/ZnS NPs (curve a) showed characteristic peaks at 3425 cm^{-1} (N–H stretching vibration), 1625 cm^{-1}(N–H bend), and 1400 cm^{-1}(C–N stretching vibration), revealing that an amide bond was formed. The results suggested that the ZnS: Mn^{2+} QDs@MEA and $FeO_x@COOH$ were successfully combined through an EDC/NHS reaction process.

To study the influence of surface modification on the magnetic behavior of $FeO_x@COOH$ MNPs, FeO_x/ZnS NPs and $FeO_x/ZnS@MIPs$, the VSM magnetization curves of the as-prepared magnetic materials were compared in Figure 4. As shown in Figure 4, all of the magnetic hysteresis loops of the magnetic materials displayed a typical super-paramagnetic characteristic and high magnetization, and the magnetization saturation values of $FeO_x@COOH$ MNPs, FeO_x/ZnS NPs, and $FeO_x/ZnS@MIPs$ were about 57.348, 45.1033, 24.5796 emu g^{-1}, respectively. The magnetization saturation value of prepared NPs was lower than that of pure Fe_3O_4 magnetite (about 80 emu g^{-1}), which is due to some mixed Fe_2O_3 that lowers the magnetization saturation. The as-prepared magnetic NPs still demonstrated a strong magnetic response and FeO_x was used to represent the mixture to describe the magnetic behavior in this paper. The magnetization saturation values of $FeO_x/ZnS@MIPs$ were lower than that of $FeO_x@COOH$ MNPs, which may be attributed to the MIP shell on the surfaces of FeO_x/ZnS NPs. Additionally, the as-prepared magnetic nanocomposites also showed a rapid magnetic response to the applied magnetic field. Once the magnetic field was removed, the magnetic nanocomposites homogeneously and quickly redispersed with a slight shake. This demonstrates that the $FeO_x/ZnS@MIPs$ possess rapid magnetic responsivity and good dispersibility, which enables them to be practically used for the rapid separation or enrichment of analyte in complex samples.

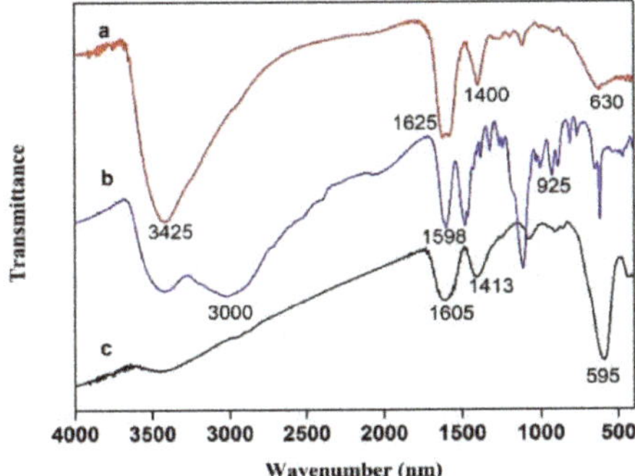

Figure 3. FT-IR spectra of (**a**) FeO_x/ZnS NPs, (**b**) ZnS: Mn^{2+} QDs@MEA and (**c**) $FeO_x@COOH$ nanoparticles.

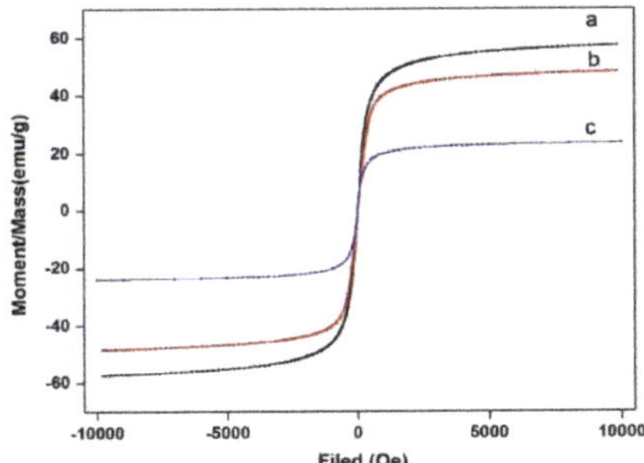

Figure 4. The magnetic hysteresis loops of (**a**) $FeO_x@COOH$ magnetic nanoparticles, (**b**) FeO_x/ZnS NPs, and (**c**) $FeO_x/ZnS@MIPs$.

The optical properties of ZnS: Mn^{2+} QDs@MEA and FeO_x/ZnS NPs were investigated by UV–Vis spectra and fluorescence spectra. As shown in Figure S2a,b, the QDs and FeO_x/ZnS NPs exhibit semblable absorption spectrum, indicating the QDs successfully combined with the FeO_x MNPs. Compared with ZnS: Mn^{2+} QDs@MEA, the absorption peak of FeO_x/ZnS NPs is less pronounced, which attributed to the broad and strong absorption of the combined Fe_3O_4 MNPs [51]. The ZnS: Mn^{2+} QDs@MEA and FeO_x/ZnS NPs show well-resolved emission spectra, with both of the maximum emission peaks located at 586 nm. The characteristic emission peak of FeO_x/ZnS NPs corresponded to the ZnS: Mn^{2+} QDs@MEA, implying that the fluorescence property was not significantly affected in the EDC/NHS reaction process. Both samples possess a narrow and symmetrical emission peak and large Stokes shift, which is more suitable for fluorescence labeling in vivo or biosensing target analytes in complex biological samples.

3.3. Fluorescence Response to Time and Adsorption Kinetics

The effect of reaction time on the fluorescence intensity and the adsorption kinetics curves of BPA onto $FeO_x/ZnS@MIPs$ are presented in Figures S3 and S4, respectively. As shown in Figure S3, the fluorescence intensity of $FeO_x/ZnS@MIPs$ showed a rapid decrease in the first 2 min and achieved stable fluorescence intensity after being incubated for 5 min, which corresponded closely to the adsorption equilibrium time (Figure S4). The adsorption equilibrium for the $FeO_x/ZnS@NIPs$ emerged at 4 min in the experiment; however, the fluorescence quenching and adsorption amount of

FeO$_x$/ZnS@NIPs was lower than that of the MIP ones. This is due to the non-specific binding cavities that are formed in the FeO$_x$/ZnS@NIPs' synthesis process, the non-specific adsorption was dominant in the FeO$_x$/ZnS@NIPs, which resulted in a lower binding capacity and lower fluorescence quenching. These results verified that the optimal reaction time of the FeO$_x$/ZnS@MIPs for detecting BPA was 5 min.

3.4. Binding Performance

The adsorption isotherm was investigated through a batch affinity adsorption experiment of FeO$_x$/ZnS@MIPs. A series of BPA standard samples were used to evaluate the adsorption isotherms of BPA on the FeO$_x$/ZnS@MIPs and FeO$_x$/ZnS@NIPs. As shown in Figure 5, the adsorption capacity of FeO$_x$/ZnS@MIPs increased quickly with the increasing concentration of BPA. The FeO$_x$/ZnS@MIPs exhibited a higher binding amount of BPA than that of the FeO$_x$/ZnS@NIPs. Because of the 3D-imprinted cavities in FeO$_x$/ZnS@MIPs, which possessed better chemical and structure matching with the template BPA, a large number of empty recognition cavities were available in the FeO$_x$/ZnS@MIPs, which enabled BPA molecules to easily enter, and resulted in excellent adsorption capacity. To further study the adsorption performance of FeO$_x$/ZnS@MIPs, the experimental data were fitted with Langmuir and Freundlich isotherm models. Fitting results showed that the binding properties were best described by the Langmuir isotherm model (R = 0.9930), which revealed that the adsorption behavior of FeO$_x$/ZnS@MIPs was basically monolayer adsorption onto a surface with a homogeneous system. The Langmuir isotherm equation of FeO$_x$/ZnS@MIPs for BPA was $C_e/Q = 0.2302 + 0.01964\,C_e$, and the maximum adsorption capacity calculated by the Langmuir isotherm model was 50.92 mg g^{-1}.

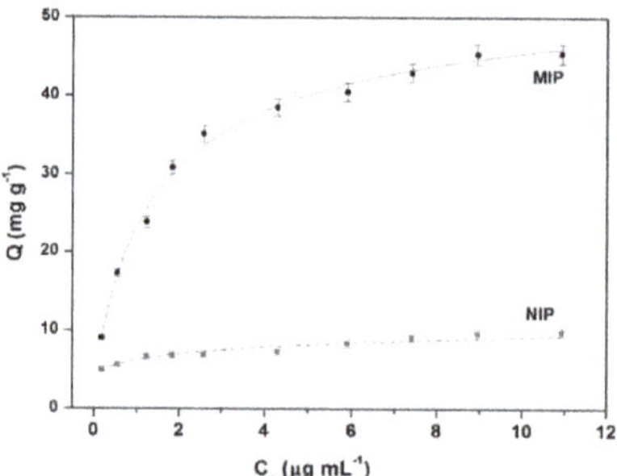

Figure 5. The adsorption isotherm and Langmuir fit of FeO$_x$/ZnS@MIPs. Experimental conditions: citrate buffer solution (0.02 mol L^{-1}, pH = 6.4), room temperature.

3.5. Fluorescence Sensing of BPA

To maintain the fluorescence stability of the FeO$_x$/ZnS@MIPs, all experiments were performed in citrate buffer solution at pH 6.4. The typical fluorescence quenching of imprinted FeO$_x$/ZnS NPs in the concentration of BPA ranged from 1.0 to 80 ng mL^{-1}, as shown in Figure 6. It can be seen that the FeO$_x$/ZnS@MIPs showed noticeable fluorescence emission responses to different concentrations of BPA. As a control, the FeO$_x$/ZnS@NIP was slightly quenched by BPA because of there being no specific recognized site on the surface of the NIPs; thus, fewer BPA molecules were bound by non-specific interactions. In contrast, the fluorescence emission of the FeO$_x$/ZnS@MIPs quenched gradually with the increasing concentration of BPA. The fluorescence intensity of FeO$_x$/ZnS@MIPs in the presence of BPA (80 ng mL^{-1}) was only 32.7% compared to that in the absence of BPA. The experiment results showed that the fluorescence quenching of the FeO$_x$/ZnS@MIPs depended on the specific binding with template BPA. Therefore, FeO$_x$/ZnS@MIPs can be used for the detection of trace BPA. Under optimal conditions, the plot of fluorescence intensity change (F_0/F-1) versus the concentration of BPA (ng mL^{-1}) showed a good linear relationship. The linear relationship between fluorescence intensity and the BPA concentrations was in the range of 0 to 80 ng mL^{-1} with a correlation coefficient of 0.9968 (n = 11).

The corresponding limit of detection (LOD) following the IUPAC criteria ($3\sigma/S$) was calculated as 0.3626 ng mL^{-1}. This value of LOD was lower than the permitted maximum residue limits of BPA in drinking water (10 ng mL^{-1}, GB 5749-2006) [40], indicating that the as-prepared FeO$_x$/ZnS@MIPs can be employed for environmental and drinking water safety monitoring. The comparison of this performance with other reported analytical methods for BPA detection is shown in Table 1. The results showed that the strategy presented is more rapid, selective and sensitive than the traditional methods. The proposed magnetic/fluorescence molecularly imprinted polymer coated FeO$_x$/ZnS nanocomposites not only have the merits of convenience and low cost, but also have high selectivity and a comparable or lower limit of detection, which makes them a promising fluorescent sensor for the selective and sensitive detection of target molecules.

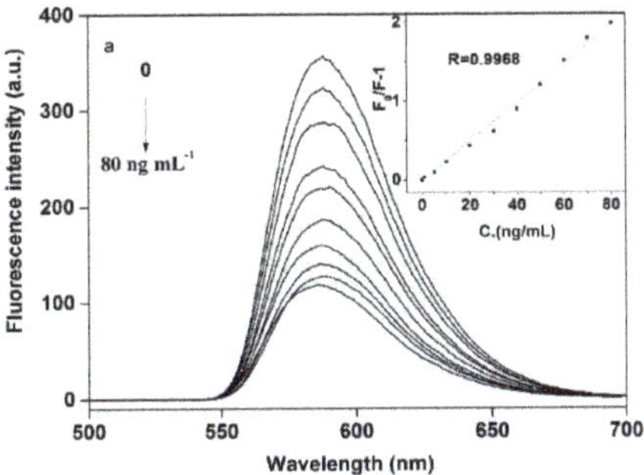
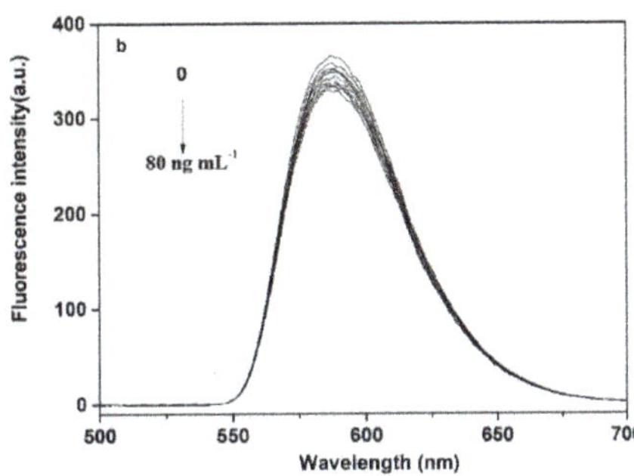

Figure 6. Fluorescence emission spectra of (**a**) FeO$_x$/ZnS@MIPs and (**b**) FeO$_x$/ZnS@NIPs with increasing concentrations of BPA. Inset graphs: the linear calibration of the fluorescence intensity change (F_0/F) versus BPA concentration. Experimental conditions: BPA (0, 5, 10, 20, 30, 40, 50, 60, 70, 80 ng mL^{-1}), citrate buffer solution (0.02 mol L^{-1}, pH = 6.4), room temperature.

Table 1. The comparison of the results of different analytical techniques for the detection of BPA.

Detection Technique	Liner Range (ng mL^{-1})	LOD (ng mL^{-1})	Imprinting Factor (IF)	References
MIP/SERS	5.0×10^2–2.28×10^4	120	1.90	[32]
MIP-coated Au NCs	0–2.98×10^3	22.8	2.25	[33]
Colorimetric method	35–1.40×10^2	0.11	/	[34]
Fluorescent aptasensor	10–80	1.86	/	[35]
MIP-coated CDs	22.8–9.60×10^2	6.84	/	[36]
Voltammetric sensor	11.4–2.28×10^3	1.82	/	[37]
Imprinting SiO$_2$-coated CdTe NPs	11.4–2.28×10^2	1.368	5.2	[38]
Ferrocenyl-based MIP	1.07–1.8	0.7296	1.84	[39]
Colorimetric sensor-MIP	2.28–2.28×10^2	1.41	/	[40]
Electrochemical sensor-MIP	18.4–2.28×10^4	8.66	/	[41]
MIP-SPE/HPLC	0.11–22.8	0.11	1.97	[42]
GC-MS	1–200	1.0	/	[43]
FeO$_x$/ZnS@MIPs	0–80	0.3626	11.19	This Work

SERS: surface enhanced Raman scattering; GC-MS: gas chromatography-mass spectrometer; SPE: solid-phase extraction; Au NCs: gold nanoclusters; CDs: carbon dots; LOD: limit of detection.

3.6. Rebinding Selectivity

The specific recognition ability of FeO$_x$/ZnS@MIPs was further investigated. The imprinting factor (IF) and selectivity coefficient (SC) were used to evaluate the selectivity of the FeO$_x$/ZnS@MIPs towards the template BPA and structural analogs (Figure S5). The imprinting factor is the ratio of the K_{MIP} and K_{NIP} (K is the slope of the linear equation), and the selectivity coefficient is the ratio of IF for the template molecule and structural analogs (Table S1).

As shown in Figure 7, the BPA molecule exhibited a significant fluorescence quenching effect on the FeO$_x$/ZnS@MIPs. Table S1 shows that the imprinting factor for BPA was 11.19, which is much larger than that of BPZ, BP, and PTBP (1.47, 1.66, and 1.39, respectively). This was because the template BPA had more access to the recognition cavities of $FeO_x/ZnS@MIPs$, which were formed in the imprinting process. Conversely, there are no complementary binding sites formed between $FeO_x/ZnS@NIPs$ and analyte molecules. The BPA and its analogs bound on the $FeO_x/ZnS@NIPs$ are mainly bound through non-specific interactions. Therefore, the K_{NIP} for the BPA and structural analogs (BPZ, BP and PTBP) was almost the same, and a lower fluorescence quenching effect on the $FeO_x/ZnS@NIPs$ was observed (Figure 7). The experimental results showed that the selectivity of the MIPs for template BPA was much higher than that of structural analogs, indicating that a molecular imprinting process can greatly enhance the selectivity of $FeO_x/ZnS@NPs$.

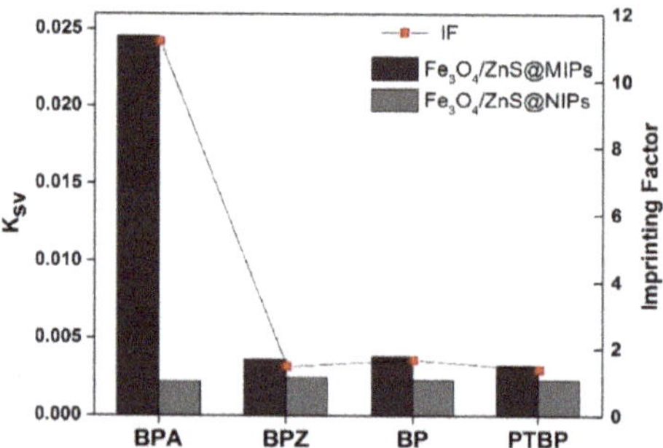

Figure 7. The quenching constant (Ksv) and the imprinting factor (IF) of BPA, 1.1-Bis (4-hydroxyphenyl) cyclohexane (BPZ), 4,4′-bisphenol (BP) and 4-tert-butylphenol (PTBP).

3.7. Analysis of Real Samples

In order to demonstrate the practical applicability of $FeO_x/ZnS@MIPs$ in real samples, different water samples (drinking water, tap water, and lake water) were used to evaluate the separation effectiveness and detection accuracy. The detection strategy is shown in Figure S1.

As shown in Table 2, the detected BPA values in the lake sample were 1.52, 7.39 and 6.54 ng mL^{-1}, respectively. BPA was not found in drinking water and tap water at the detection limit level of 0.36 ng mL^{-1}. The recovery rates of the present $FeO_x/ZnS@MIPs$ ranged from 90.8% to 103.1%, and the relative standard deviation (RSD) values were between 2.7% and 5.4% (n = 3). The above results proved that the prepared $FeO_x/ZnS@MIPs$ could be successfully applied to the magnetic separation and fluorescence detection of the target molecule in practical applications.

Table 2. Detection of BPA in real water samples using $FeO_x/ZnS@MIPs$ sensor.

Sample	Detected (ng mL^{-1})	Added (ng mL^{-1})	Measured (ng mL^{-1}) [a]	Recovery (%)	RSD (n = 3, %)
Drinking water	n.d [a]	2.28	2.20	96.5	3.5
		22.80	22.44	98.4	2.7
Tap water	n.d	2.28	2.12	93.0	4.1
		22.80	21.15	92.8	3.3
Lake water sample1	1.52	2.28	2.07	90.8	4.2
		22.80	20.73	90.9	4.9
Lake water Sample2	7.39	2.28	2.35	103.1	5.3
		22.80	22.25	97.6	4.5
Lake water Sample3	6.54	2.28	2.31	101.3	5.4
		22.80	21.98	96.4	3.8

[a] Average value; RSD: relative standard deviation.

3.8. Recyclability and Stability

The recyclability test was done by observing the changes in fluorescence intensity with six rebind/elution cycles. As shown in Figure S6a, the fluorescence properties of the FeO_x/ZnS@MIPs performance slightly decreased in five regeneration cycles. However, the fluorescence intensity decreased 25.5% and 31.1% in the next two regeneration cycles, respectively. This indicates that too much binding/removing template will affect the structure of the FeO_x/ZnS@MIPs. The stability of FeO_x/ZnS@MIPs was evaluated by measuring the initial fluorescence intensities and those after 1–35 days of storage under dark conditions (Figure S6b). It can be seen that the fluorescence intensity showed no significant change after being stored for a long time. These results suggest that FeO_x/ZnS@MIPs have good regeneration capacity and stability.

4. Conclusions

In this study, a novel approach for the preparation FeO_x/ZnS@MIPs has been successfully developed. The fabricated core-shell nanocomposites integrated selective magnetic separation and fluorescence analysis, therefore, they can be used for direct magnetic separation and the selective detection of trace BPA in complex environmental or biological matrices. A series of rebinding experiments showed that the proposed FeO_x/ZnS@MIP-based sensor has better selectivity and sensitivity for target BPA than its analogues. The FeO_x/ZnS@MIPs also showed good reusability and stability in practical applications. Moreover, by changing the template molecule, this novel approach to multifunctional sensor preparation can be expanded to other organic molecules or protein biomolecules. Such superior optical and physical merits endow the FeO_x/ZnS@MIPs with promising potential in food, environmental, clinical diagnostic, and biomedical research.

Supplementary Materials: The supplementary materials are available online at http://www.mdpi.com/2073-4360/11/7/1210/s1.

Author Contributions: X.Z. designed the experiments; data processing, analysis, and interpretation were performed by X.Z., S.Y., W.C., and X.Z., W.C. performed the experiments. Some tests and suggestions were provided by A.L., Y.W. and X.Z., Y.L. wrote the paper. All authors agree on the order of contribution and have given approval to the final version of the manuscript.

Funding: We would like to acknowledge the financial support from the scientific research projects of Nanyang Normal University.

Acknowledgments: We are very grateful to the research groups of the biological analysis laboratory at the Beijing Institute of Technology. We would also like to show our gratitude to Aiqin Luo.

Conflicts of Interest: The authors declare no conflict of interest.

References

1. Yiu, H.H.P.; Niu, H.; Biermans, E.; Tendeloo, G.V.; Rosseinsky, M.J. Designed multifunctional nanocomposites for biomedical applications. *Adv. Funct. Mater.* **2010**, *20*, 1599–1609. [CrossRef]

2. Shrivastava, S.; Jadon, N.; Jain, R. Next-generation Polymer Nanocomposite-based Electro-chemical Sensors and Biosensors: A review. *TrAC Trend Anal. Chem.* **2016**, *82*, 55–67. [CrossRef]

3. Tammari, E.; Nezhadali, A.; Lotfi, S.; Veisi, H. Fabrication of an electrochemical sensor based on magnetic nanocomposite Fe3O4/β-alanine/Pd modified glassy carbon electrode for determination of nanomolar level of clozapine in biological model and pharmeutical samples. *Sens. Actuators B Chem.* **2017**, *241*, 879–886. [CrossRef]

4. Gan, T.; Zhang, X.; Gu, Z.; Zhao, Y. Recent Advances in Upconversion Nanoparticles–Based multifunctional nanocomposites for combined cancer therapy. *Adv. Mater.* **2015**, *27*, 7692–7712.

5. Bruchez, M.J.; Moronne, M.; Gin, P.; Weiss, S.; Alivisatos, A.P. Semiconductor nanocrystals as fluorescent biological labels. *Science* **1998**, *281*, 2013–2016. [CrossRef] [PubMed]

6. Gua, W.; Gong, S.; Zhou, Y.; Xia, Y. Ratiometric sensing of metabolites using dual–emitting ZnS: Mn^{2+} quantum dots as sole luminophore via surface chemistry design. *Biosens. Bioelectron.* **2017**, *90*, 487–493. [CrossRef] [PubMed]

7.   Ulbrich, K.; Holá, K.; Šubr, V.; Bakandritsos, A.; Tuček, J.; Zbořil, R. Targeted drug delivery with polymers and magnetic nanoparticles: Covalent and noncovalent approaches, release control, and clinical studies. *Chem. Rev.* **2016**, *116*, 5338–5431. [CrossRef] [PubMed]

8.   Huang, J.; Li, Y.; Orza, A.; Lu, Q.; Guo, P.; Wang, L.; Yang, L.; Mao, H. Magnetic nanoparticle facilitated drug delivery for cancer therapy with targeted and image–guided approaches. *Adv. Funct. Mater.* **2016**, *26*, 3818–3836. [CrossRef]

9.   Shylesh, S.; Schünemann, V.; Thiel, W.R. Magnetically separable nanocatalysts: Bridges between homogeneous and heterogeneous catalysis. *Angew. Chem. Int. Ed.* **2010**, *49*, 3428–3459. [CrossRef]

10. Rocha-Santos, T.A.P. Sensors and biosensors based on magnetic nanoparticles. *TrAC Trends Anal. Chem.* **2014**, *62*, 28–36. [CrossRef]

11. Dallas, P.; Bourlinos, A.B.; Niarchos, D.; Petridis, D. Synthesis of tunable sized capped magnetic iron oxide nanoparticles highly soluble in organic solvents. *J. Mater. Sci.* **2007**, *42*, 4996–5002. [CrossRef]

12. Shi, D.; Ni, M.; Zeng, J.; Ye, J.; Ni, P.; Liu, X.; Chen, M. Simultaneous detection and removal of metal ions based on a chemosensor composed of a rhodamine derivative and cyclodextrin–modified magnetic nanoparticles. *J. Mater. Sci.* **2015**, *50*, 168–175. [CrossRef]

13. Arvand, M.; Hemmati, S. Magnetic nanoparticles embedded with graphene quantum dots andmultiwalled carbon nanotubes as a sensing platform for electrochemical detection of progesterone. *Sens. Actuators B Chem.* **2017**, *238*, 346–356. [CrossRef]

14. Dehghani, M.; Nasirizadeh, N.; Yazdanshenas, M.E. Determination of cefixime using a novel electrochemical sensor produced with gold nanowires/graphene oxide/electropolymerized molecular imprinted polymer. *Mater. Sci. Eng. C* **2019**, *96*, 654–660. [CrossRef]

15. Xu, Z.; Deng, P.; Li, J.; Tang, S.; Cui, Y. Modification of mesoporous silica with molecular imprinting technology: A facile strategy for achieving rapid and specific adsorption. *Mater. Sci. Eng. C* **2019**, *94*, 684–693. [CrossRef]

16. Leng, Y.; Wu, W.; Li, L.; Lin, K.; Sun, K.; Chen, X.; Li, W. Magnetic/Fluorescent barcodes based on cadmium–free near–infrared–emitting quantum dots for multiplexed detection. *Adv. Funct. Mater.* **2016**, *26*, 7581–7589. [CrossRef]

17. Wen, C.; Xie, H.; Zhang, Z.; Wu, L.; Hu, J.; Tang, M.; Wu, M.; Pang, D. Fluorescent/magnetic micro/nano–spheres based on quantum dots and/or magnetic nanoparticles: Preparation, properties, and their applications in cancer studies. *Nanoscale* **2016**, *8*, 12406–12429. [CrossRef]

18. Chowdhuri, A.R.; Singh, T.; Ghosh, S.K.; Sahu, S.K. Carbon dots embedded magnetic nanoparticles @chitosan @metal organic framework as a nanoprobe for pH sensitive targeted anticancer drug delivery. *ACS Appl. Mater. Interfaces* **2016**, *8*, 16573–16583. [CrossRef]

19. Ha, S.; Li, X.; Wang, Y.; Chen, S. Multifunctional imprinted polymers based on CdTe/CdS and magnetic graphene oxide for selective recognition and separation of p–t–octylphenol. *Chem. Eng. J.* **2015**, *271*, 87–95.

20. Li, X.; Jiao, H.F.; Shi, X.Z.; Sun, A.; Wang, X.; Chai, J.; Li, D.X.; Chen, J. Development and application of a novel fluorescent nanosensor based on FeSe quantum dots embedded silica molecularly imprinted polymer for the rapid optosensing of cyfluthrin. *Biosens. Bioelectron.* **2018**, *99*, 268–273. [CrossRef]

21. Lahcen, A.A.; Baleg, A.A.; Baker, P.; Iwuoha, E.; Amine, A. Synthesis and electrochemical characterization of nanostructured magnetic molecularly imprinted polymers for 17-β-Estradiol determination. *Sens. Actuators B Chem.* **2017**, *241*, 698–705. [CrossRef]

22. Gao, R.; Hao, Y.; Zhang, L.; Cui, X.; Liu, D.; Zhang, M.; Tang, Y.; Zheng, Y. A facile method for protein imprinting on directly carboxyl–functionalized magnetic nanoparticles using non–covalent template immobilization strategy. *Chem. Eng. J.* **2016**, *284*, 139–148. [CrossRef]

23. Haupt, K.; Linares, A.V.; Bompart, M.; Bui, B.T.S. *Molecular Imprinting;* Springer: New York, NY, USA, 2012; pp. 1–28.

24. Cieplak, M.; Kutner, W. Artificial Biosensors: How can molecular imprinting mimic biorecognition? *Trends Biotechnol.* **2016**, *34*, 922–941. [CrossRef]

25. Zhang, W.; He, X.; Li, W.; Zhang, Y. Thermo-sensitive imprinted polymer coating CdTe quantum dots for target protein specific recognition. *Chem. Commun.* **2012**, *48*, 1757–1759. [CrossRef]

26. Zhao, Y.; Ma, Y.; Li, H.; Wang, L. Composite QDs@MIP nanospheres for specific recognition and direct fluorescent quantification of pesticides in aqueous media. *Anal. Chem.* **2012**, *84*, 386–395. [CrossRef]

27. Zhao, X.; Cui, Y.; Wang, J.; Wang, J. Preparation of fluorescent molecularly imprinted polymers via pickering emulsion interfaces and the application for visual sensing analysis of Listeria Monocytogenes. *Polymers* **2019**, *11*, 984. [CrossRef]

28. Chantada-Vázquez, M.P.; Sánchez-González, J.; Peña-Vázquez, E.; Tabernero, M.J.; Bermejo, A.M.; Bermejo-Barrera, P.; Moreda-Piñeiro, A. Synthesis and characterization of novel molecularly imprinted polymer–coated Mn-doped ZnS quantum dots for specific fluorescent recognition of cocaine. *Biosens. Bioelectron.* **2016**, *75*, 213–222.

29. Vom Saal, F.S.; Hughes, C. An Extensive new literature concerning low–dose effects of bisphenol A shows the need for a new risk assessment. *Environ. Health Persp.* **2005**, *113*, 926–933. [CrossRef]

30. Keri, R.A.; Ho, S.; Hunt, P.A.; Knudsen, K.E.; Soto, A.M.; Prinsf, G.S. An evaluation of evidence for the carcinogenic activity of bisphenol A. *Reprod. Toxicol.* **2007**, *24*, 240–252. [CrossRef]

31. Horan, T.S.; Pulcastro, H.; Lawson, C.; Gerona, R.; Martin, S.; Gieske, M.C.; Sartain, C.V.; Hunt, P.A. Replacement Bisphenols Adversely Affect Mouse Gametogenesis with Consequences for Subsequent Generations. *Curr. Biol.* **2018**, *28*, 2948–2954. [CrossRef]

32. Xue, J.; Li, D.; Qu, L.; Long, Y. Surface-imprinted core–shell Au nanoparticles for selective detection of bisphenol A based on surface-enhanced Raman scattering. *Anal. Chim. Acta* **2013**, *777*, 57–62. [CrossRef]

33. Wu, X.; Zhang, Z.; Li, J.; You, H.; Li, Y.; Chen, L. Molecularly imprinted polymers-coated gold nanoclusters for fluorescent detection of bisphenol A. *Sens. Actuators B Chem.* **2015**, *211*, 507–514. [CrossRef]

34. Xu, J.; Li, Y.; Bie, J.; Jiang, W.; Guo, J.; Luo, Y.; Shen, F.; Sun, C. Colorimetric method for determination of bisphenol A based on aptamer-mediated aggregation of posotively charged gold nanoparticles. *Microchim. Acta* **2015**, *182*, 2131–2138. [CrossRef]

35. Li, Y.; Xu, J.; Wang, L.; Huang, Y.; Guo, J.; Cao, X.; Shen, F.; Luo, Y.; Sun, C. Aptamer-based fluorescent detection of bisphenol A using nonconjugated gold nanoparticles and CdTe quantum dots. *Sens. Actuators B Chem.* **2016**, *222*, 815–822. [CrossRef]

36. Liu, G.; Chen, Z.; Jiang, X.; Feng, D.; Zhao, J.; Fan, D.; Wang, W. In-situ hydrothermal synthesis of molecularly imprinted polymerscoated carbon dots for fluorescent detection of bisphenol A. *Sens. Actuators B Chem.* **2016**, *228*, 302–307. [CrossRef]

37. Su, B.; Shao, H.; Li, N.; Chen, X.; Cai, Z.; Chen, X. A sensitive bisphenol A voltammetric sensor relying on AuPd nanoparticles/graphene compositesmodified glassy carbon electrode. *Talanta* **2017**, *166*, 126–132. [CrossRef]

38. Qiu, C.; Xing, Y.; Yang, W.; Zhou, Z.; Wang, Y.; Liu, H.; Xu, W. Surface molecular imprinting on hybrid SiO_2-coated CdTe nanocrystals for selective optosensing of bisphenol A andits optimal design. *App. Surf. Sci.* **2015**, *345*, 405–417. [CrossRef]

39. Rebocho, S.; Cordas, C.M.; Viveiros, R.; Casimiro, T. Development of a ferrocenyl-based MIP in supercritical carbon dioxide: Towards an electrochemical sensor for bisphenol A. *J. Supercrit. Fluids* **2018**, *135*, 98–104. [CrossRef]

40. Kong, Q.; Lin, Y.; Ge, S.; Yu, J. A novel microfluidic paper-based colorimetric sensor based on molecularly imprinted polymer membranes for highly selective and sensitive detection of bisphenol A. *Sens. Actuators B Chem.* **2017**, *243*, 130–136. [CrossRef]

41. Chai, R.; Kan, X. Au-polythionine nanocomposites: A novel mediator for bisphenol A dual-signal assay based on imprinted electrochemical sensor. *Anal. Bional. Chem.* **2019**, *411*, 3839–3847. [CrossRef]

42. Wu, Y.T.; Zhang, Y.H.; Zhang, M.; Liu, F.; Wan, Y.C.; Huang, Z.; Ye, L.; Zhou, Q.; Shi, Y.; Lu, B. Selective and simultaneous determination of trace bisphenol A and tebuconazole in vegetable and juice samples by membrane–based molecularly imprinted solid–phase extraction and HPLC. *Food Chem.* **2014**, *164*, 527–535. [CrossRef]

43. Deceuninck, Y.; Bichon, E.; Durand, S.; Bemrah, N.; Zendong, Z.; Morvan, M.L.; Marchand, P.; Dervilly-Pinel, G.; Antignac, J.P.; Leblanc, J.C.; et al. Development and validation of a specific and sensitive gas chromatography tandem mass spectrometry method for the determination of bisphenol A residues in a large set of food items. *J. Chromatogr. A* **2014**, *1362*, 241–249. [CrossRef]

44. Viñas, P.; Campillo, N.; Martínez-Castillo, N.; Hernández-Córdoba, M. Comparison of two derivatization–based methods for solid–phase microextraction–gas chromatography–mass spectrometric determination of bisphenol A, bisphenol S and biphenol migrated from food cans. *Anal. Bioanal. Chem.* **2010**, *397*, 115–125. [CrossRef]

45. Wang, H.; He, Y.; Ji, Y.; Yan, X.P. Surface Molecular imprinting on Mn-Doped ZnS quantum dots for room-temperature phosphorescence optosensing of pentachlorophenol in water. *Anal. Chem.* **2009**, *81*, 1615–1621. [CrossRef]

46. Zhang, X.; Yang, S.; Jiang, R.; Sun, L.; Pang, S.; Luo, A. Fluorescent molecularly imprinted membranes as biosensor for the detection of target protein. *Sens. Actuators B Chem.* **2018**, *254*, 1078–1086. [CrossRef]

47. Yang, S.; Zhang, X.; Zhao, W.; Sun, L.Q.; Luo, A.Q. Preparation and evaluation of Fe_3O_4 nanoparticles incorporated molecularly imprinted polymers for protein separation. *J. Mater. Sci.* **2016**, *51*, 937–949. [CrossRef]

48. Lakowicz, J.R. Introduction to Fluorescence. In *Principles of Fluorescence Spectroscopy*, 3rd ed.; Lakowicz, J.R., Ed.; Springer: Boston, MA, USA, 2006; pp. 1–26.

49. Zhu, R.; Zhao, W.; Zhai, M.; Wei, F.; Cai, Z.; Sheng, N.; Hu, Q. Molecularly imprinted layer-coated silica nanoparticles for selective solid-phase extraction of bisphenol A from chemical cleansing and cosmetics samples. *Anal. Chim. Acta* **2010**, *658*, 209–216. [CrossRef]

50. Li, X.; Li, C.; Chen, L. Preparation of multifunctional magnetic–fluorescent nanocomposites for analysis of tetracycline hydrochloride. *New J. Chem.* **2015**, *39*, 9976–9982. [CrossRef]

51. Sathe, T.R.; Agrawal, A.; Nie, S.M. Mesoporous silica beads embedded with semiconductor quantum dots and iron oxide nanocrystals: Dual-function microcarriers for optical encoding and magnetic separation. *Anal. Chem.* **2006**, *78*, 5627–5632. [CrossRef]

Article

Synthesis of Silane-Based Poly(thioether) via Successive Click Reaction and Their Applications in Ion Detection and Cell Imaging

Zhiming Gou, Xiaomei Zhang, Yujing Zuo and Weiying Lin *

Institute of Fluorescent Probes for Biological Imaging, School of Chemistry and Chemical Engineering, School of Materials Science and Engineering, University of Jinan, Shandong 250022, China
* Correspondence: weiyinglin2013@163.com; Fax: +86-531-8276-9031

Received: 19 June 2019; Accepted: 3 July 2019; Published: 25 July 2019

Abstract: A series of poly(thioether)s containing silicon atom with unconventional fluorescence were synthesized via successive thiol click reaction at room temperature. Although rigid π-conjugated structure did not exist in the polymer chain, the poly(thioether)s exhibited excellent fluorescent properties in solutions and showed visible blue fluorescence in living cells. The strong blue fluorescence can be attributed to the aggregation of lone pair electron of heteroatom and coordination between heteroatom and Si atom. In addition, the responsiveness of poly(thioether) to metal ions suggested that the selectivity of poly(thioether) to Fe^{3+} ion could be enhanced by end-modifying with different sulfhydryl compounds. This study further explored their application in cell imaging and studied their responsiveness to Fe^{3+} in living cells. It is expected that the described synthetic route could be extended to synthesize novel poly(thioether)s with superior optical properties. Their application in cell imaging and ion detection will broaden the range of application of poly(thioether)s.

Keywords: poly(thioether); ion detection; cell imaging; organosilicone; fluorescence quenching

---

1. Introduction

Poly(thioether) are important base polymers that exhibit excellent physical and chemical properties, such as good chemical and weather resistance, high impermeability, and high heat insulation properties [1]. These excellent properties made it to become a promising functional polymer for the application of optical materials, medical impression materials ion detection probes, energy storage and conversion materials [2–4]. Thus far, many methods have been developed for the ion detection, including microelectrode, absorption spectroscopy [5]. However, these methods are difficult to dynamically monitor cellular ion changes. By contrast, fluorescence spectroscopy has become a powerful tool for fluorescent labeling, ion sensing and cell imaging due to its high sensitivity, excellent selectivity and dynamic monitoring [6,7]. However, as one kind of functional polymer, poly(thioether)s are rarely applied in cell imaging, especially for Fe ion detection in living cells.

At present, two synthetic routes were adopted to prepare poly(thioether). One is the ring-opening polymerization of episulfide [8,9], and the other is thiol click polymerization between dithiols and dienes or dithiols and alkyne [10–12]. As one of most widely used and important reactions, the thiol click reaction has served as useful tool to fabricated dendrimer and block polymer [13,14], and also has been used for end-/side-group functionalization and surface modification [15,16].

Trimethylsilylacetylene (TMSA) is a versatile precursor to prepare important intermediates or unsaturated compounds with unique properties, and also to be used as a protecting group for functional modification. NaKa group and Miura group reported the synthesis of silyl-substituted thiophene derivatives and silyl-substituted fulvene derivatives based on TMSA, respectively [17,18]. Yokozawa et al. described the synthesis of ethynyl-functionalized poly(3-hexylthiophene) with one terminal ethynyl group, which has been very suitable for subsequent preparation of block copolymers via a click reaction [19]. Grirrane et al. synthesized a novel family of dipropargylamines through a double catalytic A^3-coupling of primary amine, formaldehyde and TMSA, and subsequent deprotection of two terminal trimethylsilyl (TMS) groups endow them with high values of post-modification [20]. Furthermore, Alejandro and Pilar reported metal-monolayer-metal molecular electronic devices based on oligoynes $(Me_3Si-(C \equiv C)_4-SiMe_3)$ through fluoride-induced deprotection of terminal trimethylsilyl (TMS) groups [21,22], and this nascent surface modification technique provided new perspectives for the fabrication of molecular electrochemical sensors.

Among the above reports, studies on poly(thioether)s using TMSA are rare. Here, this study reported the synthesis of functionalized poly(thioether)s containing silane via a two-step successive addition reaction (Scheme 1). Firstly, the addition of trimethylsilylacetylene (TMSA) to a dithiol with different molar ratio yielded a series of poly(thioether)s. Secondly, vinyl poly(thioether) (**P1** and **P2**) was selected as intermediates, and reserved vinyl groups further reacted with thiol and finally gained the functionalized poly(thioether)s. The authors further explored their fluorescence properties and used them in cell imaging and ion detection. It was found that the post-functionalization of poly(thioether)s obviously enhanced their responsiveness to Fe^{3+} and avoided the disturbance of Fe^{2+}. The obtained poly(thioether)s and post-functionalized products provided a simple route to synthesize sulfur-containing organosilicone polymer with superior optical properties and extended their application in the field of cell imaging and ion detection.

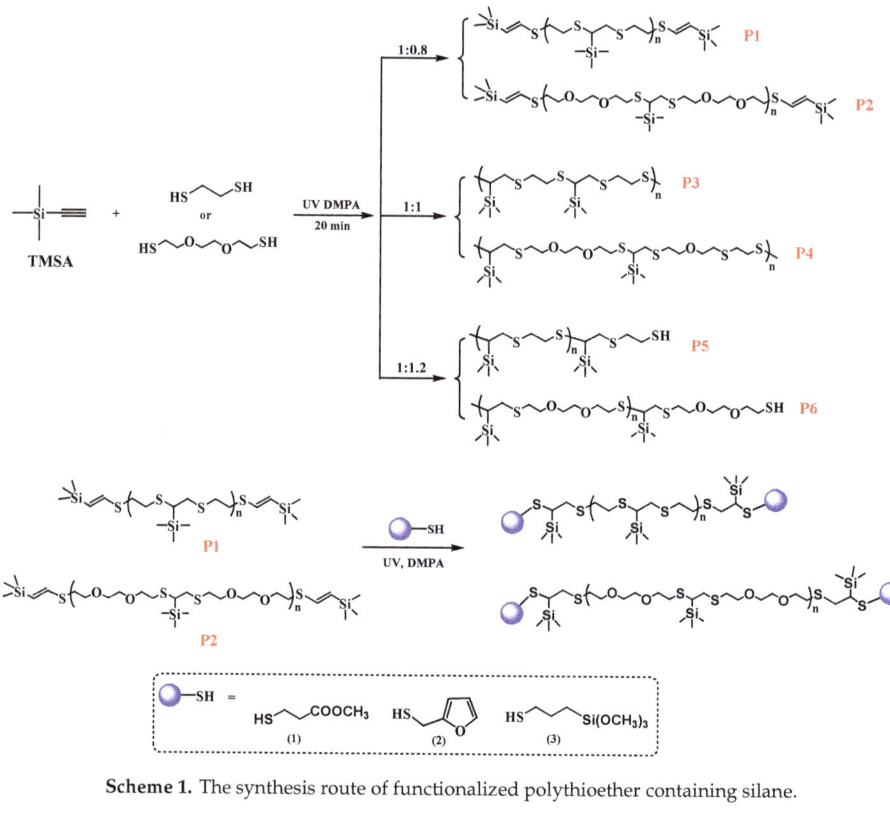

Scheme 1. The synthesis route of functionalized polythioether containing silane.

2. Materials and Experiments

2.1. Materials

Trimethylsilylacetylene (TMSA), 2,2-dimethoxy-2-phenylacetophenone (DMPA), 1,2-ethanedithiol (EDT), and 3,6-dioxa-1,8-octanedithiol (DODT) were purchased from Aladdin Co. (Shanghai, China) and used directly. Furfuryl mercaptan, trimethoxysilylpropanethiol, and methyl 3-mercaptopropionate were supplied by Macklin biochemical (Shanghai, China) Co., Ltd. and used as received. All solvents were bought from Chemical Technology (Shanghai, China) Co., Ltd. and directly used without further purification.

2.2. Characterization and Measurements

NMR (^{1}H and ^{13}C) spectra were recorded by Bruker AVANCE 400 MHz using CDCl$_3$ as solvents, and without tetramethylsilane as the internal standard. Thermogravimetric analysis (TGA) was measured using SDTQ 600 at 10 °C·min^{-1} under N$_2$. The luminescence emission spectra of the poly(thioether)s were recorded by using a Hitachi F–4500 fluorescence spectrophotometer with Xe lamp as an excitation source. The data of molecular weights was measured by gel permeation chromatography (GPC) using a Waters 515 liquid chromatograph with a refractive index detector 2414 and using tetrahydrofuran (THF) as the elution solvent.

2.3. Synthesis of Poly(thioether)

The synthetic route and structure of poly(thioether)s have been shown in Scheme 1. The detailed procedure for **P1** was as follows: TMSA (0.98 g, 10.0 mmol), EDT (0.75 g, 8.0 mmol) and DMPA (1 wt%) were placed in a glass vessel. The above mixture was exposed to a UV light source (365 nm, 100 w) with stirring for 20 min at room temperature. The crude product was washed three times with cold methanol and then reduced pressure distillation. The obtained light-yellow solid was **P1**. Yield: 96%. A series of poly(thioether)s (**P2**, **P3**, **P4**, **P5**, and **P6**) were synthesized according to the same above reaction condition and the treating process.

The data of **P1**: ^{1}H-NMR (400 MHz, CDCl$_3$, ppm): δ = 0.05–0.20 (-SiC*H*$_3$), 2.05 (-SiCHC*H*$_2$S-), 2.69–3.14 (-SiC*H*C*H*$_2$S-, -SC*H*$_2$C*H*$_2$S-), 5.72 and 5.77 (s, -SiC*H*$_2$=C*H*$_2$S-), 6.47 and 6.52 (s, -SiC*H*$_2$=C*H*$_2$S-). ^{13}C-NMR (400 MHz, CDCl$_3$, ppm): δ = −3.0 (SiC*H*$_3$), 31.6 (-SiCHC*H*$_2$S-), 32.8 and 36.5 (-SC*H*$_2$C*H*$_2$S-, -SiCHC*H*$_2$S-), 110.7 and 133.0 (-SiC*H*$_2$=C*H*$_2$S-).

The data of **P2**: ^{1}H-NMR (400 MHz, CDCl$_3$, ppm): δ = 0.07–0.13 (-SiC*H*$_3$), 2.07 (-SiCHC*H*$_2$S-), 2.75–2.95 (-SiCHC*H*$_2$S-, -SC*H*$_2$C*H*$_2$O-), 3.03–3.08 (-OC*H*$_2$C*H*$_2$O-), 3.61–3.70 (-SC*H*$_2$C*H*$_2$O-), 6.49 and 6.53 (-SiC*H*$_2$=C*H*$_2$S-). ^{13}C-NMR (400 MHz, CDCl$_3$, ppm): 2.06–2.07 (SiC*H*$_3$), 29.7 and 32.89 (-OC*H*$_2$C*H*$_2$S-), 31.7 (-SiCHC*H*$_2$S-), 36.7 (-SiCHC*H*$_2$S-), 70.7 (-OC*H*$_2$C*H*$_2$O-), 70.9 (-SC*H*$_2$C*H*$_2$O-), 110.9 and 133.1 (-SiC*H*$_2$=C*H*$_2$S-). Yield: 95%.

The data of **P3**: ^{1}H-NMR (400 MHz, CDCl$_3$, ppm): δ = 0.06–0.19 (m, -SiC*H*$_3$), 2.07 (-SiCHC*H*$_2$S-), 2.79–3.09 (-SiC*H*C*H*$_2$S-, -SC*H*$_2$C*H*$_2$S-). ^{13}C-NMR (400 MHz, CDCl$_3$, ppm): −3.0 (SiC*H*$_3$), 24.6 (-SC*H*$_2$C*H*$_2$SH), 31.6 (-SiCHC*H*$_2$S-), 32.9 (-SC*H*$_2$C*H*$_2$S-), 33.7 (-SC*H*$_2$C*H*$_2$SH), 36.5 (-SiCHC*H*$_2$S-). Yield: 97%.

The data of **P4**: ^{1}H-NMR (400 MHz, CDCl$_3$, ppm): δ = 0.07–0.13 (-SiC*H*$_3$), 2.07 (-SiCHC*H*$_2$S-), 2.73–2.91 (-SiCHC*H*$_2$S-, -SC*H*$_2$C*H*$_2$O-), 3.02–3.06 (-OC*H*$_2$C*H*$_2$O-), 3.62–3.68 (-SC*H*$_2$C*H*$_2$O-). ^{13}C-NMR (400 MHz, CDCl$_3$, ppm): 2.06–2.07 (SiC*H*$_3$), 29.8 and 32.9 (-OC*H*$_2$C*H*$_2$S-), 31.8 (-SiCHC*H*$_2$S-), 36.9 (-SiCHC*H*$_2$S-), 70.8 (-OC*H*$_2$C*H*$_2$O-), 70.9 (-SC*H*$_2$C*H*$_2$O-). Yield: 93%.

The data of **P5**: ^{1}H-NMR (400 MHz, CDCl$_3$, ppm): δ = 0.05–0.16 (m, -SiC*H*$_3$), 1.35 (-SC*H*$_2$C*H*$_2$S*H*), 2.05 (-SiCHC*H*$_2$S-), 2.69–3.05 (-SiCHC*H*$_2$S-, -SC*H*$_2$C*H*$_2$S-). ^{13}C-NMR (400 MHz, CDCl$_3$, ppm): −1.76 (SiC*H*$_3$), 24.6 (-SC*H*$_2$C*H*$_2$SH), 31.6 (-SiCHC*H*$_2$S-), 32.9 (-SC*H*$_2$C*H*$_2$S-), 33.7 (-SC*H*$_2$C*H*$_2$SH), 36.5 (-SiCHC*H*$_2$S-). Yield: 92%.

The data of **P6**: ^{1}H-NMR (400 MHz, CDCl$_3$, ppm): δ = 0.07–0.16 (-SiC*H*$_3$), 1.34 (-OC*H*$_2$C*H*$_2$S*H*), 2.09 (-SiCHC*H*$_2$S-), 2.74–3.07 (-SiCHC*H*$_2$S-, -SC*H*$_2$C*H*$_2$O-), 3.62–3.64 (-OC*H*$_2$C*H*$_2$O-), 3.70–3.75

($-SCH_2CH_2O-$, $-OCH_2CH_2SH$). ^{13}C-NMR (400 MHz, $CDCl_3$, ppm): 2.07 ($SiCH_3$), 21.0 ($-OCH_2CH_2SH$), 29.6 and 32.9 ($-OCH_2CH_2S-$), 67.1–71.3 ($-OCH_2CH_2O-$, $-SCH_2CH_2O-$, $-OCH_2CH_2SH$). Yield: 94%.

Methyl-mercaptopropionate, trimethoxysilylpropanethiol and furfuryl mercaptan were selected and used to the functionalization of **P1** and **P2**. Taking methyl-mercaptopropionate as an example, the above obtained **P1**, methyl-mercaptopropionate (4.0 mmol) and DMPA (1 wt%) were dissolved in THF (2 mL) and reacted following the above-mentioned procedure. The crude product was washed three times with cold methanol and then reduced pressure distillation. The obtained light-yellow solid was **P1-1**. Yield: 93%. A series of functionalized poly(thioether)s (**P1-1**, **P1-2**, **P1-3**, **P2-1**, **P2-2**, and **P2-3**) were synthesized according to the same procedure.

The data of **P1-1**: ^1H NMR (400 MHz, $CDCl_3$): 0.03–0.09 ($SiCH_3$), 2.21 ($SiCHCH_2S$), 2.47–2.51($SCH_2CH_2COOCH_3$), 2.60–2.99 ($SiCHCH_2S$, SCH_2CH_2S), 3.61 ($COOCH_3$). ^{13}C-NMR (400 MHz, $CDCl_3$, ppm): −3.01 ($SiCH_3$), 29.6 ($SCH_2CH_2COCH_3$), 31.61 (SCH_2CH_2O), 32.7 ($SiCHCH_2S$), 33.4 ($SCH_2CH_2COCH_3$), 36.5 and 38.1 ($SiCHCH_2S$), 69.8 (OCH_2CH_2S), 172.1 ($COOCH_3$).

The data of **P1-2**: ^1H NMR (400 MHz, $CDCl_3$): 0.01–0.10 ($SiCH_3$), 2.23 ($SiCHCH_2S$), 2.51 and 3.18 ($SiCHCH_2S$), 2.58–3.01 (SCH_2CH_2S), 3.33–3.43 (SCH_2CO), 6.27 and 6.40 ($C=CH_2CH_2CH_2O$), 7.58 ($C=CH_2CH_2CH_2O$). ^{13}C NMR (400 MHz, $CDCl_3$): −1.75 ($SiCH_3$), 30.0 (SCH_2CO), 31.0 and 31.6 (SCH_2CO), 32.6 and 34.6 ($SiCHCH_2S$), 36.4($SiCHCH_2S$), 106.9–111.1 ($-C=CHCH=CHO-$), 142.3 ($-C=CHCH=CHO-$), 151.4 ($-C=CHCH=CHO-$). Yield: 94%.

The data of **P1-3**: ^1H NMR (400 MHz, $CDCl_3$): 0.01–0.15 ($SiCH_3$), 0.56–0.74 ($OSiCH_2CH_2CH_2S$), 1.53–1.66 ($OSiCH_2CH_2CH_2S$), 2.17–2.26 ($SiCHCH_2S$), 2.43-2.52 and 3.18 ($SiCHCH_2S$), 2.63–3.04 (SCH_2CH_2S), 3.41–3.50 (OCH_3). ^{13}C NMR (400 MHz, $CDCl_3$): −2.0 ($SiCH_3$), 8.35 ($SiCH_2CH_2CH_2S$), 9.65 ($SiCH_2CH_2CH_2S$), 31.4 and 31.7 (SCH_2CH_2S, $SiCHCH_2S$), 33.1 ($SiCH_2CH_2CH_2S$), 36.7 ($SiCHCH_2S$), 48.9 and 50.2 (OCH_3). Yield: 93%.

The data of **P2-1**: ^1H NMR (400 MHz, $CDCl_3$): 0.07–0.10 ($SiCH_3$), 2.23 ($SiCHCH_2S$), 2.59–2.65 (SCH_2CH_2O), 2.66–2.76 ($SiCHCH_2S$), 2.78–2.85 ($SCH_2CH_2CO,$), 3.52 (OCH_2CH_2O), 3.54–3.60 (SCH_2CH_2O), 3.61 (OCH_3). ^{13}C-NMR (400 MHz, $CDCl_3$, ppm): −1.89 and −0.53 ($SiCH_3$), 29.8 ($SCH_2CH_2COCH_3$), 31.61 (SCH_2CH_2O), 32.9 ($SiCHCH_2S$), 33.6 ($SCH_2CH_2COCH_3$), 36.7 and 38.3 ($SiCHCH_2S$), 51.9 ($COOCH_3$), 69.8 (OCH_2CH_2O), 70.9 (OCH_2CH_2S), 172.3 ($COOCH_3$). Yield: 92%.

The data of **P2-2**: ^1H NMR (400 MHz, $CDCl_3$): 0.01–0.09 ($SiCH_3$), 2.24 ($SiCHCH_2S$), 2.67–2.76 (SCH_2CH_2O), 2.51 and 2.76–2.92 ($SiCHCH_2S$), 3.52–3.54 (SCH_2CO), 3.55–3.59 (OCH_2CH_2O), 3.75–3.95 (SCH_2CH_2O), 6.26 and 6.38 ($-C=CHCH=CHO-$), 7.57 ($-C=CHCH=CHO-$). ^{13}C NMR (400 MHz, $CDCl_3$): −1.74 ($SiCH_3$), 29.7 (SCH_2CO), 31.0 and 31.8 (SCH_2CH_2O), 32.8 and 34.7 ($SiCHCH_2S$), 36.6($SiCHCH_2S$), 70.0 (OCH_2CH_2O), 70.8 (SCH_2CH_2O), 106.9–111.0 ($-C=CHCH=CHO-$), 142.7 ($-C=CHCH=CHO-$), 151.8 ($-C=CHCH=CHO-$). Yield: 94%.

The data of **P2-3**: ^1H NMR (400 MHz, $CDCl_3$): 0.07–0.10 ($SiCH_3$), 0.58–0.72 ($OSiCH_2CH_2CH_2S$), 1.55–1.62 ($SiCH_2CH_2CH_2S$), 2.24 ($SiCHCH_2S$), 2.43–2.52 and 2.99 ($SiCHCH_2S$), 3.42 (SCH_2CO), 3.48–3.59 (OCH_2CH_2O), 3.50–3.54 (OCH_3), 3.42–3.59 (SCH_2CH_2O). ^{13}C NMR (400 MHz, $CDCl_3$): −1.9 ($SiCH_3$), 8.33 ($SiCH_2CH_2CH_2S$), 9.67 ($SiCH_2CH_2CH_2S$), 31.3 and 31.7 (SCH_2CH_2O, $SiCHCH_2S$), 32.9 ($SiCH_2CH_2CH_2S$), 36.7 ($SiCHCH_2S$), 48.9 and 50.1 (OCH_3), 70.0 (OCH_2CH_2O), 70.9 (SCH_2CH_2O). Yield: 91%.

2.4. Fluorescence Property and Ion Detection

Poly(thioether)s were dissolved in ethanol (0.3 mg/mL) and their fluorescent emission spectra were measured under excitation wavelengths of 330 nm.

The responsiveness of poly(thioether)s to ions were detected by adding corresponding chlorine salt (100 μL, 1 mmol/L) to poly(thioether)s solution (900 μL, 0.2 mg/mL).

The HeLa cells were incubated with 4.0 μg of poly(thioether)s for 20 min at 37 °C and washed with phosphate buffer solution (PBS). The cell images of poly(thioether)s in HeLa cells were observed under a confocal microscope after adding new culture medium. In addition, the responsiveness of poly(thioether)s to Fe^{3+} in HeLa cells were measured by successively adding 10 μL, 30 μL, 50 μL and 70 µL Fe^{3+} ion (1 mmol/L), respectively. The response process was observed in situ under confocal microscopy.

3. Results and Discussion

3.1. Synthesis of Poly(thioether)s and Post-Functionalization

A series of poly(thioether)s were synthesized using different molar ratio (1:0.8, 1:1.0, and 1:1.2) between TMSA and dithiol (Scheme 1). In view of the same reaction condition and purification step, 1,2-ethanedithiol based poly(thioether)s (**P1**, **P3**, and **P5**) was selected to be analyzed and their ^{1}H NMR have been shown in Figure 1. The characteristic peak of alkenyl groups (-*CH=CH*-) appeared at approximately 5.75 ppm and 6.5 ppm only existed in the polymer when the molar ratio was 1:0.8. By the same token, the characteristic peak of sulfhydryl groups (-*SH*) appeared at approximately 1.7 ppm and was presented when the molar ratio was 1:1.2. The appearance of the peak at 1.7 ppm when the molar ratio was 1:1 could be attribute to the low boiling point of TMSA, which resulted in the volatilization of part of TMSA in the reaction process. In addition, the thiol-alkyne reaction proceeded under a radical step-growth mechanism involving a two step addition. Therefore, the dithiol radicals can react with alkynyl groups from different active sites and generate two types of product structures (α-addition and β-addition) which exist simultaneously in the polymer backbone. The appearance of peak at 1.35 ppm belonged to α-addition products. However, a small amount of α-addition could not have had a significant influence on the fluorescent property of poly(thioether)s. This is because the integral structure of linear polymer chain is likely to remain unchanged. ^{1}H-NMR data indicates that the thiol-alkyne click reaction is a simple and an efficient approach to prepare poly(thioether)s with a special functional group by simply changing the molar ratio. Further, the remained end functional group could be used for further modification.

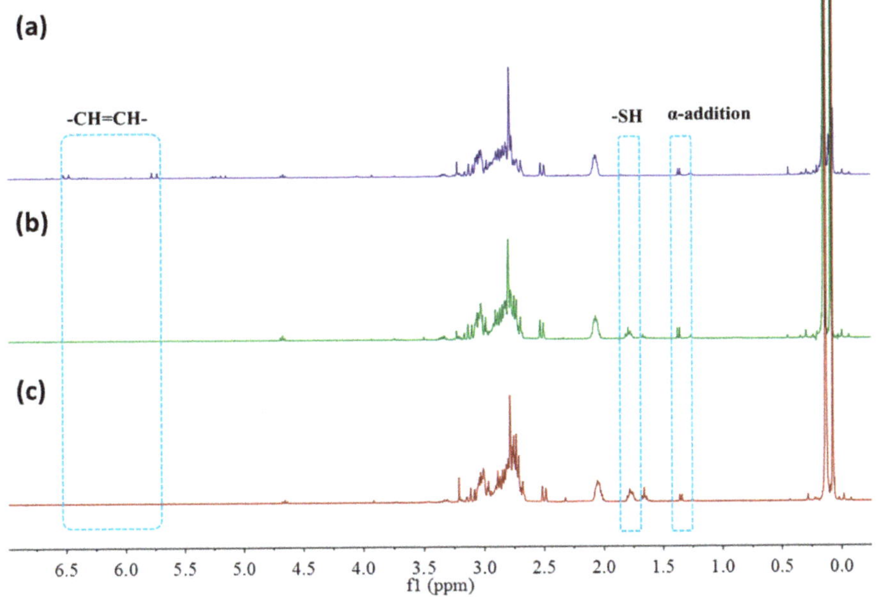

Figure 1. ^{1}H NMR data of (**a**) **P1**, (**b**) **P3**, and (**c**) **P5**.

Theoretically, the reaction with a molar ratio of 1:1 is likely to react completely and maximize the molecular weight of polymers. However, the obtained polymers in this situation did not have reactive sites. When one monomer is excessive, the non-stoichiometric alkyne-to-dithiol molar ratio may result in a handful of functional groups in chain end of polymers. The polymers can possess short polymer chains and low molecular weight. In addition, these maintained functional groups could be used for post-functionalization and naturally various polymers with interesting properties. Carbon-carbon double bonds were maintained in chain end when the molar ratio was 1:0.8, and unreacted sulfhydryl groups were maintained when the molar ratio was 1:1.2. Moreover, the excessive dithiol can lead to the polymers possessing shorter chain lengths than excessive alkyne. Therefore, among the three different molar ratios, the sequence of their molecular weight may be as follows: $M_{w\ (1:1)} > M_{w\ (1:0.8)} > M_{w\ (1:1.2)}$. This speculation has been confirmed by the test outcome. The molecular weight of polymers was measured by GPC and the detailed data have been depicted in Table 1.

Table 1. Gel permeation chromatography (GPC) data for the poly(thioether)s (**P1** to **P6**).

Sample	M_n (g.mol^{-1})	M_w (g.mol^{-1})	M_z (g.mol^{-1})	PDI	M_z/M_w
P1	3210	4000	4910	1.24	1.23
P2	3400	4320	5380	1.27	1.24
P3	9420	14950	21020	1.58	1.40
P4	22800	44460	78360	1.95	1.76
P5	1860	1990	2150	1.07	1.08
P6	2340	2800	3440	1.19	1.23

The thiol-alkyne reaction proceeded via radical step-growth involving two successive addition processes. The first addition generates an intermediate that contains alkenyl group, and subsequent thiol-alkene reaction yields poly(thioether)s. When one monomer is excessive, the non-stoichiometric alkyne-to-dithiol molar ratio can result in a handful of functional groups in the chain end of polymers. Poly(thioether)s containing alkenyl (**P1** and **P2**) were selected to further modification. A series of functionalized poly(thioether)s (**P1-1**, **P1-2**, **P1-3**, **P2-1**, **P2-2**, and **P2-3**) were synthesized with different sulfhydryl compounds via thiol-alkene click reaction (Scheme 1), and a detailed synthetic procedure was described in the experimental section. The modified sulfhydryl compound can bring new and interesting properties to poly(thioether)s, although the number of modified functional groups is very small.

3.2. Fluorescent Properties

Poly(thioether)s (**P1**, **P2**, **P3**, **P4**, **P5**, and **P6**) and functionalized poly(thioether)s (**P1-1**, **P1-2**, **P1-3**, **P2-1**, **P2-2**, and **P2-3**) emit a blue color in the solution and their emission spectra excited by 330 nm as shown in Figure 2. **P1**, **P3**, and **P5** were synthesized with different molar ratios between TMSA and EDT, their fluorescence intensity (Figure 2a) decreased sharply as the molar ratio of dithiol increased. While **P2**, **P4**, and **P6** were synthesized with different molar ratios between TMSA and DODT, their fluorescence intensity (Figure 2b) slightly decreased as the molar ratio of dithiol increased. The unconventional fluorescence can be attributed to the aggregation of long pair electrons of heteroatom (O atom and S atom). Meanwhile, the coordination bonds of O → Si and S → Si were both conducive to cause strong blue photoluminescence in poly(thioether)s [23,24]. The aggregation and coordination in poly(thioether)s containing the O atom (**P2**, **P4**, and **P6**) were stronger than poly(thioether)s without the O atom (**P1**, **P3**, and **P5**). The interaction between heteroatom and silicon atom resulted in the significant difference of fluorescence intensity among poly(thioether)s (**P1**, **P3**, and **P5**) and negligible difference of fluorescence intensity among poly(thioether)s (**P2**, **P4**, and **P6**).

The introduction of sulfhydryl compounds can bring new properties to poly(thioether)s (**P1** and **P2**). The end modified poly(thioether)s demonstrated different trends of fluorescent intensity. The functionalized poly(thioether)s (**P1-1**, **P1-2**, and **P1-3**) based on **P1** showed obvious fluorescence quenching after the end-modification (Figure 2c). The introduction of three sulfhydryl compounds all resulted in the decreasing of fluorescence intensity, especially to methyl-mercaptopropionate. The functionalized poly(thioether)s (**P2-1**, **P2-2**, and **P2-3**) based on **P2** showed slight fluorescence enhancement after the end-modification (Figure 2d). The introduction of three sulfhydryl compounds did not show a significant difference to the fluorescence intensity of **P1**. The difference between two types of functionalized poly(thioether)s indicated that the end modified functional groups had an obvious influence on the fluorescent property of poly(thioether)s, although the number of modified functional groups was very small. The different effect of the end modified groups to **P1** and **P2** could be attributed to the structure of the polymer chain. The oxygen atom in the main chain enhanced the structural stability of **P2** and weakened the influence of the end modified groups on the fluorescent properties of poly(thioether)s. The above results indicated that the fluorescent properties of functionalized poly(thioether)s were influenced by the end modified groups and the structure of the polymer chain.

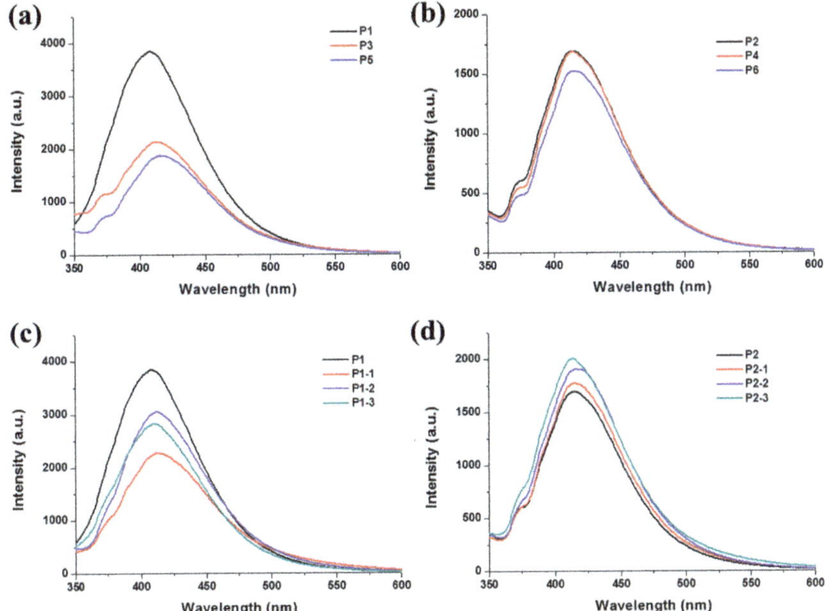

Figure 2. Fluorescence spectra of poly(thioether)s (**a**) **P1**, **P3** and **P5**, (**b**) **P2**, **P4** and **P6**, and functionalized poly(thioether)s (**c**) **P1-1** to **P1-3** and (**d**) **P2-1** to **P2-3** at 0.2 mg/mL final concentration in ethanol solution (slits = 10 nm).

3.3. Ion Detection and Cell Imaging

As one types of most important bioactive substances in biological systems, metal cations are vital to the stability and effectivity of the immune system. To study the responsiveness of poly(thioether)s to various cations, a series of chloride salts (Al^{3+}, Ba^{2+}, Ca^{2+}, Co^{2+}, Cu^{2+}, Fe^{2+}, Fe^{3+}, Mg^{2+}, Ni^{2+}, and Sn^{2+}) were added to the solutions of poly(thioether)s (**P1** and **P2**) and their end-modified derivatives. As shown in Figure 3a,e, the fluorescence intensity of **P1** and **P2** both decreased at different levels after adding metal ions, and especially for Fe^{2+} and Fe^{3+}. This behavior of fluorescence quenching of poly(thioether)s after adding metal ions can be attributed to the coordination between heteroatom (O atom and S atom) and metal ion, especially for Fe^{3+}. This coordination destroyed the interaction between heteroatom and Si atom, and subsequently decreased the aggregation extent of the long pair electrons of heteroatom. These two factors described above resulted in the fluorescence quenching of poly(thioether)s.

In addition, the end-modified with different sulfhydryl compounds resulted in significant differences between their responsiveness to Fe^{2+} and Fe^{3+}. The introduction of methyl 3-mercaptopropionate did not bring obvious changes of **P1-1** to iron ions (Fe^{2+} and Fe^{3+}) compared to **P1** (Figure 3b), while **P2-1** had negligible selectivity to Fe^{2+} and Fe^{3+} compared to **P2** (Figure 3f). However, the introduction of furfuryl mercaptan and trimethoxysilylpropanethiol significantly enhanced the selective responses of **P1** and **P2** derivates to Fe^{3+}. It can be observed that **P1** derivates (Figure 3c,d) and **P2** derivates (Figure 3g,h) all exhibited good selective to Fe^{3+}. The above results suggested that end-modified functional groups with a very small amount could obviously change the fluorescent properties of poly(thioether)s.

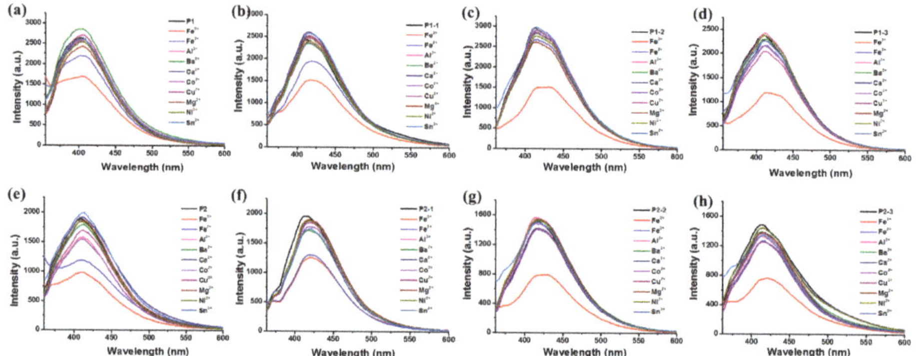

Figure 3. Fluorescence intensity response of functionalized poly(thioether)s (**a**) **P1**, (**b**) **P1-1**, (**c**) **P1-2**, (**d**) **P1-3**, (**e**) **P2**, (**f**) **P2-1**, (**g**) **P2-2**, and (**h**) **P2-3** to different metal cations (0.1 mM) in ethanol solution. (λ_{ex} = 330 nm, slits = 10 nm).

Regarding the good selectivity of functionalized poly(thioether)s to Fe^{3+}, this study further explored their fluorescence behaviors in living cells. First, the HeLa cells were stained with poly(thioether)s (**P1** to **P6**) for 20 min at 37 °C and washed with PBS solution. Their confocal fluorescence images were observed and depicted in Figure 4. The images show visible blue fluorescence in the dark filed and mainly distributed in the cytoplasm. The distribution was due to the way the poly(thioether)s entered into the cell. The polymers enter the cell by simple diffusion. The blue fluorescence of poly(thioether)s belonged to unconventional fluorescence because rigid π-conjugated structures and heterocycle structures did not exist in the obtained poly(thioether)s. Furthermore, silicon atoms in the polymer chain could enhance their fluorescent properties due to their empty 3d orbital [25]. The fluorescence intensity of poly(thioether)s containing the O atom (**P2**, **P4** and **P6**) in cells were stronger than poly(thioether)s without containing the O atom (**P1**, **P3** and **P5**). This may relate to the structure of the polymer chain. The O atom in the main chain enhanced the coordination in polymers, therefore poly(thioether)s containing O atom exhibited stronger fluorescence in living cells. In addition, it can be observed that some poly(thioether)s absorbed on the surface of the cells in the bright filed. The results suggest that poly(thioether)s could easily enter living cells and could be absorbed on the surface of the cells.

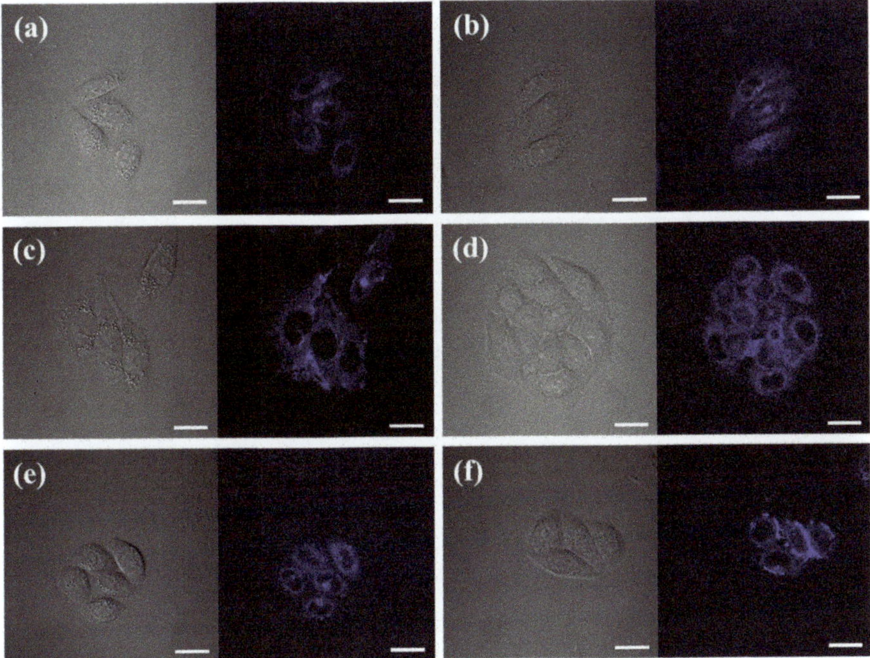

Figure 4. Confocal fluorescence images of (**a**) P1, (**b**) P2, (**c**) P3, (**d**) P4, (**e**) P5, and (**f**) P6 in HeLa cells. bar = 20 µm.

To further investigate the responsiveness of poly(thioether)s to Fe^{3+} in living cells, **P1** and **P2** derivates were selected as a sample and their fluorescent quenching behaviors were observed in situ under confocal microscopy. HeLa cells were stained only with poly(thioether)s and their cell images were observed under confocal microscopy. The responsiveness of poly(thioether)s to Fe^{3+} in HeLa cells were measured by successively adding Fe^{3+} ions. The fluorescence intensity of HeLa cells in the blue field was calculated by confocal microscopy. Confocal fluorescence images of **P1** and their derivates were observed as shown in Figure 5. With the increasing Fe^{3+} concentrations, the fluorescence intensity of **P1** remained roughly unchanged. Meanwhile, the end-modification with different sulfhydryl compounds brought obvious responsiveness to Fe^{3+}. **P1-1** showed slightly fluorescence quenching after adding successive concentrations of Fe^{3+}. **P1-2** and **P1-3** both exhibited obvious fluorescence quenching with the increasing Fe^{3+} ions, especially for **P1-3**. Similarly, the fluorescence intensity of **P2** remained roughly the same with the increasing Fe^{3+} concentrations (Figure 6). Three types of end-modified poly(thioether)s show a similar trend in the change of fluorescence quenching.

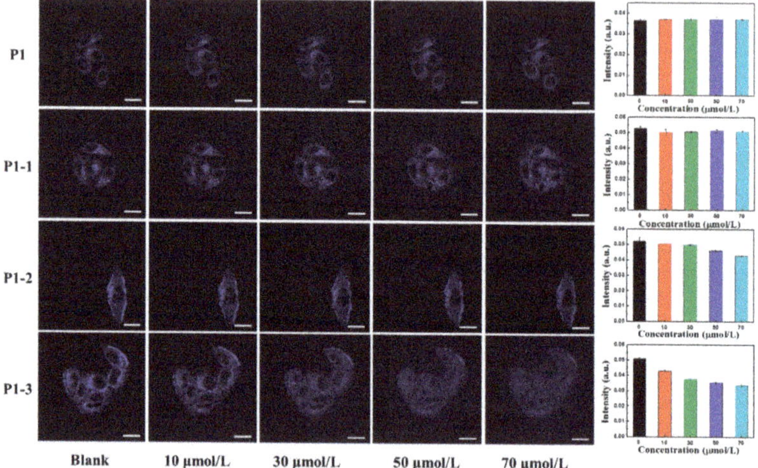

Figure 5. Confocal fluorescence images and quantified relative fluorescence intensity of **P1** and their derivates after adding successive concentrations of Fe^{3+} ions in HeLa cells. bar = 20 μm. The statistical analyses were performed with student's *t*-test ($n = 4$). *$3.37 \times 10^{-8} < p < 1.57 \times 10^{-4}$ and the error bars represent standard deviations (± S. D.).

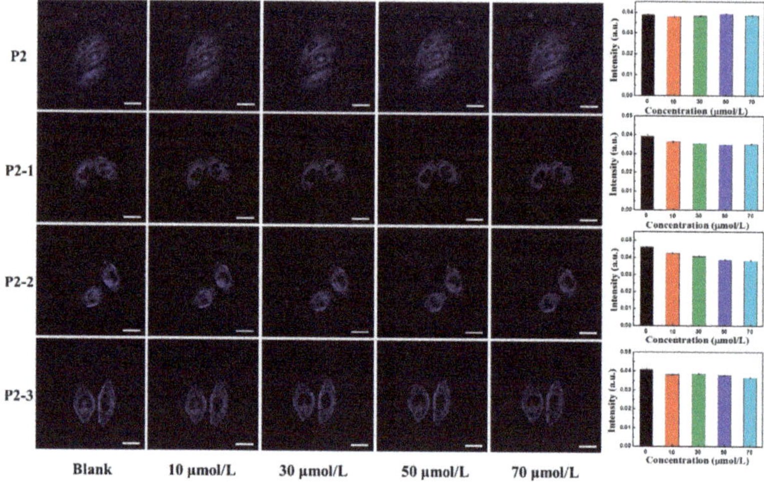

Figure 6. Confocal fluorescence images and quantified relative fluorescence intensity of **P2** and their derivates after adding successive concentrations of Fe^{3+} ion in HeLa cellss. bar = 20 μm. The statistical analyses were performed with student's *t*-test ($n = 4$). *$5.2 \times 10^{-9} < p < 2.39 \times 10^{-6}$ and the error bars represent standard deviations (± S. D.).

3.4. Thermogravimetric Analysis (TGA)

To evaluate the thermal stability of poly(thioether)s, **P1** and **P2** were selected and their thermodynamics properties were measured by TGA analysis at 10 °C·min^{-1} in N_2. The TGA data indicated that the poly(thioether)s exhibited high thermal stability (Figure 7). The thermal decomposition temperature (T_d) of **P1** and **P2** were at approximately 290 °C and 320 °C, respectively. **P2** possessed higher T_d than **P1**, which indicated that the presence of oxygen atoms in the polymer chain increased the thermal performance of poly(thioether)s. In addition, **P1** exhibited two degradation steps.

The first weight loss step of **P1** before 250 °C could be ascribed to the sublimation of low molecular weight of poly(thioether)s. The second weight loss of **P1** can be attributed to the decomposition and sublimation of the main chain. The high T_d of **P1** and **P2** suggested that the good thermal stability of poly(thioether)s, which was prepared by simple thiol-alkyne click reaction at room temperature.

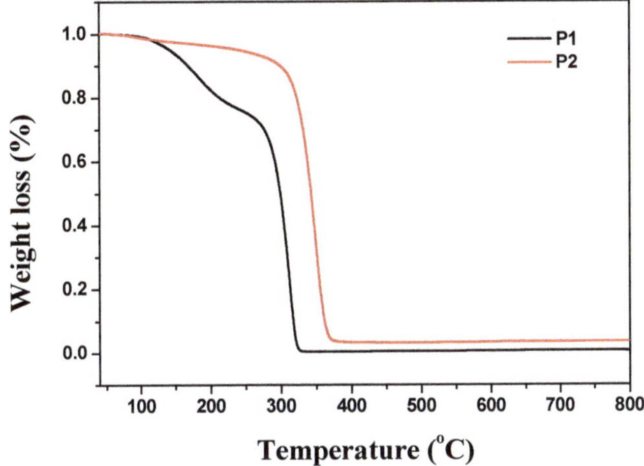

Figure 7. Thermogravimetric analysis (TGA) curves of **P1** and **P2**.

4. Conclusions

A series of silane-based poly(thioether)s and functionalized poly(thioether)s were synthesized via successive thiol click reaction. These poly(thioether)s exhibited excellent fluorescent properties in solutions and showed visible blue fluorescence in living cells. The responsiveness of poly(thioether)s to metal ions can be adjusted by end-modifying with different sulfhydryl compounds. The functionalized poly(thioether)s showed high selectivity to Fe^{3+} detection and still exhibited responsiveness to Fe^{3+} ions in living cells. The synthetic route of poly(thioether)s described herein could be extended to synthesize novel organosilicone fluorescent materials, and their application in ion detection and cell imaging can extend the application field of poly(thioether)s.

Author Contributions: W.L. directed the research, Z.G. and Y.Z. designed the experiments, X.Z. and Z.G. performed the experiments, Z.G., Y.Z. and W.L. wrote the paper.

Funding: This work was financially supported by NSFC (21877048, 21472067, 21672083), Taishan Scholar Foundation (TS201511041), the startup fund of University of Jinan (309-10004, 160100332, 140200321), and Natural Science Foundation of Shandong Province (ZR2018BB022).

Conflicts of Interest: There are no conflict to declare.

References

1. Hearon, K.; Nash, L.D.; Rodriguez, J.N.; Lonnecker, A.T.; Raymond, J.E.; Wilson, T.S.; Wooley, K.L.; Maitland, D.J. A high-performance recycling solution for polystyrene achieved by the synthesis of renewable poly (thioether) networks derived from D-limonene. *Adv. Mater.* **2014**, *26*, 1552–1558. [CrossRef] [PubMed]
2. Lim, J.; Pyun, J.; Char, K. Recent approaches for the direct use of elemental sulfur in the synthesis and processing of advanced materials. *Angew. Chem. Int. Ed.* **2015**, *54*, 3249–3258. [CrossRef] [PubMed]
3. He, L.; Zhao, H.; Theato, P. No heat, no light—The future of sulfur polymers prepared at room temperature is bright. *Angew. Chem. Int. Ed.* **2018**, *57*, 13012–13014. [CrossRef] [PubMed]
4. Zhang, C.-J.; Zhu, T.-C.; Cao, X.-H.; Hong, X.; Zhang, X.-H. Poly (thioether) s from Closed-System One-Pot Reaction of Carbonyl Sulfide and Epoxides by Organic Bases. *J. Am. Chem. Soc.* **2019**, *141*, 5490–5496. [CrossRef] [PubMed]

5. Bansod, B.; Kumar, T.; Thakur, R.; Rana, S.; Singh, I. A review on various electronchemical techniques for heavy metal ions detection with different sensing platforms. *Biosens. Bioelectron.* **2019**, *94*, 443–455. [CrossRef] [PubMed]

6. Van Der Vlist, E.J.; Nolte, E.N.; Stoorvogel, W.; Arkesteijn, G.J.; Wauben, M.H. Fluorescent labeling of nano-sized vesicles released by cells and subsequent quantitative and qualitative analysis by high-resolution flow cytometry. *Nat. Protoc.* **2012**, *7*, 1311–1326. [CrossRef] [PubMed]

7. Mousseau, F.; Berret, J.-F.; Oikonomou, E.K. Design and Applications of a Fluorescent Labeling Technique for Lipid and Surfactant Preformed Vesicles. *ACS Omega* **2019**, *4*, 10485–10493. [CrossRef]

8. Napoli, A.; Tirelli, N.; Kilcher, G.; Hubbell, A.; Hubbell, J. New Synthetic Methodologies for Amphiphilic Multiblock Copolymers of Ethylene Glycol and Propylene Sulfide. *Macromolecules* **2001**, *34*, 8913–8917. [CrossRef]

9. Raffa, P.; Wever, D.A.Z.; Picchioni, F.; Broekhuis, A.A. Polymeric Surfactants: Synthesis, Properties, and Links to Applications. *Chem. Rev.* **2015**, *115*, 8504–8563. [CrossRef]

10. Sarapas, J.M.; Tew, G.N. Thiol-ene step-growth as a versatile route to functional polymers. *Angew. Chem. Int. Ed.* **2016**, *55*, 15860–15863. [CrossRef]

11. Türünç, O.; Meier, M.A.R. A novel polymerization approach via thiol-yne addition. *J. Polym. Sci. Polym. Chem.* **2012**, *50*, 1689–1695. [CrossRef]

12. Lowe, A.B. Thiol-yne 'click'/coupling chemistry and recent applications in polymer and materials synthesis and modification. *Polymer* **2014**, *55*, 5517–5549. [CrossRef]

13. Banuls, M.J.; Gonzalez-Martinez, M.A.; Sabek, J.; Garcia-Ruperez, J.; Maquieira, A. Thiol-click photochemistry for surface functionalization applied to optical biosensing. *Anal. Chim. Acta* **2019**, *1060*, 103–113. [CrossRef] [PubMed]

14. Tunca, U. Click and multicomponent reactions work together for polymer chemistry. *Macromol. Chem. Phys.* **2018**, *219*, 1800163. [CrossRef]

15. Dai, Y.; Zhang, X.; Xia, F. Click chemistry in functional aliphatic polycarbonates. *Macromol. Rapid Commun.* **2017**, *38*, 1700357. [CrossRef] [PubMed]

16. Li, P.-Z.; Wang, X.-J.; Zhao, Y. Click chemistry as a versatile reaction for construction and modification of metal-organic frameworks. *Coordin. Chem. Rev.* **2019**, *380*, 484–518. [CrossRef]

17. Naka, A.; Mihara, T.; Ishikawa, M. Platinum-catalyzed reactions of 3, 4-bis (dimethylsilyl)-and 2, 3, 4, 5-tetrakis (dimethylsilyl) thiophene with alkynes and alkenes. *J. Organomet. Chem.* **2019**, *879*, 1–6. [CrossRef]

18. Suzuki, S.; Kinoshita, H.; Miura, K. Palladium-catalyzed regio-and stereoselective synthesis of (E)-1, 3-bissilyl-6-arylfulvenes from aryl iodides and silylacetylenes. *Org. Lett.* **2019**, *21*, 1612–1616. [CrossRef]

19. Zhang, G.; Ohta, Y.; Yokozawa, T. Exclusive synthesis of poly (3-hexylthiophene) with an ethynyl group at only one end for effective block copolymerization. *Macromol. Rapid. Commun.* **2018**, *39*, 1700586. [CrossRef]

20. Grirrane, A.; Alvarez, E.; García, H.; Corma, A. Double A3—Coupling of Primary Amines Catalysed by Gold Complexes. *Chem. A Eur. J.* **2018**, *24*, 16356–16367. [CrossRef]

21. Moneo, A.; González-Orive, A.; Bock, S.; Fenero, M.; Herrer, I.L.; Milan, D.C.; Lorenzoni, M.; Cea, P.; Pérez-Murano, F.; Low, P.J.; et al. Towards molecular electronic devices based on 'all-carbon' wires. *Nanoscale* **2018**, *10*, 14128–14138. [CrossRef] [PubMed]

22. Herrer, L.; González-Orive, A.; Marques-Gonzalez, S.; Martín, S.; Nichols, R.J.; Serrano, J.L.; Low, P.J.; Cea, P.; Solans, S.M. Electrically transmissive alkyne-anchored monolayers on gold. *Nanoscale* **2019**, *11*, 7976–7985. [CrossRef] [PubMed]

23. Feng, S.; Lu, H.; Zhang, J. Controllable photophysical properties and self-assembly of siloxane-poly (amidoamine) dendrimers. *Phys. Chem. Chem. Phys.* **2015**, *17*, 26783–26789.

24. Lu, H.; Feng, L.; Li, S.; Zhang, J.; Lu, H.; Feng, S. Unexpected Strong Blue Photoluminescence Produced from the Aggregation of Unconventional Chromophores in Novel Siloxane–Poly(amidoamine) Dendrimers. *Macromolecules* **2015**, *48*, 476–482. [CrossRef]

25. Gou, Z.; Zuo, Y.; Tian, M.; Lin, W. Siloxane-Based Nanoporous Polymers with Narrow Pore-size Distribution for Cell Imaging and Explosive Detection. *ACS Appl. Mater. Interfaces* **2018**, *10*, 28979–28991. [CrossRef] [PubMed]

Article

Application of PEG-Covered Non-Biodegradable Polyelectrolyte Microcapsules in the Crustacean Circulatory System on the Example of the Amphipod *Eulimnogammarus verrucosus*

Ekaterina Shchapova [1,2], Anna Nazarova [1], Anton Gurkov [1,2], Ekaterina Borvinskaya [3], Yaroslav Rzhechitskiy [1], Ivan Dmitriev [1], Igor Meglinski [4,5,6] and Maxim Timofeyev [1,2,*]

1   Institute of Biology, Irkutsk State University, 664025 Irkutsk, Russia
2   Baikal Research Centre, 664003 Irkutsk, Russia
3   Institute of Biology, Karelian Research Center of the Russian Academy of Sciences, 185910 Petrozavodsk, Russia
4   Optoelectronics and Measurement Techniques Laboratory, University of Oulu, 90570 Oulu, Finland
5   Aston Institute of Materials Research, School of Engineering & Applied Science, Aston University, Birmingham B4 7ET, UK
6   School of Life & Health Sciences, Aston University, Birmingham B4 7ET, UK
*   Correspondence: m.a.timofeyev@gmail.com; Tel.: +7-3952-243-077

Received: 15 May 2019; Accepted: 24 July 2019; Published: 27 July 2019

Abstract: Layer-by-layer assembled microcapsules are promising carriers for the delivery of various pharmaceutical and sensing substances into specific organs of different animals, but their utility in vivo inside such an important group as crustaceans remains poorly explored. In the current study, we analyzed several significant aspects of the application of fluorescent microcapsules covered by polyethylene glycol (PEG) inside the crustacean circulatory system, using the example of the amphipod *Eulimnogammarus verrucosus*. In particular, we explored the distribution dynamics of visible microcapsules after injection into the main hemolymph vessel; analyzed the most significant features of *E. verrucosus* autofluorescence; monitored amphipod mortality and biochemical markers of stress response after microcapsule injection, as well as the healing of the injection wound; and finally, we studied the immune response to the microcapsules. The visibility of microcapsules decreased with time, however, the central hemolymph vessel was confirmed to be the most promising organ for detecting the spectral signal of implanted microencapsulated fluorescent probes. One million injected microcapsules (sufficient for detecting stable fluorescence during the first hours after injection) showed no toxicity for six weeks, but in vitro amphipod immune cells recognize the PEG-coated microcapsules as foreign bodies and try to isolate them by 12 h after contact.

Keywords: biocompatibility; immunity; implantable sensors; invertebrate; layer-by-layer; primary cell cultures

---

1. Introduction

Various implantable nano- and micro-scaled structures have great potential for becoming the foundation of novel technologies for monitoring and manipulating the state of diverse organisms. Such structures are, for example, polyelectrolyte microcapsules produced by the layer-by-layer (LbL) adsorption technique [1,2]. This technique is based on the deposition of some polymers onto various colloidal microparticles [3–5], including porous structures, such as calcium carbonate in the form of vaterite that can be preloaded with certain substances and further dissolved to obtain soft microcapsules with the substances inside [6,7]. These microcapsules can be easily prepared, using either biodegradable or non-biodegradable polymers to control their stability inside an organism for different applications [3]. Strongly charged poly(allylamine hydrochloride) (PAH) and poly(sodium 4-styrenesulfonate) (PSS) are among the most popular non-biodegradable polyelectrolytes used as sequential building blocks for microcapsule shells [8–10]. These two polymers form bilayers with a structure close to lamellar and can be used to prepare shells of given thickness [9].

For increasing the overall biocompatibility of the microcapsules, a polyethylene glycol (PEG) coating was suggested [11]. PEG can be attached to polyelectrolyte microcapsules in the form of a graft copolymer, with the charged main polymer sticking to the microcapsule surface electrostatically, while backbones of PEG obduce the microcapsule and reduce their aggregation and friction, which simplifies their distribution in the circulatory system [12,13], as well as decreasing protein adsorption to the surface, which can lower or at least postpone the immune response [11,13].

Crustaceans are an important source of aquatic food protein and popular research objects in ecophysiology [14,15]. However, possibilities for the application of microcapsules as an implantable tool for monitoring and adjusting the physiological status of crustaceans in vivo remain poorly explored. Previously, we applied the PEG-covered polyelectrolyte microcapsules to deliver the pH-sensitive fluorescent dye SNARF-1 into the main hemolymph vessel of the amphipod *Eulimnogammarus verrucosus* (Figure 1) and monitor pH changes in vivo [16]. The same approach is perspective for measuring other parameters, such as the concentration of different ions and metabolites [17], directly in the circulatory system of various crustaceans with translucent exoskeletons. The microcapsules used were composed of non-biodegradable PAH and PSS, to increase their stability and prolong the possibility of sensing.

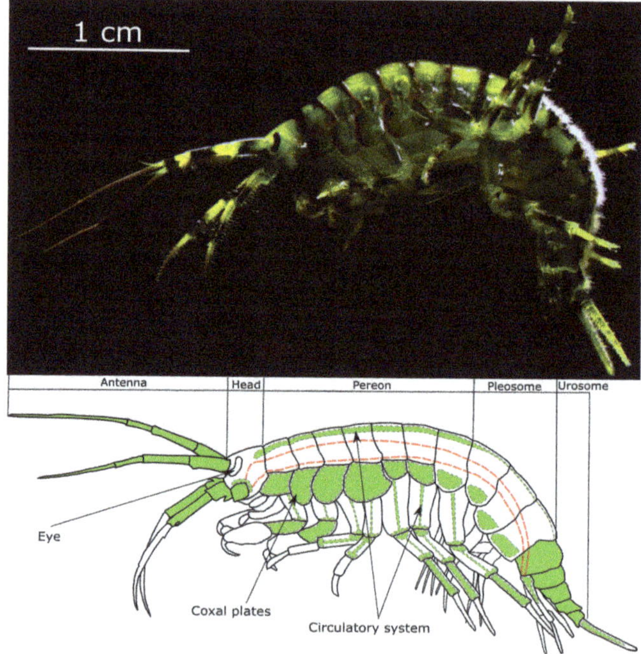

Figure 1. A photo of the amphipod *Eulimnogammarus verrucosus* and schematic designation of all body parts where the microcapsules were observed (colored by green) after injection into the circulatory system. The red dashed line indicates the approximate position of the digestive system. The microcapsules contained fluorescein (in the form of FITC-albumin) and were monitored in the green channel.

In the present study, we analyze various nuances arising from the application of microencapsulated fluorescent dyes in the crustacean circulatory system, using the example of the same amphipod *E. verrucosus* (Figure 1), endemic to Lake Baikal [18]. In particular, we monitor the visibility of the PEG-coated PAH/PSS microcapsules after injection into the main hemolymph vessel and analyze the autofluorescence of *E. verrucosus* to highlight the most promising body parts for the application of microencapsulated molecular probes. Then, we assess the survival and stress response of the amphipods after injection of the microcapsules. Finally, we study the wound healing after injection and the amphipod immune response to the microcapsules.

2. Materials and Methods

2.1. Materials

All the chemicals used for the preparation of polyelectrolyte microcapsules and further procedures were of analytical grade and were applied without additional purification. The conjugate of fluorescein isothiocyanate with bovine serum albumin (FITC-albumin; #A9771) was purchased from Sigma-Aldrich (St. Louis, MO, USA), the conjugate of seminaphtharhodafluor-1 with dextran (SNARF-1-dextran; #D-3304) was bought from Thermo Fisher Scientific (Eugene, OR, USA). Polymers poly(allylamine hydrochloride) (PAH; #283215) and poly(sodium 4-styrenesulfonate) (PSS; #243051) were provided by Sigma-Aldrich (produced in USA and Belgium, respectively). The poly(L-lysine)-graft-poly(ethylene glycol) copolymer (PLL-g-PEG; #SZ34-67) was purchased from SuSoS (Dübendorf, Switzerland).

2.2. Preparation of Microcapsules

Fluorescent dyes were encapsulated at room temperature using the layer-by-layer adsorption of oppositely charged polyelectrolytes, as described previously [19]. The conjugates FITC-albumin and SNARF-1-dextran were co-precipitated in porous $CaCO_3$ microcores by mixing 2 mL of 4 mg/mL conjugate solution with 0.615 mL of 1 M $CaCl_2$ and 1 M Na_2CO_3 solutions. The cores were then covered with 12 layers of oppositely charged polyelectrolytes PAH and PSS and with the final layer of PLL-g-PEG. After dissolving $CaCO_3$ cores in 0.1 M ethylenediaminetetraacetic acid solution (pH 7.1), the microcapsules had the following structure: dye conjugate/(PAH/PSS)$_6$/PLL-g-PEG.

2.3. Animal Sampling and Maintenance

Adult individuals (~25–35 mm-long) of *Eulimnogammarus verrucosus* (Gerstfeldt, 1858) were caught with a hand net (kick sampling) in the Baikal littoral zone near the village of Listvyanka (51°52′13.3″ N 104°49′41.9″ E), at depths of 0–1.2 m. The amphipod species is not endangered or protected; no specific permission was required for sampling. During acclimation, amphipods were kept in well-aerated 3 L aquaria at a water temperature of approximately 6 °C, at a density of 15–20 individuals per aquarium. Amphipods were given one week to acclimatize to laboratory conditions before any experiments. The high levels of activity, feeding, and absence of mortality during the acclimation period confirmed that laboratory maintenance was not stressful for the amphipods.

All experimental procedures with amphipods were conducted in accordance with the EU Directive 2010/63/EU for animal experiments and were approved by the Animal Subjects Research Committee of Institute of Biology at Irkutsk State University.

2.4. Injection into the Circulatory System of Amphipods

The number of microcapsules being injected was one million per animal. The 2 μL suspension of microcapsules in the isotonic solution (0.9% aqueous solution of NaCl) was injected into the central hemolymph vessel of amphipods between the sixth and seventh segments of pereon using the IM-9B microinjector (Narishige, Tokyo, Japan). The injected microcapsules contained FITC-albumin, unless otherwise specified. During the injection, the individual was immobilized inside a wet polyurethane sponge with a temperature of approximately 6 °C.

2.5. Fluorescent Microscopy and Spectroscopy of Amphipods

Microcapsules and autofluorescence of *E. verrucosus* were visualized under the Mikmed-2 fluorescence microscope (LOMO, Saint Petersburg, Russia), with either the EOS 1200 camera (Canon, Taiwan) or the QE Pro spectrometer (Ocean Optics, Largo, FL, USA; acquisition range 350–1100 nm, INTSMA-200 optical slit, integration time 1–5 s) attached. The spectrometer was connected to the microscope, as described previously [19]. The animals were restrained under the 10× microscope objective using a thermostatic cell containing circulating Baikal water with a temperature of 5–8 °C, according to previous recommendations [20].

Fluorescent microscopy was performed in either the green channel (the peak excitation wavelength at 496 nm; long-pass emission filter from ~510 nm) or the red channel (the peak excitation wavelength at 546 nm; long-pass emission filter from ~585 nm). The parameters of the green channel were standard for visualization of FITC and similar dyes; the illumination conditions of the red channel were optimized to excite SNARF-1 in the used optical system [16,21]. Data analysis was performed using the Scilab package (www.scilab.org).

2.6. Determination of Biochemical Stress Response Markers

The biochemical measurements were performed in whole individuals frozen in liquid nitrogen. For the quantitative determination of 70 kDa heat shock proteins (HSP70), we used the method of Bedulina et al. [22]. The HSP70 content was determined by western blot. SDS electrophoresis was performed in polyacrylamide gel blocks using the Mini-PROTEAN II electrophoretic cell (Bio-Rad, USA). Relative molecular weights of proteins were estimated using the low molecular weight marker kit (Biomol, UK). Proteins were transferred to a polyvinylidene fluoride membrane, as described in Bedulina et al. [23]. For HSP70 assessment, the blots were incubated with primary antibody (anti-HSP70 antibody produced in mice; Sigma-Aldrich, #H5147) and then with secondary antibody (anti-mouse IgG:AP Conj.; Sigma-Aldrich #A3562). Bovine HSP70 (Sigma, #H9776) was used as a positive control. HSP70 levels were measured by the semiquantitative analysis of grey values on scanned western blot membranes, using the ImageJ software within the Fiji package [24].

Lactate levels were determined using the Lactate-Vital express kit (Vital–Diagnostics, St. Petersburg, Russia) according to Axenov-Gribanov et al. [25] Spectrophotometry was performed using the Cary 50 spectrophotometer (Varian, Palo Alto, CA, USA) at 505 nm (absorbance, here and further).

The activity of the enzyme glutathione S-transferase (GST; EC 2.5.10.13) was measured after extraction, according to Shchapova et al. [26]. The ratio of tissue homogenization buffer to amphipod biomass was 3:1 (v:w). Specimens were homogenized in 0.1 M sodium phosphate buffer (pH 6.5) and centrifuged at $10,000\times g$ for 3 min. Glutathione S-transferase activity was measured using 0.97 mM 1-chloro-2,4-dinitrobenzene as substrate at 340 nm (pH 6.5) according to Habig et al. [27]. The Bradford assay was used to evaluate protein concentrations [28]. The assay was based on the binding of coomassie brilliant blue G-250 to residues of certain amino acids (mostly arginine and lysine), leading to an increase in absorbance at 595 nm [29]. Enzyme activities are expressed in nkat/mg protein.

2.7. Monitoring the Healing of the Injection Wound

The process of exoskeleton healing after the injection was monitored for four weeks. We defined three stages of repair based on the observed wound melanization [30]: no healing (no black coloration); healing is ongoing (indicated by the black coloration of the wound border); the exoskeleton is repaired. The external examination of the chitin state was followed by an internal checkup after the dissection of four–five individuals (euthanized with clove oil suspension) at each time point. Amphipods were examined under the SPM0880 stereomicroscope (Altami, Saint Petersburg, Russia).

2.8. Determination of the Phenoloxidase Activity

Phenoloxidase (PO) activity was measured in the amphipod hemolymph in response to microcapsule injection. Additionally, we tested the sensitivity of the PO system to yeast *Saccharomyces cerevisiae* as a positive control. Yeast cells resuspended in isotonic solution were injected as described for the microcapsules at concentrations of one million or six million cells per animal.

Hemolymph extracts were taken between the seventh and eighth dorsal segment using a sterile needle at 6 h, 12 h, 1 day, and 3 days after injection. Seventy microliters of hemolymph were collected into a sterile, pre-chilled glass capillary and placed into a 0.5 mL microcentrifuge tube containing 0.07 mL cold phosphate buffered saline (PBS; 150 mM NaCl, 10 mM Na_2HPO_4, pH 8) with 7 mg/mL phenylmethanesulfonyl fluoride, as modified from [31]. All samples were frozen in liquid nitrogen and stored at $-80\,°C$ until later measurements. After thawing, the samples were centrifuged for 10 min at 500 g and 4 °C to pellet the cell fraction.

About 10 µL hemolymph extract was mixed with 40 µL PBS, 280 µL distilled water, and 40 µL 3,4-dihydroxy-L-phenylalanine (L-DOPA; 4 mg/mL of distilled water) to measure the PO activity. The determination of PO activity was based on the catalytic conversion of L-DOPA to dopachrome, with the solution staining in red-brown color [32]. Measurements were performed with the Cary 50 spectrophotometer at a wavelength of 490 nm (absorbance) for 40 min. Enzyme activity was assessed as the slope of the reaction curve during the linear reaction phase [31].

2.9. Primary Culture of Amphipod Hemocytes

We established the primary culture of hemocytes isolated from the hemolymph of amphipods *E. verrucosus*. Before the hemolymph extraction, the dorsal side of the pereon surface was sterilized with 70% ethanol. The central hemolymph vessel was punctured with a sterile needle, and hemolymph was collected with a sterile glass capillary. The amphipod hemolymph was mixed (1:1) with the isotonic anticoagulant solution (150 mM NaCl, 5 mM Na_2HPO_4, 30 mM sodium citrate, 10 mM EDTA, pH 8.0; filtered through a 0.45 µm syringe filter) on ice to avoid undesired degranulation of granulocytes. Cells were separated by low-speed centrifugation (23 g for 2 min at 4 °C); the humoral fraction was discarded, and hemocytes were washed with a sterile buffer solution (150 mM NaCl, 5 mM Na_2HPO_4, pH 8.0), based on the available data on the hemolymph composition [33,34]. The washing procedure was repeated twice. The viability of amphipod hemocytes was assessed using vital staining with 0.4% trypan blue in a hemocytometer under the Mikmed-2 microscope. The used hemocyte extraction procedure resulted in at least 90% hemocyte viability, a median cell loss of approximately one-third of hemocytes, no increase in degranulation (as compared to the granule concentration in hemolymph), and aggregation of no more than 20% of the cells (aggregates were defined as a cluster of at least three hemocytes).

Hemocytes were suspended in sterile Leibovitz's L-15 medium with L-glutamine (#1840516, Life Technologies, Belfast, UK), containing 15% fetal bovine serum (#FB-1001, Biosera, South America). The average volume of *E. verrucosus* hemolymph was 50–100 µL, and the used medium volume was 50 µL. The hemocyte concentration in the media was about 4000 cells/µL, which is approximately equal to the upper natural concentration in amphipod hemolymph. Cells were kept in 0.5 mL microtubes at 6–8 °C, with regular tube rotation to imitate the hemolymph circulation in the circulatory system of the animal. The viability of hemocytes in the primary culture did not drop below 90% for the three days.

2.10. Analysis of Hemocyte Aggregation after Contact with Microcapsules

The obtained primary culture of amphipod hemocytes was incubated with polyelectrolyte microcapsules or a suspension of *S. cerevisiae* cells, used as a positive control, to test the sensitivity of the culture to foreign bodies. Before the test, microcapsules or yeast cells were washed in sterile buffer (0.23 g for 5 min) and resuspended in the L-15 medium. Fifty microliter aliquots of the primary hemocyte culture, with approximately 0.2 million cells, were mixed with 2 µL aliquots containing either 0.2 million yeast cells or one million microcapsules (identical to the number injected during the other experiments). The recognition of the foreign bodies by hemocytes was assessed by measuring the hemocyte aggregation, defined as the proportion of free hemocytes compared to the initial total number of hemocytes in the sample. The aggregation process was examined at 1, 5, 12, 17, and 24 h after the start of incubation.

The staining of the hemocyte aggregates was performed with a dichlorotris(1,10-phenanthroline)ruthenium (II) hydrate (RuPhen$_3$; Sigma-Aldrich, USA, #343714) and visualized in the green fluorescent channel. RuPhen$_3$ contains aromatic components and may bind to hydrophobic structures of the cell, such as membrane lipids. Compounds with similar structure were used for the staining of nuclear components [35]. In the case of *E. verrucosus*, RuPhen$_3$ was able to emphasize nuclei and granules of a significant part of hemocytes (but not all). Part of the microcapsules was also stained by RuPhen$_3$ due to the interaction with PSS in the microcapsule shell [17].

2.11. Preparation of the Histological Section

For histological analysis, individual amphipods were sacrificed with deep anesthesia (in clove oil suspension) one week post injection of microcapsules containing SNARF-1-dextran. About 0.5 mL of Davidson's solution was injected into the central vessel. Then, the urosome was removed, and crustaceans were completely immersed into Davidson's solution for one day [36]. On the next day, the samples were rinsed, dissected to 0.5 mm-thick pieces and stored in 70% methyl alcohol. The specimens were embedded in paraffin, routinely using an STP 120 Spin Tissue Processor (Thermo Fisher Scientific, Walldorf, Germany), according to the protocol for tissues [37]. Tissues were cut into 6 µm-thick-slices using the HM-440 microtome (Thermo Fisher Scientific, Walldorf, Germany). The resulting sections were dewaxed and examined in the red fluorescent channel to detect microcapsules. Stained histological sections (hematoxylin and eosin) were studied in the bright field.

2.12. Statistical Analysis

The statistical significance of differences between the experimental and control (parallel if available or initial) groups was analyzed using the Mann–Whitney U test, with the Holm correction for multiple comparisons in R (www.r-project.org). The differences were considered significant with $p < 0.05$.

3. Results and Discussion

3.1. Visibility of the Microcapsules inside Amphipods: Distribution and Autofluorescence

The possibility of visualizing fluorescent microcapsules is of primary importance for such applications as physiological sensing in loco and contrasting various features of the crustacean circulatory system. Previously, we visualized the microcapsules only in the central hemolymph vessel of amphipods and did not analyze their behavior in other body parts [16]. The current study of the microcapsule visibility in the *E. verrucosus* body was carried out for six weeks, after the injection of one million microcapsules into the circulatory system. This number was chosen to obtain a stable, high concentration of microcapsules in the central hemolymph vessel for most individuals right after the delivery and was used throughout the whole study. The microcapsules contained the bright dye fluorescein (in the form of FITC-albumin) and were visualized in the green fluorescent channel.

Figure 1 demonstrates all body parts of *E. verrucosus* where the microcapsules were observed. Since chitin and tissues of *E. verrucosus* have sufficient translucency and mostly low autofluorescence intensity in the green channel, at least some microcapsules could be visualized in the circulatory system in all segments of the amphipod body. Figure 2 provides quantitative information on the distribution of visible microcapsules. The color key indicates the median concentrations of microcapsules per the area of the respective parts of the circulatory system.

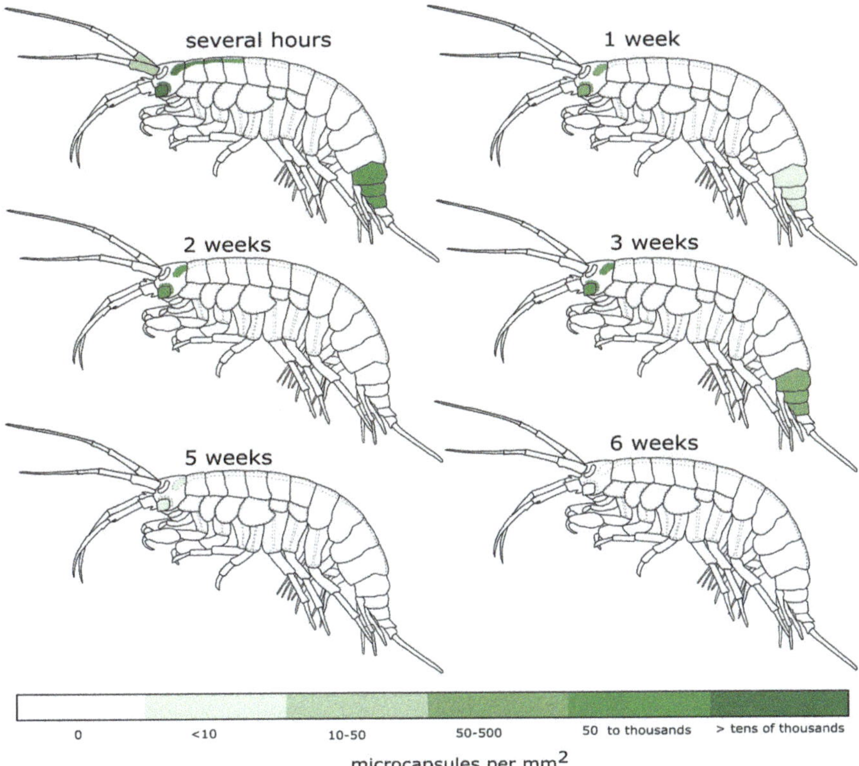

Figure 2. Monitoring of the visible microcapsules in different segments of *E. verrucosus* during the six weeks after injection. The microcapsules containing FITC-albumin were injected into the main hemolymph vessel (1 million per animal) and monitored in the green channel. The intensity of color indicates the median number of recognized microcapsules per area of the corresponding body part (n = 11–15 animals).

During the first hours after injection, the observed median concentrations of microcapsules were among the highest in the central hemolymph vessel inside the first three segments of the amphipod pereon, but they decreased drastically already one week after the injection, and the microcapsules were not visible in more than half of the individuals (Figure 2). Thus, acquiring the spectral signal from microencapsulated optical probes inside the central hemolymph vessel is possible for the majority of injected individuals only during the first hours or days after injection. It is an important limitation that should be considered in experimental design involving the probe-carrying microcapsules. A similar decrease was observed in the case of the first segment of upper antennae, another potentially perspective organ for physiological sensing. However, the concentration of microcapsules in these segments was substantially lower than in the central hemolymph vessel; moreover, as the antennae are close to the amphipod eye, direct illumination for fluorescence excitation leads to anxiety for the animal.

Throughout the six weeks of the experiment, the highest concentration of visible microcapsules and the lowest rate of its decrease were observed in the amphipod head, and the median concentration in this area reached zero only at the end of the experiment (Figure 2). This fact opens intriguing possibilities for long-term measurements of various hormones in close vicinity to the central nervous system of crustaceans using implanted microsensors, as well as targeted delivery of some substances in this region using resolving microcapsules. However, similar to antennae, physiological sensing in the head will require the application of either animal anesthesia or luminescent sensors, which do not require any exciting illumination. The last body part where high microcapsule concentrations were found for most individuals is the amphipod urosome. However, this part of *E. verrucosus* has significant armor that limits the application of fluorescent sensors (see below), but it may be of interest for other species. Additionally, the immobilization of amphipods and other crustaceans by the end of the body can be less convenient than by the central body area [20], which also makes this segment less favorable.

Variability in the concentration of microcapsules in other body parts was shown to be too unstable to use at least half of individuals with injected microcapsules for physiological sensing. Nevertheless, special attention should be given to amphipod coxal plates (Figure 1), despite the median number of microcapsules in the plates never exceeding zero, for example, several dozens of microcapsules were visible in the first coxal plate of one third of individuals right after the injection. The distinctive feature of this body part is proximity of the hemolymph flow to the body surface. Thus, microcapsules in coxal plates can be visualized with the highest magnification possible among all non-moving parts of the amphipod body. This feature may allow local measurements of hemolymph parameters using the spectral signal from even single fluorescent microcapsule, while at least many dozens of them are required in the central hemolymph vessel [16].

The reasons for the gradual decrease in visibility of the PEG-covered microcapsules in the amphipod body remain unexplored. Despite the microcapsules being assembled of non-biodegradable polymers, they potentially could be decomposed to the individual polymers, for example, under extreme pH [38,39], which may be produced by the amphipod hemocytes [40] participating in the immune response. However, previous studies showed that cells of vertebrates seem to be unable to disintegrate the phagocytosed PAH/PSS-based microcapsules [21,41]. Additionally, the dissection of *E. verrucosus* after the six-week experiment showed plenty of intact fluorescent microcapsules around internal organs. So, they could simply migrate to deeper tissues and become invisible. The possibility of excretion of the microcapsules from the amphipod organism was not studied. It should also be noted that the microcapsules could become invisible due to the deposition of melanin pigments around them during the immune response [42]. In this case, they could stay at the same place in the circulatory system during the whole experiment, but become optically isolated.

We also studied the autofluorescence of the amphipods *E. verrucosus*, which may prevent the effective visualization of various fluorescent microsensors. The autofluorescence was visualized in the green and red channels, since excitation by short-wavelength light may be phototoxic [43] and is thus less useful for physiological studies. As mentioned already, the smooth parts of the amphipod body mostly have a relatively low background autofluorescence. Additionally, the autofluorescence spectra, especially in the green channel, have relatively low variability between individuals. Examples of such spectra for the first two pereon segments containing the central vessel of the circulatory system are depicted in Figure 3. The stability of the autofluorescence spectra (at least in the range 600–650 nm for the red channel) makes it possible to perform their subtraction from the spectral signal of the fluorescent microsensors, as has been shown previously [16].

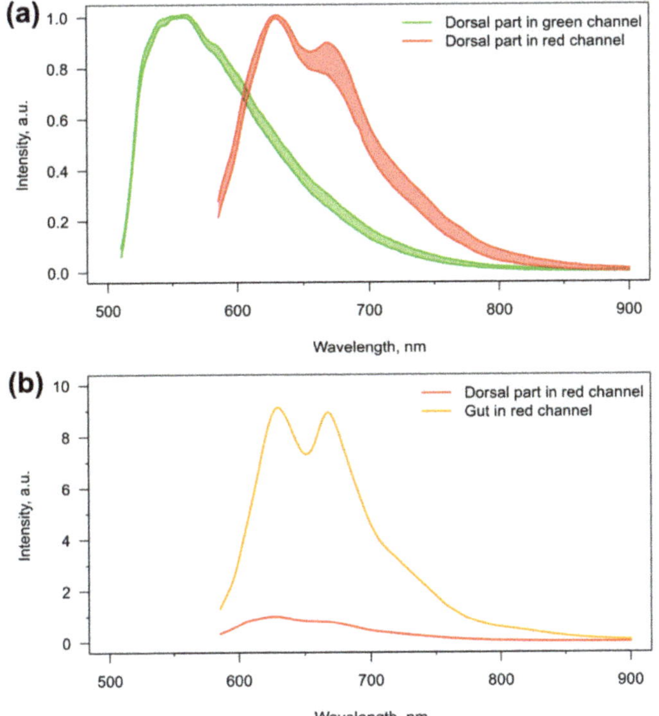

Figure 3. Autofluorescence of *E. verrucosus*. (**a**) Variability ($n = 5$) in autofluorescence spectra of the dorsal parts of the first two pereon segments in green and red channels, and (**b**) representative comparison of autofluorescence intensity between the dorsal and central (containing gut) parts of third pereon segment of the same individual in red channel. Spectra on the upper panel are aligned at the regions of peak intensity.

However, there are two factors that significantly increase the intensity of autofluorescence in some parts of the amphipod body. First of all, sclerotized armor, such as spines and setae, has similar spectra, but enhanced intensity of autofluorescence in both channels. This is the case especially for the highly armored urosome of *E. verrucosus* (Figure 1), which makes separation of autofluorescence from the spectrum of microsensors difficult. Additionally, in the red channel, we observed pronounced autofluorescence in the area of the digestive system (Figure 1), which is approximately ten-fold higher than for the area of the central hemolymph vessel of the same pereon segment (Figure 3). The intense autofluorescence of the intestine in the red channel can contaminate the spectra of microsensors located both in the central hemolymph vessel and the coxal plates, and requires careful positioning of the amphipod under the objective.

The performed analyses allow us to conclude that despite the visibility of the PEG-coated polyelectrolyte microcapsules in the central hemolymph vessel of amphipods significantly decreasing within several days after injection, this organ is indeed the most promising for physiological monitoring using different implanted fluorescent microsensors inside the hemolymph of crustaceans. This is due to the combination of a number of factors: the convenience of injection into the vessel, easiness of fixation of the animal by this body part, high microcapsule concentration, sufficient remoteness from the eye and low level of autofluorescence.

3.2. Reaction of the E. verrucosus Organism to the Injection of Microcapsules

Another important aspect of the application of any substance being injected into the organism is toxicity. This is especially significant for non-biodegradable implants, which is most probably the case for the PAH/PSS-based microcapsules. The injection of microcapsules in the isotonic solution or saline alone revealed no effect on *E. verrucosus* survival in comparison to the parallel control group, without any injections, during six weeks (Figure 4). Similarly, no distinct effect of the same microcapsules was previously observed on the survival of the fish *Danio rerio* [21].

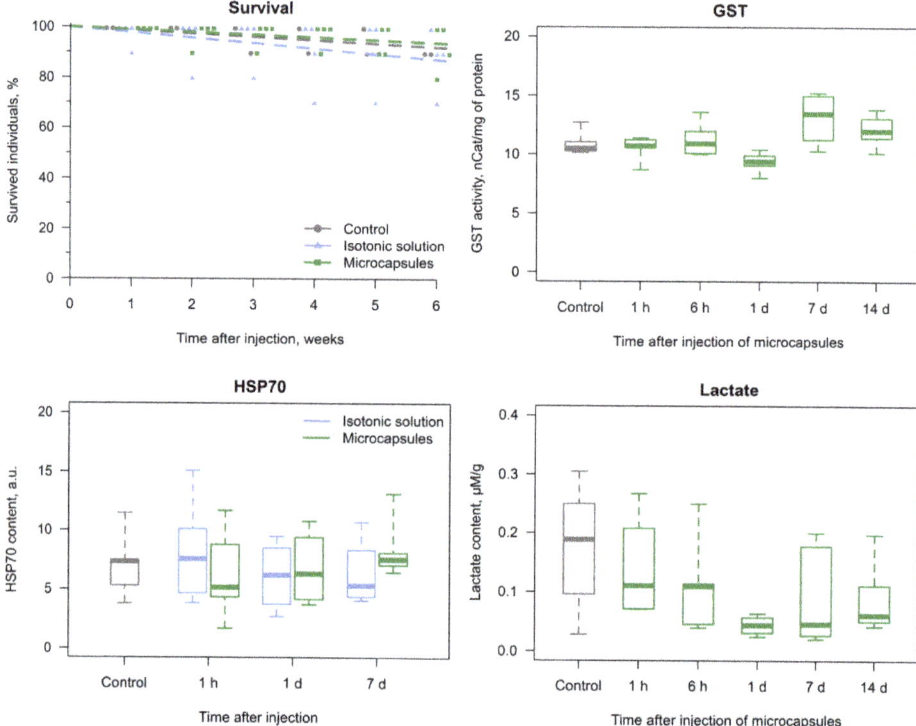

Figure 4. The reaction of the organism of amphipods *E. verrucosus* to the injection of 1 million microcapsules. The analyses included monitoring of mortality during six weeks ($n = 10$ per replicate; each point is a separate replicate); glutathione S-transferase activity (GST; $n = 4$–5) and lactate content ($n = 5$–7) for two weeks; and the content of heat shock proteins HSP70 ($n = 4$–6) for one week. No statistically significant differences from respective control groups (initial or parallel) were observed for the biochemical parameters.

Additionally, we evaluated three biochemical markers of stress response after the injections (Figure 4): concentrations of lactate and heat shock proteins of the family HSP70, as well as activity of glutathione S-transferase (GST). GST activity is a widely used biomarker of intoxication, as it responds to various xenobiotics in many aquatic animals [44,45]. It was measured to assess the possible influence of microcapsule components after their potential disintegration (see above). Lactate is a product of anaerobic glycolysis activated under deficient oxygen supply [25], and its concentration was determined to disclose the possible local occlusions of the hemolymph flow caused by the microcapsules. HSP70 are considered by many authors as a sensitive biomarker that responds to a variety of stressors, including toxic effects and hypoxia [46–48]. GST activity and lactate concentrations monitored for two weeks, as well as HSP70 concentration evaluated for one week, showed no reaction (all $p > 0.15$) to the injections (Figure 4).

So, according to the results of our study, the chosen concentration of PEG-covered polyelectrolyte microcapsules demonstrates no lethal or significant sublethal effects on the amphipods *E. verrucosus*. The diameter of polyelectrolyte microcapsules was less than the size of crustacean hemocytes, and the appearance of currently unobserved toxic effects due to occlusions of the hemolymph flow is unexpected also for smaller crustaceans, unless a higher number of the microcapsules per hemolymph volume is used.

Since the integrity of amphipod integuments is an important factor in toxicological studies, we also studied the process of exoskeleton repair after the injections between the sixth and seventh segments of *E. verrucosus* pereon (Figure S1). The wound healing begins by the 10th day after the injection, and complete restoration of the chitin integrity was observed in 28 days from the day of the procedure. Since the visibility of microcapsules inside amphipods decreases faster than the exoskeleton can be restored, any toxicological experiments involving the implantation of the microencapsulated optical probes into the amphipod circulatory system will require artificial repair of the injection wound using, for example, some tissue adhesives.

3.3. Amphipod Immune Response to the Microcapsules

Finally, an important factor for the use of implantable sensors is the potential stimulation of the immune response, but the information on the possibility of immune reaction to implants with the PEG coating is fairly absent for invertebrates. The immune response of crustaceans is globally divided into the humoral and cellular branches [49]. The main humoral immune reaction that can be dangerous for the contact of any sensor with the internal environment (and thus its functionality) is melanization of the foreign body involving the enzyme phenoloxidase. Hemocytes are the cells circulating in crustacean hemolymph; they participate in the immune response by phagocytosis and aggregation around different foreign bodies [50], and thus can also isolate the sensor from the hemolymph.

The cells of the yeast *Saccharomyces cerevisiae* (similar in size to the microcapsules) were used as a positive control to test the sensitivity of *E. verrucosus* PO to foreign objects. Indeed, the median PO activity increased after the injection of yeast cells into the main hemolymph vessel, but a statistically significant difference from the parallel control group was observed only after the introduction of six million cells ($p = 0.04$) and not one million (Figure 5a). In the case of microcapsules (one million per animal; see above), there were no statistically significant differences from the control group (all $p > 0.3$) for three days after injection (Figure 5b).

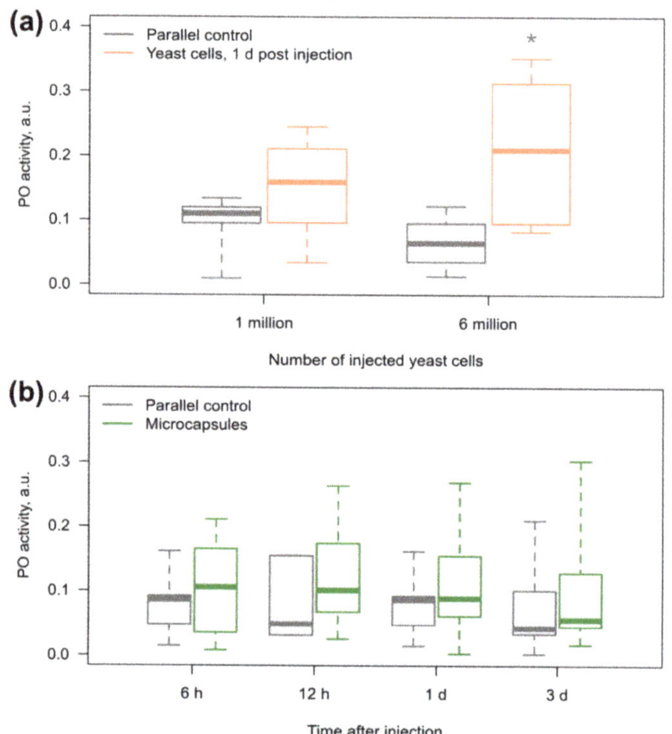

Figure 5. Hemolymph phenoloxidase (PO) activity of amphipods in response to injection of (**a**) yeast and (**b**) microcapsules. The injection of yeast cells (n = 6–8) was used as a positive control to verify the sensitivity of *E. verrucosus* PO to foreign bodies before monitoring of PO response to 1 million of the microcapsules (n = 5–11). * designates statistically significant difference from the parallel control group that did not receive any injections with $p < 0.05$.

The sensitivity of *E. verrucosus* hemocytes to the microcapsules was investigated in the primary culture in vitro, as was previously suggested for the red palm weevil *Rhynchophorus ferrugineus* [51]. It should be mentioned that the current study provides the first description of an established primary culture of hemocytes for amphipods in general. Again, yeast cells were used as the positive control and caused a statistically significant elevation in the hemocyte aggregation already in five hours ($p = 0.01$; all further $p < 0.001$) after introduction into the culture media (Figure 6a). The median proportion of hemocyte aggregation in the presence of yeast cells remained nearly the same from 5 to 24 h of the experiment.

Contact of the amphipod hemocytes with the microcapsules also triggered their aggregation, but at a lower rate (Figure 6a). In particular, a statistically significant difference from the parallel control group was observed only at 12 h ($p < 0.003$ here and further), while the median level of hemocyte aggregation reached the median level caused by yeast even later, at 24 h, although the introduced concentration of microcapsules was five times higher than the concentration of yeast cells. The microcapsules were observed inside the aggregates (Figure 6b), and their concentration in the culture media clearly decreased when the hemocyte aggregates appeared (data not shown). In several cases, we found a black coloration of the hemocyte aggregates that probably indicated the reaction of melanization (Figure S2). In order to partially visualize the membrane structures of hemocytes inside the aggregates, we applied the hydrophobic dye RuPhen3 with bright orange fluorescence. This staining clearly demonstrated that a significant proportion of hemocyte nuclei were located outside the cells (Figure 6b), which indicates the death of many cells during the immune reaction to the microcapsules.

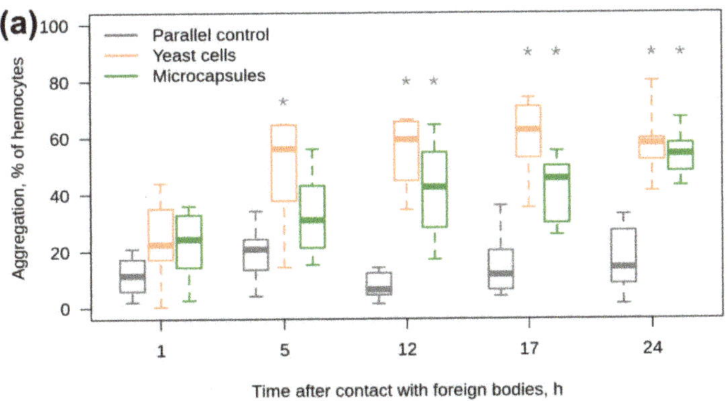

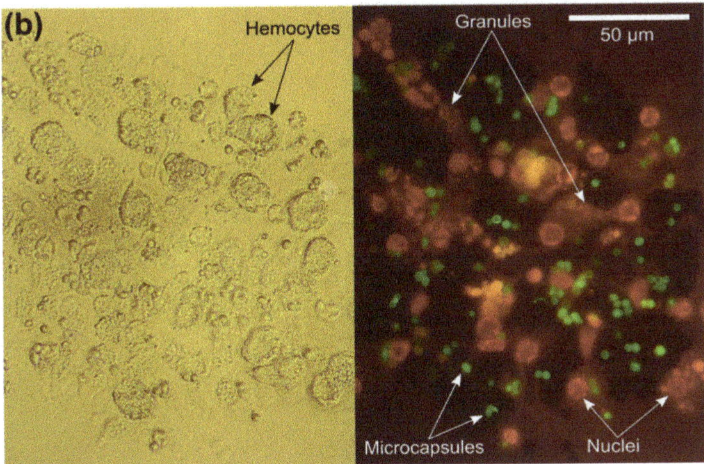

Figure 6. The reaction of primary culture of *E. verrucosus* hemocytes to yeast and microcapsules. (a) Monitoring of hemocyte aggregation after the introduction of yeast cells used as a positive control (1:1 of hemocytes) and microcapsules (5:1 of hemocytes) to the hemocyte media during 24 h ($n = 8$). (b) A representative example of a squashed aggregate of hemocytes with the microcapsules after the end of incubation in brightfield channel (left) and green fluorescent channel with the orange RuPhen$_3$ staining (right). Microcapsules contained green FITC-albumin but partially obtained the yellow coloration due to contact with RuPhen$_3$. * designates statistically significant difference from the parallel control group with $p < 0.05$.

In order to also verify the recognition of the microcapsules by *E. verrucosus* hemocytes in vivo, we performed a histological analysis of the amphipod body part containing the central hemolymph vessel for three individuals one week post injection. Indeed, we were able to find aggregates of microcapsules tightly enclosed by some cells, probably hemocytes, in two out of three samples (Figure S3). This fact, together with the observed in vitro melanization of hemocyte aggregates, also supports the hypothesis of decreasing microcapsule visibility due to immune reaction, at least to some extent. Additionally, it is more mild evidence against the possibility of decomposition of the PAH/PSS/PLL-g-PEG microcapsules to individual polymers inside crustaceans.

So, our results unambiguously demonstrate the existence of a cellular immune reaction of amphipods to the polyelectrolyte microcapsules, despite the PEG coating. Due to this effect, such microcapsules and similar probe-carrying microstructures composed of non-biodegradable polymers may become a promising tool for studying the behavior of immune cells in vivo. However, the stability of microcapsules inside the hemocyte aggregates makes them less suitable for prolonged (more than several hours) measurements of various hemolymph parameters, since it is hardly possible to distinguish the optical signals from free microcapsules and microcapsules within the hemocyte aggregates. Additionally, the stable microcapsules may attract more hemocytes and more resources for immune response than the biodegradable ones. Thus, the application of biodegradable polyelectrolyte microcapsules might be more recommended for physiological measurements inside crustaceans in vivo.

4. Conclusions

This study provides the first complex analysis of application of LbL-assembled polyelectrolyte microcapsules in the crustacean organism on the example of *E. verrucosus*. It is also the first study assessing the invertebrate immune response to PEG-covered implants. According to our results, the central hemolymph vessel should be proposed as the most convenient organ for visualization of fluorescent microsensors in most small crustaceans with translucent integument. The tested number of microcapsules (that was sufficient for stable visualization in the central hemolymph vessel) did not show lethal or sublethal toxic effects. Although the cellular immune response to the PEG-coated microcapsules was delayed in comparison to microorganisms, it also was the case. Nevertheless, during the first hours the proportion of involved hemocytes was low and could not affect a significant number of microcapsules. This makes the application of such microcapsules for physiological sensing possible, at least for the first several hours after injection.

Supplementary Materials: The following are available online at http://www.mdpi.com/2073-4360/11/8/1246/s1, Figure S1: Checkup of *E. verrucosus* exoskeleton integrity after injection between the sixth and seventh segments of pereon (area of the central hemolymph vessel), Figure S2: Melanization (dark area in the center) inside an aggregate of hemocytes with microcapsules two days post introduction of microcapsules into the culture media, Figure S3: Example of hemocyte accumulation around an aggregate of microcapsules one week post injection into the central hemolymph vessel of *E. verrucosus*.

Author Contributions: Conceptualization, A.G., E.S., E.B., A.N. and M.T.; Data curation, E.S., A.G., A.N. and E.B.; Formal analysis, E.S., A.G. and A.N.; Funding acquisition, M.T.; Investigation, E.S., A.N., I.D. and Y.R.; Methodology, E.S., A.G., A.N. and I.M.; Project administration, M.T. and I.M.; Supervision, A.G., E.B., M.T. and I.M.; Validation, E.S., A.G. and A.N.; Visualization, E.S., A.G. and A.N.; Writing—original draft, E.S., A.G. and A.N.; Writing—review & editing, E.S., A.G., E.B., I.D., Y.R., M.T. and I.M.

Funding: The study was supported by the Russian Science Foundation (#17-14-01063), the Russian Foundation for Basic Research (#19-54-04008) and Lake Baikal Foundation (#02-3/14). E.S. also acknowledges the partial support from the Ministry of Science and Higher Education of Russia and Deutscher Akademischer Austauschdienst ("Goszadanie": #6.13468.2019/13.2).

Acknowledgments: We acknowledge Lidia Yu. Raevskaya and Irina A. Zaravnyaeva (Irkutsk Regional Diagnostic Center, Pathology Department) for help in the preparation of the histological section, as well as Polina Drozdova (Irkutsk State University) for the photo of *E. verrucosus* and help in the preparation of the manuscript.

Conflicts of Interest: The authors declare no conflict of interest. The funders had no role in the design of the study; in the collection, analyses, or interpretation of data; in the writing of the manuscript, or in the decision to publish the results.

References

1. Donath, E.; Sukhorukov, G.B.; Caruso, F.; Davis, S.A.; Möhwald, H. Novel hollow polymer shells by colloid-templated assembly of polyelectrolytes. *Angew. Chem. Int. Ed.* **1998**, *37*, 2201–2205. [CrossRef]
2. Timin, A.S.; Gould, D.J.; Sukhorukov, G.B. Multi-layer microcapsules: Fresh insights and new applications. *Expert Opin. Drug Deliv.* **2017**, *14*, 583–587. [CrossRef] [PubMed]
3. Gentile, P.; Carmagnola, I.; Nardo, T.; Chiono, V. Layer-by-layer assembly for biomedical applications in the last decade. *Nanotechnol.* **2015**, *26*, 422001. [CrossRef] [PubMed]

4.   Pham, T.; Bui, T.; Nguyen, V.; Bui, T.; Tran, T.; Phan, Q.; Hoang, T. Adsorption of polyelectrolyte onto nanosilica synthesized from rice husk: Characteristics, mechanisms, and application for antibiotic removal. *Polymers* **2018**, *10*, 220. [CrossRef] [PubMed]

5.   Pham, T.D.; Tran, T.T.; Pham, T.T.; Dao, T.H.; Le, T.S. Adsorption characteristics of molecular oxytetracycline onto alumina particles: The role of surface modification with an anionic surfactant. *J. Mol. Liq.* **2019**, *287*, 110900. [CrossRef]

6.   Trushina, D.B.; Bukreeva, T.V.; Antipina, M.N. Size-controlled synthesis of vaterite calcium carbonate by the mixing method: Aiming for nanosized particles. *Cryst. Growth Des.* **2016**, *16*, 1311–1319. [CrossRef]

7.   Svenskaya, Y.I.; Fattah, H.; Inozemtseva, O.A.; Ivanova, A.G.; Shtykov, S.N.; Gorin, D.A.; Parakhonskiy, B.V. Key parameters for size-and shape-controlled synthesis of vaterite particles. *Cryst. Growth Des.* **2017**, *18*, 331–337. [CrossRef]

8.   Kreft, O.; Javier, A.M.; Sukhorukov, G.B.; Parak, W.J. Polymer microcapsules as mobile local pH-sensors. *J. Mater. Chem.* **2007**, *17*, 4471–4476. [CrossRef]

9.   Guzmán, E.; Mateos-Maroto, A.; Ruano, M.; Ortega, F.; Rubio, R.G. Layer-by-Layer polyelectrolyte assemblies for encapsulation and release of active compounds. *Adv. Colloid Interface Sci.* **2017**, *249*, 290–307. [CrossRef]

10. Pham, T.D.; Do, T.U.; Pham, T.T.; Nguyen, T.A.H.; Nguyen, T.K.T.; Vu, N.D.; Le, T.S.; Vu, C.M.; Kobayashi, M. Adsorption of poly (styrenesulfonate) onto different-sized alumina particles: Characteristics and mechanisms. *Colloid Polym. Sci.* **2019**, *297*, 13–22. [CrossRef]

11. Wattendorf, U.; Kreft, O.; Textor, M.; Sukhorukov, G.B.; Merkle, H.P. Stable stealth function for hollow polyelectrolyte microcapsules through a poly (ethylene glycol) grafted polyelectrolyte adlayer. *Biomacromolecules* **2008**, *9*, 100–108. [CrossRef] [PubMed]

12. Sadovoy, A.; Teh, C.; Korzh, V.; Escobar, M.; Meglinski, I. Microencapsulated bio-markers for assessment of stress conditions in aquatic organisms in vivo. *Laser Phys. Lett.* **2012**, *9*, 542. [CrossRef]

13. Knop, K.; Hoogenboom, R.; Fischer, D.; Schubert, U.S. Poly (ethylene glycol) in drug delivery: Pros and cons as well as potential alternatives. *Angew. Chem. Int. Ed.* **2010**, *49*, 6288–6308. [CrossRef] [PubMed]

14. Bondad-Reantaso, M.G.; Subasinghe, R.P.; Josupeit, H.; Cai, J.; Zhou, X. The role of crustacean fisheries and aquaculture in global food security: Past, present and future. *J. Invertebr. Pathol.* **2012**, *110*, 158–165. [CrossRef] [PubMed]

15. Callaghan, N.I.; MacCormack, T.J. Ecophysiological perspectives on engineered nanomaterial toxicity in fish and crustaceans. *Comp. Biochem. Physiol. Part C Toxicol. Pharmacol.* **2017**, *193*, 30–41. [CrossRef] [PubMed]

16. Gurkov, A.; Shchapova, E.; Bedulina, D.; Baduev, B.; Borvinskaya, E.; Meglinski, I.; Timofeyev, M. Remote in vivo stress assessment of aquatic animals with microencapsulated biomarkers for environmental monitoring. *Sci. Rep.* **2016**, *6*, 36427. [CrossRef] [PubMed]

17. Kazakova, L.I.; Shabarchina, L.I.; Anastasova, S.; Pavlov, A.M.; Vadgama, P.; Skirtach, A.G.; Sukhorukov, G.B. Chemosensors and biosensors based on polyelectrolyte microcapsules containing fluorescent dyes and enzymes. *Anal. Bioanal. Chem.* **2013**, *405*, 1559–1568. [CrossRef] [PubMed]

18. Takhteev, V.V.; Berezina, N.A.; Sidorov, D.A. Checklist of the Amphipoda (Crustacea) from continental waters of Russia, with data on alien species. *Arthropoda Sel.* **2015**, *24*, 335–370. [CrossRef]

19. Borvinskaya, E.; Gurkov, A.; Shchapova, E.; Karnaukhov, D.; Sadovoy, A.; Meglinski, I.; Timofeyev, M. Simple and effective administration and visualization of microparticles in the circulatory system of small fishes using kidney injection. *J. Vis. Exp.* **2018**, *136*, e57491. [CrossRef]

20. Gurkov, A.; Borvinskaya, E.; Shchapova, E.; Timofeyev, M. Restraining small decapods and amphipods for in vivo laboratory studies. *Crustaceana* **2018**, *91*, 517–525. [CrossRef]

21. Borvinskaya, E.; Gurkov, A.; Shchapova, E.; Baduev, B.; Meglinski, I.; Timofeyev, M. Distribution of PEG-coated hollow polyelectrolyte microcapsules after introduction into the circulatory system and muscles of zebrafish. *Biol. Open* **2018**, *7*, bio030015. [CrossRef] [PubMed]

22. Bedulina, D.S.; Evgen'ev, M.B.; Timofeyev, M.A.; Protopopova, M.V.; Garbuz, D.G.; Pavlichenko, V.V.; Luckenbach, T.; Shatilina, Z.M.; Axenov-Gribanov, D.V.; Gurkov, A.N.; et al. Expression patterns and organization of the hsp70 genes correlate with thermotolerance in two congener endemic amphipod species (*Eulimnogammarus cyaneus* and *E. verrucosus*) from Lake Baikal. *Mol. Ecol.* **2013**, *22*, 1416–1430. [CrossRef] [PubMed]

23. Bedulina, D.S.; Zimmer, M.; Timofeyev, M.A. Sub-littoral and supra-littoral amphipods respond differently to acute thermal stress. *Comp. Biochem. Physiol. Part B Biochem. Mol. Biol.* **2010**, *155*, 413–418. [CrossRef] [PubMed]

24. Schindelin, J.; Arganda-Carreras, I.; Frise, E.; Kaynig, V.; Longair, M.; Pietzsch, T.; Preibisch, S.; Rueden, C.; Saalfeld, S.; Schmid, B.; et al. Fiji: An open-source platform for biological-image analysis. *Nat. Methods* **2012**, *9*, 676. [CrossRef] [PubMed]

25. Axenov-Gribanov, D.; Bedulina, D.; Shatilina, Z.; Jakob, L.; Vereshchagina, K.; Lubyaga, Y.; Gurkov, A.; Shchapova, E.; Luckenbach, T.; Lucassen, M.; et al. Thermal preference ranges correlate with stable signals of universal stress markers in Lake Baikal endemic and Holarctic amphipods. *PLoS ONE* **2016**, *11*, e0164226. [CrossRef] [PubMed]

26. Shchapova, E.P.; Axenov-Gribanov, D.V.; Lubyaga, Y.A.; Shatilina, Z.M.; Vereshchagina, K.P.; Madyarova, E.V.; Timofeyev, M.A. Crude oil at concentrations considered safe promotes rapid stress-response in Lake Baikal endemic amphipods. *Hydrobiologia* **2018**, *805*, 189–201. [CrossRef]

27. Habig, W.H.; Pabst, M.J.; Jakoby, W.B. Glutathione S-transferases the first enzymatic step in mercapturic acid formation. *J. Biol. Chem.* **1974**, *249*, 7130–7139.

28. Bradford, M.M. A rapid and sensitive method for the quantitation of microgram quantities of protein utilizing the principle of protein-dye binding. *Anal. Biochem.* **1976**, *72*, 248–254. [CrossRef]

29. Kruger, N.J. The Bradford method for protein quantitation. In *The Protein Protocols Handbook*; Walker, J.M., Ed.; Humana Press: Totowa, NJ, USA, 2009; pp. 17–24.

30. Bilandžija, H.; Abraham, L.; Ma, L.; Renner, K.J.; Jeffery, W.R. Behavioural changes controlled by catecholaminergic systems explain recurrent loss of pigmentation in cavefish. *Proc. R. Soc. B* **2018**, *285*, 20180243. [CrossRef]

31. Cornet, S.; Biard, C.; Moret, Y. Variation in immune defence among populations of *Gammarus pulex* (Crustacea: Amphipoda). *Oecologia* **2009**, *159*, 257–269. [CrossRef]

32. Laughton, A.M.; Siva-Jothy, M.T. A standardised protocol for measuring phenoloxidase and prophenoloxidase in the honey bee, Apis mellifera. *Apidologie* **2011**, *42*, 140–149. [CrossRef]

33. Jakob, L.; Axenov-Gribanov, D.V.; Gurkov, A.N.; Ginzburg, M.; Bedulina, D.S.; Timofeyev, M.A.; Luckenbach, T.; Lucassen, M.; Sartoris, F.J.; Pörtner, H.O. Lake Baikal amphipods under climate change: Thermal constraints and ecological consequences. *Ecosphere* **2016**, *7*, 1–15. [CrossRef]

34. George, S.K.; Dhar, A.K. An improved method of cell culture system from eye stalk, hepatopancreas, muscle, ovary, and hemocytes of Penaeus vanname. *In Vitro Cell. Dev. Biol. Anim.* **2010**, *46*, 801–810. [CrossRef] [PubMed]

35. Rajendiran, V.; Palaniandavar, M.; Periasamy, V.S.; Akbarsha, M.A. Ru(phen)2(dppz)]2+ as an efficient optical probe for staining nuclear components. *J. Inorg. Biochem.* **2010**, *104*, 217–220. [CrossRef] [PubMed]

36. Presnell, J.K.; Schreibman, M.P.; Humason, G.L. *Humason's Animal Tissue Techniques Humason. Humason's Animal Tissue Techniques*; Johns Hopkins University Press: Baltimore, MD, USA, 1997.

37. Mikodina, E.V.; Sedova, M.A.; Chmilevsky, D.A.; Mikulin, A.E.; Pianova, S.V.; Poluektova, O.G. *Histology for Ichthyologists: Experience and Advices*; VNIRO: Moscow, Russia, 2009.

38. Antipov, A.A.; Sukhorukov, G.B.; Leporatti, S.; Radtchenko, I.L.; Donath, E.; Möhwald, H. Polyelectrolyte multilayer capsule permeability control. *Colloids Surf.* **2002**, *198*, 535–541. [CrossRef]

39. Déjugnat, C.; Sukhorukov, G.B. pH-responsive properties of hollow polyelectrolyte microcapsules templated on various cores. *Langmuir* **2004**, *20*, 7265–7269. [CrossRef]

40. Tuan, V.V.; Dantas-Lima, J.J.; Thuong, K.V.; Li, W.; Grauwet, K.; Bossier, P.; Nauwynck, H.J. Differences in uptake and killing of pathogenic and non-pathogenic bacteria by haemocyte subpopulations of penaeid shrimp, Litopenaeus vannamei, (Boone). *J. Fish Dis.* **2016**, *39*, 163–174. [CrossRef]

41. De Geest, B.G.; Vandenbroucke, R.E.; Guenther, A.M.; Sukhorukov, G.B.; Hennink, W.E.; Sanders, N.N.; Demeester, J.; De Smedt, S.C. Intracellularly degradable polyelectrolyte microcapsules. *Adv. Mater.* **2006**, *18*, 1005–1009. [CrossRef]

42. Christensen, B.M.; Li, J.; Chen, C.C.; Nappi, A.J. Melanization immune responses in mosquito vectors. *Trends Parasitol.* **2005**, *21*, 192–199. [CrossRef]

43. Marek, V.; Mélik-Parsadaniantz, S.; Villette, T.; Montoya, F.; Baudouin, C.; Brignole-Baudouin, F.; Denoyer, A. Blue light phototoxicity toward human corneal and conjunctival epithelial cells in basal and hyperosmolar conditions. *Free Radic. Biol. Med.* **2018**, *126*, 27–40. [CrossRef]

44. Domingues, I.; Agra, A.R.; Monaghan, K.; Soares, A.M.; Nogueira, A.J. Cholinesterase and glutathione-S-transferase activities in freshwater invertebrates as biomarkers to assess pesticide contamination. *Environ. Toxicol. Chem.* **2010**, *29*, 5–18. [CrossRef] [PubMed]

45. Dong, M.; Zhu, L.; Shao, B.; Zhu, S.; Wang, J.; Xie, H.; Wang, J.H.; Wang, F.H. The effects of endosulfan on cytochrome P450 enzymes and glutathione S-transferases in zebrafish (Danio rerio) livers. *Ecotoxicol. Environ. Saf.* **2013**, *92*, 1–9. [CrossRef] [PubMed]

46. Rajak, P.; Roy, S. Heat shock proteins and pesticide stress. In *Regulation of Heat Shock Protein Responses*; Springer: Cham, Switzerland, 2018.

47. Wong, L.L.; Do, D.T. The role of heat shock proteins in response to extracellular stress in aquatic organisms. In *Heat Shock Proteins in Veterinary Medicine and Sciences*; Springer: Cham, Switzerland, 2017.

48. Mohanty, B.P.; Mahanty, A.; Mitra, T.; Parija, S.C.; Mohanty, S. *Heat Shock Proteins in Stress in Teleosts*; Springer: Cham, Switzerland, 2018.

49. Oliver, J.D.; Loy, J.D.; Parikh, G.; Bartholomay, L. Comparative analysis of hemocyte phagocytosis between six species of arthropods as measured by flow cytometry. *J. Invertebr. Pathol.* **2011**, *108*, 126–130. [CrossRef] [PubMed]

50. Vazquez, L.; Alpuche, J.; Maldonado, G.; Agundis, C.; Pereyra-Morales, A.; Zenteno, E. Immunity mechanisms in crustaceans. *Innate Immun.* **2009**, *15*, 179–188. [CrossRef] [PubMed]

51. Mastore, M.; Arizza, V.; Manachini, B.; Brivio, M.F. Modulation of immune responses of *Rhynchophorus ferrugineus* (Insecta: Coleoptera) induced by the entomopathogenic nematode *Steinernema carpocapsae* (Nematoda: Rhabditida). *Insect Sci.* **2015**, *22*, 748–760. [CrossRef]

 *polymers*

Article

Ratiometric Luminescent Nanoprobes Based on Ruthenium and Terbium-Containing Metallopolymers for Intracellular Oxygen Sensing

Wu-xing Zhao [1], Chao Zhou [1,2] and Hong-shang Peng [1,*]

[1]   School of Science, Minzu University of China, Beijing 100081, China
[2]   Key Laboratory of Luminescence and Optical Information, Ministry of Education, Institute of Optoelectronic Technology, Beijing Jiaotong University, Beijing 100044, China
*   Correspondence: hshpeng@bjtu.edu.cn

Received: 21 June 2019; Accepted: 24 July 2019; Published: 2 August 2019

Abstract: A collection of luminescent metal complexes have been widely used as oxygen probes in the biomedical field. However, single intensity-based detection approach usually suffered from errors caused by the signal heterogeneity or fluctuation of the optoelectronic system. In this work, respective ruthenium (II) and terbium (III) complexes were chosen to coordinate a bipyridine-branched copolymer, so that to produce oxygen-sensitive metallopolymer (Ru-Poly) and oxygen-insensitive metallopolymer (Tb-Poly). Based on the hydrophobic Ru-Poly and Tb-Poly, a ratiometric luminescent oxygen nanoprobe was facilely prepared by a nanoprecipitation method. The nanoprobes have a typical size of ~100 nm in aqueous solution, exhibiting a green-red dual-wavelength emission under the excitation of 300 nm and 460 nm, respectively. The red emission is strongly quenched by dissolved oxygen while the green one is rather stable, and the ratiometric luminescence was well fitted by a linear Stern–Volmer equation. Using the ratiometric biocompatible nanoprobes, the distribution of intracellular oxygen within three-dimensional multi-cellular tumor spheroids was successfully imaged.

Keywords: Tb-containing polymer; Ru-containing polymer; ratiometric luminescence; nanoprobe; oxygen

---

1. Introduction

Molecular oxygen is a key component for maintaining physiological activities in almost all living systems. Hence oxygen sensing is of great importance to understand related physiological and pathological processes, such as cell respiration and tumor hypoxia [1,2]. In terms of in vivo or in vitro oxygen sensing, luminescence-based approaches are much more attractive with the merits of high sensitivity, high spatiotemporal resolution, and non-invasiveness [3,4]. Various luminescent oxygen probes have been developed so far. In comparison, polymeric nanoparticles incorporated with oxygen-sensitive dyes are the most competent probes, because the porous matrix can prevent the doped dyes from interference by ions or biomolecules in complicated physiological environments, while maintaining the free diffusion of oxygen molecules [5,6]. Ruthenium (Ru) (II)-based oxygen probes are believed to be more stable than other luminescent metal complexes [7].

For most of the established luminescent oxygen nanoprobes, however, quantum yield and oxygen sensitivity are more or less undermined by the concentration quenching of indicators or obstructed diffusion of oxygen by the matrix [8,9]. Very recently we have presented a type of luminescent nanoprobe based on Ru (II)-containing metallopolymers to bypass these issues [10]. Because the oxygen probes $[Ru(bpy)_3]^{2+}$ reside on the particle surface, the nanoprobes exhibit strong red luminescence free of aggregation-induced-quenching and high oxygen sensitivity. Another common disadvantage of nanoprobes is the most widely adopted single intensity-based detection modality, which could be influenced by the incident lamp, detector, and uneven distribution of probes [11]. Although lifetime-based sensing approaches are immune from the influence of these drawbacks, the complexity and the demands on the optoelectronic components increase with the decreasing lifetimes. By contrast, luminescence ratiometric approaches (2-wavelength) allow for more accurate and robust detection with a built-in calibration. Lanthanide complexes have unique optical properties, such as narrow emission bands and large Stokes shift [12]. In particular, their luminescence is hardly influenced by oxygen due to the protection of $5s^2 5p^6$ outer-shell. Given that lanthanide ions could be chelated to polymers [13–16], a ratiometric luminescent oxygen nanoprobe thus can be constructed by using lanthanide-containing and Ru-containing metallopolymers as the reference and probe dye, respectively.

In this work, a bipyridine-branched hydrophobic copolymer was utilized to chelate Tb (III) complex and Ru (II) complex, respectively, so that to produce oxygen-insensitive metallopolymer (Tb-Poly) and oxygen-sensitive metallopolymer (Ru-Poly). Herein the green emissive Tb-Poly is chosen under the consideration that the reference signal should be distinguished from the red sensing signal of Ru-Poly. Taking advantage of the two metallopolymers, biocompatible luminescent nanoparticles (NPs) were prepared by a nanoprecipitation method. The resulting NPs give a two-wavelength emission under 300 nm and 460 nm excitation in aqueous solution, the ratio of which is highly oxygen-dependent in the experimental conditions. Based on the ratiometric luminescence, intracellular oxygen in monolayer cells and three-dimensional multi-cellular tumor spheroids were both detected.

2. Materials and Methods

2.1. Materials

Poly(styrene)-b-poly((4′-hydroxymethyl-[2,2′-bipyri-din]-4-yl) methyl acrylate) with a molecular weight of 58,800 Da (PS-PBPyA, 42,000-b-16 800; PDI (Mw/Mn), 1.18) (P16178-SBPyA, https://www.polymersource.ca/index.php?route=product/category&path=2_2190_16_105_2205_1069&product_id=11572&subtract=1&serachproduct=yes&categorystart=A-1.1) and polystyrene-graft-ethylene-oxide functionalized with carboxy (PS-PEG-COOH, molecular weight: 36,500 Da of the PS moiety; 4600 Da of the PEG-COOH; PDI, 1.3) (P15019-SEOCOOHcomb, https://www.polymersource.ca/index.php?route=product/category&path=2_23_119_2642_1067&product_id=2600&subtract=0&serachproduct=yes&categorystart=A-1.1) were purchased from Polymer Source, Inc. Terbium chloride hexahydrate (TbCl$_3$ 6H$_2$O) and acetylacetone (acac) were obtained from Aladdin and SCR, Ltd. (Shanghai), respectively. 2,2′-Bipyridine (bpy), tetrahydrofuran (THF) and N, N-dimethyl formamide (DMF) were provided by J&K Scientific, Ltd. (Beijing, China).

All solvents were prepared as analytical grade or chromatographically pure. All buffer components were of biological grade without further purification unless otherwise mentioned. Ultrapure water ($\geq$18.2 MΩ cm) was used among all experiments. The N$_2$ and O$_2$ gas (>99.999%), stored in the steel cylinder, were provided by Jinghui Gas Co., Ltd. (Beijing, China). All synthetic procedures were carried out under an inert and dry nitrogen atmosphere using standard techniques.

2.2. Preparation of Ru- and Tb-Metallopolymer Based NPs (Ru-Tb NPs)

Taking advantage of the Ru-containing polymer (Ru-Poly was synthesized according to the literature [10]) and Tb-containing polymer (Tb-Poly, whose synthesis route is showed in supporting information and schematically depicted in Figure 1, along with their respective chemical structures), Ru-Tb NPs were prepared by a modified nanoprecipitation method with the schema showed in Figure 2. A mixture (200 µL, 100 ppm) of THF solution consisting of Ru-Poly, Tb-Poly, and PS-PEG-COOH at a weight ratio of 7:3:2 was quickly injected into 8 mL of distilled water in a 25 mL vial under vigorous sonication. The NPs were formed by the conglomerating of polymer as the sudden decrease in solubility and hydrophobic interaction between polymer chains. Subsequently, the solution was purged by continuous nitrogen on a 100 °C hot plate for 60 min to remove THF and further concentrate the NPs suspension. The final step is filtration through a 0.2 μm micrometer syringe filter to eliminate larger aggregates. The resultant NP suspensions were stored at 4 °C for further use.

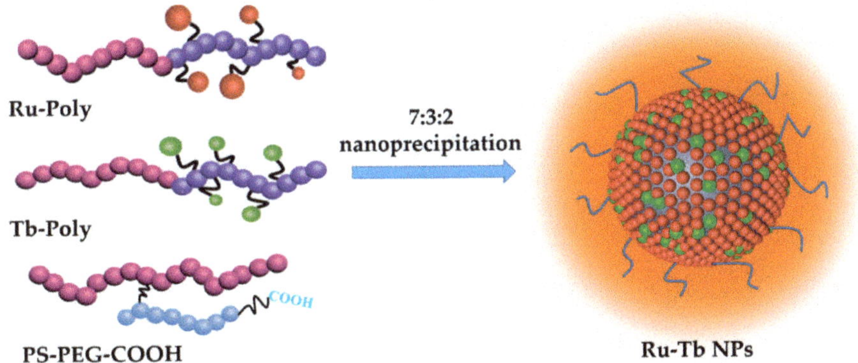

Figure 1. Schematic illustration of the synthesis of Tb-containing metallopolymer (Tb-Poly).

Figure 2. Schematic illustration of the preparation of Ru-Tb NPs.

2.3. Oxygen Calibration of Ru-Tb NPs

Oxygen calibrations for Ru-Tb NPs in aqueous solution were performed in a cuvette sealed with Parafilm (Chicago, IL, USA). A WITT gas mixer (KM60-2, WITT, Witten, Germany) was used to control the gas mixtures containing 0%, 5%, 10%, 15%, 20%, 25%, and 100% of O_2 balanced with N_2 (100%, 95%, 90%, 85%, 80%, 75%, and 0%) with an accuracy of 1% absolute, respectively. Subsequently, the O_2–N_2 mixtures were bubbled through the cuvette for about 10 min, at which point the luminescent intensity was recorded. The dissolved oxygen concentrations in aqueous solution were calculated by the equation

$$[O_2]/ppm = 43 \,(\%O_2/100\%) \tag{1}$$

where $\% O_2$ is the concentration of oxygen in the bubbling gas, and 100% is the concentration of oxygen in oxygen-saturated water (43 ppm).

2.4. Intracellular Localization of Ru-Tb NPs by Confocal Laser Scanning Microscopy

HeLa cells (cancerous cervical tumor cell line) should be incubated in a 35 mm confocal culture dish at 37 °C for 12 h firstly. After cells adhered onto the surface of the dish and reached a density of 5×10^4 cells per dish, they could be treated with 2 mL DMEM, including Ru-Tb NPs (30 μg mL^{-1})

for 6 h. The supernatant cell culture medium should be removed and the cells were washed by PBS three times before microscopic measurements. Unlike the adhered cells, the 3D multi-cellular tumor spheroids (MCTs) should be cultured for 7 days in a normal dish whose bottom was carpeted with agarose. Then the MCTs were treated with the same DMEM consisting Ru-Tb NPs for 12 h and transferred to the confocal dish before imaging. In all the experiments, luminescence and differential interference contrast (DIC) images were recorded with a Nikon A1R HD multiphoton confocal laser scanning microscope. Emission wavelengths were collected at 585–625 nm (red channel, for Ru-Poly) as the cells were excited at 488 nm, while the emissions at 525–565 nm (green channel, for Tb-Poly) were monitored under mercury lamp excitation. The same two emission channels were observed under two-photon excitation at 720 nm, while scanning Z-stack of MCTs.

2.5. Characterizations

The hydrodynamic size of NPs were tested by dynamic light scattering (DLS) using a Zetasizer Nano instrument (Malvern Instruments, Malvern, UK). The morphology of NPs was investigated by scanning electronic microscopy (SEM) taken on a JSM-7500dF (JEOL, Tokyo, Japan) at 15 kV. The UV-vis absorption and steady-state emission spectra were measured on a Model V-550 spectrophotometer (Jasco, Tokyo, Japan) and a Model F-4500 fluorescence spectrometer (Hitachi, Tokyo, Japan), respectively. The metallopolymer structure was analyzed by proton nuclear magnetic resonance (^{1}H NMR) measurements, which performed with a 600 MHz high-resolution NMR spectrometer (AVANCE 600 MHz FT NMR, Bruker, Bremen, Germany). The time-resolved photoluminescence experiments were carried out on an FLS 980 PL spectrometer (Edinburgh, UK) excited with laser diodes.

3. Results and Discussion

3.1. Preparation, Characterization and Properties of Ru-Tb NPs

A Tb-containing metallopolymer (Tb-Poly) was firstly synthesized by coordinating the precursor Tb(acac)$_3$ 3H$_2$O with bipyridine-branched polymer PS–PBPyA (see the Supporting Information for synthesis and ^{1}H NMR data, Figures S1 and S2). Figure 3a shows excitation spectra of Tb(acac)$_3$ and Tb-Poly in DMF solution. Compared to Tb(acac)$_3$, the excitation band of Tb-Poly blue shifts 33 nm due to the formation of a coordinate bond between Tb^{3+} and bpy ligand. Under the excitation of a 300 nm light, Tb-Poly exhibit the typical intra-f-f transitions of Tb^{3+} ion (Figure 3b, similar to free Tb(acac)$_3$ in Figure S3). The multiplet transitions $^5D_4 \rightarrow \,^7F_{6,5,4,3}$ are clearly observed, as well as a moderate emission at 545 nm. Time-resolved fluorescence of Tb-Poly was also measured by monitoring the 545 nm emission (Figure S4), and the phosphorescence lifetime was obtained to be 1.183 ms by a biexponential fitting. It can be derived that the lifetime of Tb-Poly is increased in comparison to that of Tb(acac)$_3$ 3H$_2$O (~0.8 ms) [17] as a result of the replacement of coordinated H$_2$O by bpy ligands. These results indicate that Tb(acac)$_3$ is successfully chelated to the side chain of PS-PBPyA. The hydrophobic metallopolymers Tb-Poly could be easily formed into NPs with a modified nanoprecipitation method [10]. After the formation of NPs, however, a weak fluorescence emerges ranging from 350 nm to 450 nm, as indicated in Figure 3b. Since free Tb-Poly in solution is nonfluorescent in this range, it is reasonable to attribute this blue emission to coordinated bpy ligands of Tb-Poly in NPs, i.e., the back donation from energy level $^5D_4$ of Tb^{3+} to chelated bpy in solid matrix [18]. This postulation can be further demonstrated by the decreased luminescence lifetime, from 1.183 to 0.94 ms (Figure S4), and lowered quantum yield, from 25.57% to 3.76% (Figure S5), of Tb-Poly in NPs.

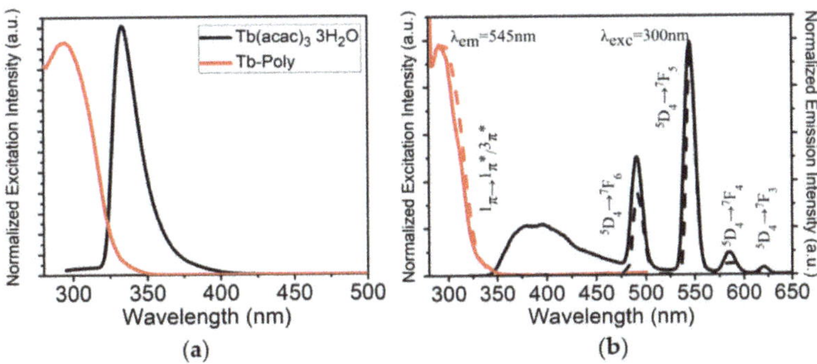

Figure 3. (**a**) Excitation spectra of Tb(acac)$_3$ 3H$_2$O (black) and Tb-Poly (red) in DMF solution. (**b**) Excitation and emission spectra of Tb-Poly in DMF (dash line) and in NPs (solid line), respectively.

The Ru-containing polymer (Ru-Poly) was similarly synthesized according to the procedures in the literature [10]. Based upon Ru-Poly and Tb-Poly, Ru-Tb NPs were prepared in combination with PS–PEG-COOH by the same nanoprecipitation method, in which the latter polymer is used to PEGylate the NPs. The resultant Ru-Tb NPs have a typical size of ~100 nm in diameter (Figure 4a). The shape and size are shown in the SEM image (Figure 4b). The doping ratio of Ru-Poly to Tb-Poly has been optimized to render the comparable emission intensity in air-saturated environments (see Figure S6 for more details). A weight ratio of 7:3:2 (Ru-Poly: Tb-Poly: PS–PEG-COOH) is finally adopted to prepare Ru-Tb NPs. It needs to point out that Ru-Tb NPs are based on a two-wavelength excitation, corresponding to the absorption of bpy ligands in Tb-Poly and metal-to-ligand charge-transfer (^{1}MLCT) in Ru-Poly, respectively. Ru-Tb NPs are very stable in various media (like water and DMEM culture) due to the stabilization of PEG, which can be long-term stored without obvious changes in size (Figure S7).

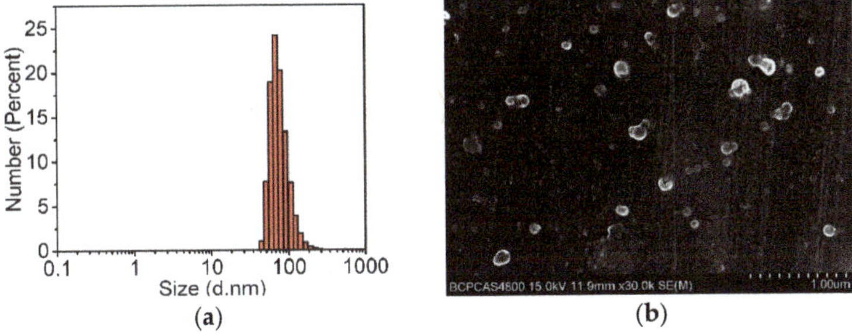

Figure 4. (**a**) Dynamic light scattering and (**b**) scanning electron microscopy image of Ru-Tb NPs.

3.2. Oxygen Sensitivity and Calibration of the Ratiometric Ru-Tb Nanoprobes

The luminescence response of the ratiometric Ru-Tb NPs toward oxygen was investigated by purging the aqueous suspension with a gas mixture of O$_2$/N$_2$. As showed in Figure 5a, the 610 nm emission of Ru-Poly is strongly quenched with the increase of dissolved oxygen, while the 545 nm emission of Tb-Poly are kept rather constant. In small-sized nanoprobes, luminescence oxygen quenching can be expressed by the linear Stern–Volmer equation, as in the case of the homogeneous system,

$$R_0/R - 1 = K_{SV} [O_2] \tag{2}$$

where R_0 is the intensity ratio (red emission at 610 nm versus green emission at 545 nm) in the absence of oxygen, R the ratio in the presence of oxygen at a given concentration, K_{SV} as the Stern–Volmer quenching constant, and $[O_2]$ as the concentration of dissolved oxygen. The Stern−Volmer plot of the intensity ratio of Ru-Tb NPs versus oxygen concentration is depicted in Figure 5b. The data were fitted quite well by the linear function (the correlation coefficient >0.997), which is important for practical applications. It needs to point out that the luminescent intensity of nanoprobes is decreased abnormally in the case of oxygen-saturated solution (Figure 5a, 43 ppm) for both the green and red emission. This may be caused by the loss of Ru-Tb NPs due to long-time purging by gas. If the detection modality is based on single intensity, this experimental data would deviate from the linear Stern−Volmer plot considerably. However, the robust ratiometric approach keeps the data still following a linear function relationship.

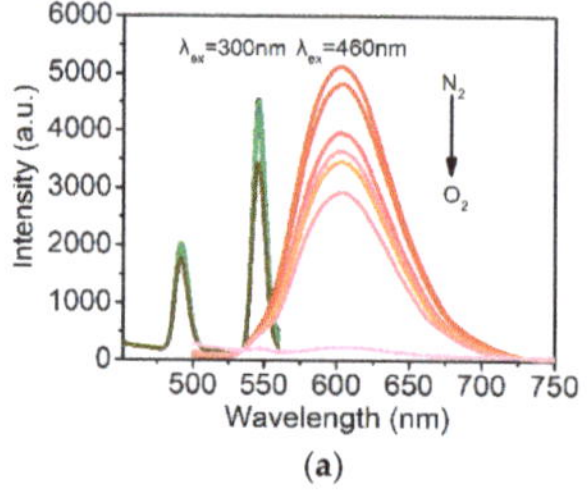
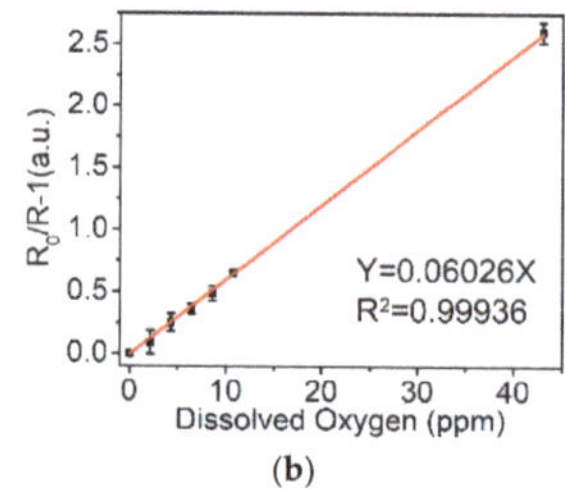

(a)   (b)

Figure 5. Oxygen sensitivity of Ru-Tb NPs in aqueous solution: (**a**) Emission spectra under 300 nm and 460 nm excitation at various oxygen concentrations (from top to bottom is 0 (nitrogen saturated), 2.15, 4.3, 8.6, 10.75, and 43 ppm (oxygen saturated) in sequence). (**b**) Ratio of red and green luminescent intensity-based Stern−Volmer plot. The experimental data (scatter) were calculated from the ratio luminescence intensities at 610 and 545 nm and linearly fitted (solid line).

3.3. Intracellular Imaging of the Ratiometric Ru-Tb Nanoprobes

The cytotoxicity of Ru-Tb NPs was tested by MTT assay on living HeLa cells (see Figure S8). The results show that a dosage of <33 μg mL^{-1} renders slight growth inhibition of cells (>85% cell viability). Afterwards, the cellular uptake and distribution of Ru-Tb NPs were explored with confocal microscopy with a dosage of 30 μg mL^{-1} (Figure 6). It can be deduced from the fluorescence images (red and green channels) and differential interference contrast (DIC) image that the PEGylated nanoprobes can be efficiently uptaken by live cells. Because of the limitation of available lasers in common confocal microscopy, the green channel is obtained under the excitation of a mercury lamp rather than a 300 nm laser. As a result, the quality of the acquired image is not very clear, and a 720 nm light is thus used in the following, based on the mechanism of two-photon excitation.

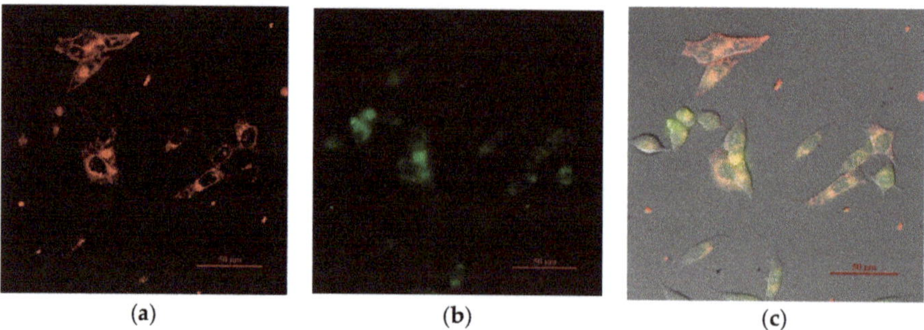

(a)   (b)   (c)

Figure 6. Images of HeLa cells treated with Ru-Tb NPs (30 μg mL^{-1}): (**a**) red channel, excitation at 488 nm and emission at 585–625 nm; (**b**) green channel, excited by mercury lamp and emission at 525–565 nm; (**c**) overlay of red and green channels, together with the DIC image. In a merged picture, colocalization of red and green signal results in orange areas. Scale bar = 50 μm.

Recently three-dimensional multi-cellular tumor spheroids (3D MCTSs) have been popularly studied to mimic the real situation of solid tumor. MCTSs are a valid intermediate between monolayer in vitro cells and in vivo tissue that are formed by heterogeneous cell aggregation [19]; the hypoxic center of MCTSs provides an ideal place for monitoring oxygen. In this experiment, 300 μm MCTSs in diameter can be obtained after 7 days' cultivation from single Hela cell. For staining experiment, 30 μg mL^{-1} Ru-Tb NPs in DMEM medium were incubated with MCTSs for 12 h. Cells were visualized by confocal microscopy immediately (Figure 7, and for the whole image, see the Figure S9). The Z-stack confocal images are taken under two-photon excitation. Interestingly, red and green channels could be monitored simultaneously under two-photon excitation at 720 nm, although the emission intensity is weaker than excited under one-photon. The emission spectra of Ru-Tb NPs were measured under excitation at 360 nm in aqueous solution, and the green and red emission peaks could be observed clearly (Figure S10).

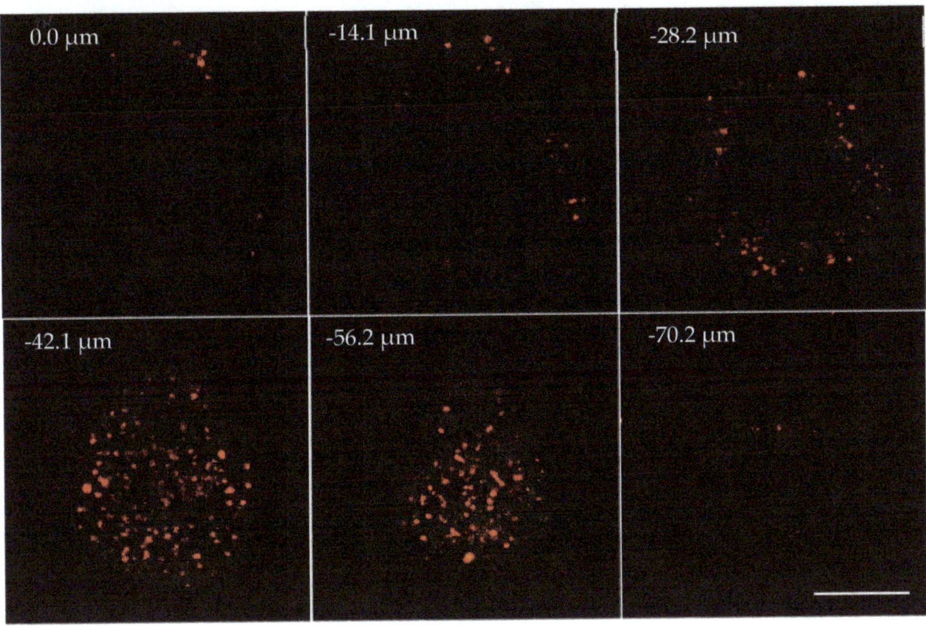

Figure 7. Z-stack of two-photon microscopy images of MCTs. The images were taken every 14.04 μm section from the top to bottom of intact MCTs. The green and red channel was collected at 525–565 nm and 585–625 nm, respectively. The scale bar = 200 μm.

From Figure 7 and Figure S9, it can be seen that the intensity of red and green emission both increase gradually with the Z-stack section depth (<~54 μm), indicating the efficient uptake of Ru-Tb NPs by MCTSs. When the section depth is above 54 μm, the intensity of red emission decreases while the green emission is rather constant. Obviously, the microenvironment of interest has a high oxygen concentration which seriously quenches the luminescence of Ru-Poly. To display the distribution of dissolved oxygen within MCTSs, the intensity ratio of red emission to green emission of Ru-Tb NPs at the depth of −36 μm is displayed in pseudocolor, and particularly, the spatial distribution of dissolved concentration is roughly quantified by recording the intensity ratio along profiles spanning the section with the confocal software (Figure 8). From the above results, the normoxic brim, and hypoxic core of MCTSs can be discerned clearly.

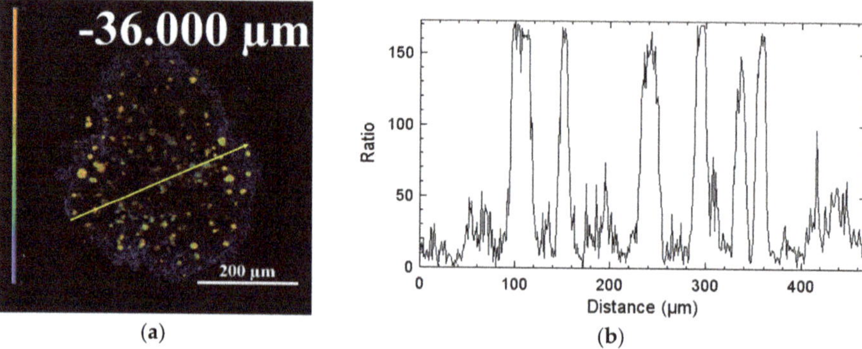

Figure 8. (**a**) Pseudocolor ratiometric intensity images (I_{F610}/I_{F545}) of sectional MCTs at depth of −36 μm and (**b**) intensity ratio along with profiles spanning labeled cells, as indicated by the lines (yellow). The color bar from blue to orange corresponds to the increase of intensity ratio from 0 to 200. Scale bar = 200 μm.

4. Conclusions

In summary, we present a type of ratiometric luminescent nanoprobe (Ru-Tb NP) for intracellular oxygen, which was based on two metallopolymers branched with oxygen probe $[Ru(bpy)^3]^{2+}$ and reference dye $[Tb(acac)_3bpy]^{3+}$, respectively. The resultant Ru-Tb NPs have a hydrodynamic size of ~100 nm in diameter, and are easily swallowed by living cells due to PEGylation. Ru-Tb NPs exhibit a two-wavelength emission (545 nm/610 nm) under the excitation of 300 nm and 460 nm, respectively. The red emission from Ru-Poly is significantly quenched by oxygen, while the green emission from Tb-Poly is rather stable. The ratiometric luminescence towards oxygen is well fitted by a linear Stern–Volmer equation. Under the two-photon excitation of a 720 nm light, distribution of dissolved oxygen in MCTSs is successfully imaged with the ratiometric nanoprobes.

Supplementary Materials: The following are available online at http://www.mdpi.com/2073-4360/11/8/1290/s1. Experimental detail on the chemical synthesis of $Tb(acac)_3$ $3H_2O$ and Tb-Poly, temperature sensitivity, figures of optical properties, and stability.

Author Contributions: Methodology, C.Z.; writing—original draft preparation, W.-x.Z.; writing—review and editing, H.-s.P.; supervision, H.-s.P.

Funding: This research was funded by Natural Science Foundation of China (Grant number 61775245) and the MUC 111 project.

Conflicts of Interest: The authors declare no conflict of interest.

References

1. Acker, T.; Acker, H. Cellular oxygen sensing need in CNS function: Physiological and pathological implications. *J. Exp. Biol.* **2004**, *207*, 3171–3188. [CrossRef] [PubMed]
2. Decker, H.; Van Holde, K.E. *Oxygen and The Evolution of Life*; Springer: Heidelberg, Germany, 2011; p. 172. [CrossRef]
3. Wang, X.D.; Wolfbeis, O.S. Optical methods for sensing and imaging oxygen: Materials, spectroscopies and applications. *Chem. Soc. Rev.* **2014**, *43*, 3666–3761. [CrossRef] [PubMed]
4. Zeng, L.; Gupta, P.; Chen, Y.; Wang, E.; Ji, L.; Chao, H.; Chen, Z.S. The development of anticancer ruthenium (ii) complexes: From single molecule compounds to nanomaterials. *Chem. Soc. Rev.* **2017**, *46*, 5771–5804. [CrossRef] [PubMed]
5. Kim, H.N.; Guo, Z.; Zhu, W.; Yoon, J.; Tian, H. Recent progress on polymer-based fluorescent and colorimetric chemosensors. *Chem. Soc. Rev.* **2011**, *40*, 79–93. [CrossRef] [PubMed]
6. Winter, A.; Schubert, U.S. Synthesis and characterization of metallo-supramolecular polymers. *Chem. Soc. Rev.* **2016**, *45*, 5311–5357. [CrossRef] [PubMed]

7. Dmitriev, R.I.; Papkovsky, D.B. Optical probes and techniques for O_2 measurement in live cells and tissue. *Cell. Mol. Life Sci.* **2012**, *69*, 2025–2039. [CrossRef] [PubMed]

8. Tian, Y.; Shumway, B.R.; Meldrum, D.R. A new cross-linkable oxygen sensor covalently bonded into poly(2-hydroxyethyl methacrylate)-co-polyacrylamide thin film for dissolved oxygen sensing. *Chem. Mater.* **2010**, *22*, 2069–2078. [CrossRef] [PubMed]

9. Wolfbeis, O.S. Luminescent sensing and imaging of oxygen: Fierce competition to the Clark electrode. *BioEssays* **2015**, *37*, 921–928. [CrossRef] [PubMed]

10. Zhou, C.; Zhao, W.X.; You, F.T.; Geng, Z.X.; Peng, H.S. Highly stable and luminescent oxygen nanosensor based on ruthenium-containing metallopolymer for real-time imaging of intracellular oxygenation. *ACS Sens.* **2019**. [CrossRef] [PubMed]

11. Hara, D.; Komatsu, H.; Son, A.; Nishimoto, S.; Tanabe, K. Water-soluble phosphorescent ruthenium complex with a fluorescent coumarin unit for ratiometric sensing of oxygen levels in living cells. *Bioconj. Chem.* **2015**, *26*, 645–649. [CrossRef] [PubMed]

12. Wang, X.; Chang, H.; Xie, J.; Zhao, B.; Liu, B.; Xu, S.; Pei, W.; Ren, N.; Huang, L.; Huang, W. Recent developments in lanthanide-based luminescent probes. *Coord. Chem. Rev.* **2014**, *273*, 201–212. [CrossRef]

13. Badiane, A.M.; Freslon, S.; Daiguebonne, C.; Suffren, Y.; Bernot, K.; Calvez, G.; Costuas, K.; Camara, M.; Guillou, O. Lanthanide-based coordination polymers with a 4,5-dichlorophthalate ligand exhibiting highly tunable luminescence: Toward luminescent bar codes. *Inorg. Chem.* **2018**, *57*, 3399–3410. [CrossRef] [PubMed]

14. Kotova, O.; Bradberry, S.J.; Savyasachi, A.J.; Gunnlaugsson, T. Recent advances in the development of luminescent lanthanide-based supramolecular polymers and soft materials. *Dalton Transact.* **2018**, *47*, 16377–16387. [CrossRef] [PubMed]

15. Beer, P.D.; Szemes, F.; Passaniti, P.; Maestri, M. Luminescent Ruthenium(II) Bipyridine–Calix[4]arene complexes as receptors for lanthanide cations. *Inorg. Chem.* **2004**, *43*, 3965–3975. [CrossRef] [PubMed]

16. Wang, Y.; Li, H.; Feng, Y.; Zhang, H.; Calzaferri, G.; Ren, T. Orienting Zeolite L Microcrystals with a functional linker. *Angewandte Chem. Int. Ed.* **2010**, *49*, 1434–1438. [CrossRef] [PubMed]

17. Parra, D.F.; Mucciolo, A.; Brito, H.F. Green luminescence system containing a Tb^{3+}-β-diketonate complex doped in the epoxy resin as sensitizer. *J. Appl. Polym. Sci.* **2004**, *94*, 865–870. [CrossRef]

18. Zhang, R.J.; Yang, K.Z.; Yu, A.C.; Zhao, X.S. Fluorescence lifetime and energy transfer of rare earth β-diketone complexes in organized molecular film. *Thin Solid Films* **2000**, *363*, 275–278. [CrossRef]

19. Huang, H.; Yang, L.; Zhang, P.; Qiu, K.; Huang, J.; Chen, Y.; Diao, J.; Liu, J.; Ji, L.; Long, J.; et al. Real-time tracking mitochondrial dynamic remodeling with two-photon phosphorescent iridium (III) complexes. *Biomaterials* **2016**, *83*, 321–331. [CrossRef] [PubMed]

Article

Cationic Fluorescent Nanogel Thermometers based on Thermoresponsive Poly(*N*-isopropylacrylamide) and Environment-Sensitive Benzofurazan

Teruyuki Hayashi [1], Kyoko Kawamoto [2], Noriko Inada [3] and Seiichi Uchiyama [2,*

[1] College of Nutrition, Koshien University, Hyogo 665-0006, Japan
[2] Graduate School of Pharmaceutical Sciences, The University of Tokyo, Tokyo 113-0033, Japan
[3] Graduate School of Life and Environmental Sciences, Osaka Prefecture University, Osaka 599-8531, Japan
* Correspondence: seiichi@mol.f.u-tokyo.ac.jp; Tel.: +81-3-5841-4768

Received: 21 June 2019; Accepted: 29 July 2019; Published: 4 August 2019

Abstract: Cationic nanogels of *N*-isopropylacrylamide (NIPAM), including NIPAM-based cationic fluorescent nanogel thermometers, were synthesized with a cationic radical initiator previously developed in our laboratory. These cationic nanogels were characterized by transmission electron microscopy (TEM), dynamic light scattering (DLS), zeta potential measurements and fluorescence spectroscopy, as summarized in the temperature-dependent fluorescence response based on the structural change in polyNIPAM units in aqueous solutions. Cellular experiments using HeLa (human epithelial carcinoma) cells demonstrated that NIPAM-based cationic fluorescent nanogel thermometers can spontaneously enter the cells under mild conditions (at 25 °C for 20 min) and can show significant fluorescence enhancement without cytotoxicity with increasing culture medium temperature. The combination of the ability to enter cells and non-cytotoxicity is the most important advantage of cationic fluorescent nanogel thermometers compared with other types of fluorescent polymeric thermometers, i.e., anionic nanogel thermometers and cationic/anionic linear polymeric thermometers.

Keywords: fluorescence; imaging; nanogel; temperature; thermometer; thermometry

---

1. Introduction

Fluorescent molecular thermometers change their fluorescence properties (e.g., fluorescence intensity, fluorescence quantum yield, fluorescence lifetime and maximum emission wavelength) with temperature [1–5]. Fluorescent polymeric thermometers based on the combination of a thermoresponsive polymer and an environment-sensitive fluorophore [6–12] are among the most sensitive fluorescent molecular thermometers. Because of their function at the molecular level in aqueous media, these sensitive fluorescent polymeric thermometers have been applied for the thermometry of small subjects, such as microfluids [13,14] and even single living cells [15–25]. Fluorescent polymeric thermometers are classified into two categories of morphology: fluorescent nanogel thermometer with crosslinking units to construct a nano-scaled particle [8,15] and fluorescent linear polymeric thermometer without crosslinking units [6,7,16]. In general, the former is more robust and less interactive with external molecules and ions, whereas the latter is capable of thermometry with a higher spatial resolution, due to its diffusivity.

In 2009, we performed the first intracellular thermometry of live mammalian COS7 (African green monkey kidney) cells using a fluorescent nanogel thermometer that was negatively charged, due to the radical initiator ammonium persulfate (APS) used in its preparation [15]. For intracellular thermometry, the microinjection technique was required for the introduction of anionic fluorescent nanogel thermometers into live COS7 cells. If the ability to spontaneously enter live cells is expected, a fluorescent nanogel thermometer should be positively charged based on the established concept that polycationic structures efficiently support the spontaneous entry of molecules into living cells [26–30]. However, the preparation of positively charged fluorescent nanogel thermometers has not been easily realized because of the lack of cationic radical initiators.

In a recent communication published in 2018 [31], we reported the use of the first cationic radical initiator, 2,2′-azobis-[2-(1,3-dimethyl-4,5-dihydro-1*H*-imidazol-3-ium-2-yl)]propane triflate (ADIP), to prepare cationic nanogels, including a cationic fluorescent nanogel thermometer (Figure 1). In the present article, we describe a detailed experimental procedure to prepare cationic NIPAM nanogels and NIPAM-based cationic fluorescent nanogel thermometers using ADIP and present the comprehensive physical and photophysical properties of these cationic nanogels (i.e., size, zeta potential and/or temperature-dependent fluorescence properties) and the functions of the cationic fluorescent nanogel thermometers in intracellular applications (i.e., ability to enter live cells, distribution in cells, cytotoxicity and response to temperature variation). These experimental protocols and new functional data will complement our earlier communication [31].

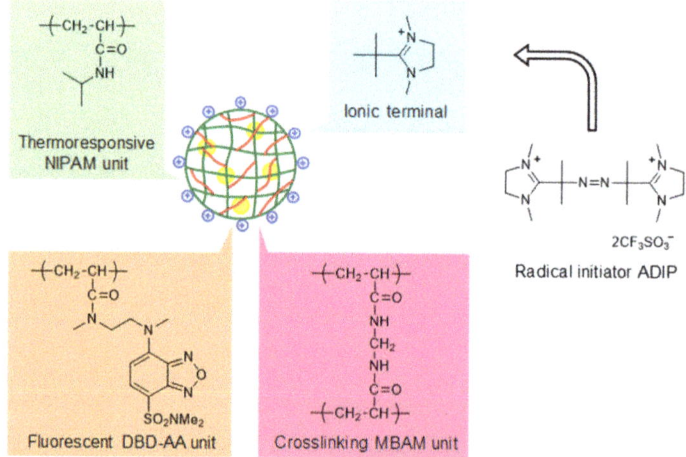

Figure 1. Cationic fluorescent nanogel thermometer prepared with cationic radical initiator ADIP (2,2′-azobis-[2-(1,3-dimethyl-4,5-dihydro-1*H*-imidazol-3-ium-2-yl)]propane triflate). NIPAM: *N*-isopropylacrylamide, DBD-AA: *N*-{2-(7-*N*,*N*-dimethylaminosulfonyl-2,1,3-benzoxadiazol-4-yl)methylamino}ethyl-*N*-methylacrylamide, MBAM: *N*,*N*′-methylenebisacrylamide.

2. Materials and Methods

Bulletized procedures corresponding to Sections 2.2 and 2.4 were described in Supplementary Materials.

2.1. Materials

NIPAM was purchased from FIJIFILM Wako Pure Chemical Corporation (98%, Osaka, Japan, catalog no. 099-03695) and was purified by recrystallization from *n*-hexane. Cetyltrimethylammonium chloride (CTAC, 95%, catalog no. 087-06032), NaCl (99%, catalog no. 196-01671) and KCl (99.9%, catalog no. 164-13122) were purchased from FIJIFILM Wako Pure Chemical Corporation and were used without further purification. MBAM (99%, catalog no. 148326) and *N*,*N*,*N*′,*N*′-tetramethylethylenediamine (TMEDA, 99%, catalog no. T22500) were purchased from Sigma-Aldrich Japan Inc. (Tokyo, Japan) and were used without further purification. D₂O was obtained from Cambridge Isotope Laboratories (D, 99.9%, Tewksbury, US, catalog no. DLM-4-100). ADIP [31] and DBD-AA [32] were synthesized and purified, as previously reported. Water was purified using a Direct-Q 3 UV system (Merck Millipore,

Burlington, MA, USA). Dialysis membranes (MEMBRA-CEL MD34, Molecular Weight Cut Off: 14,000; Pore size: 50 Å) were purchased from Viskase Companies Inc. (Willowbrook, IL, USA).

35-mm Glass dish was purchased from Iwaki/AGC Techno Glass (Shizuoka, Japan, catalog no. 3960-035). Phenol red-free DMEM (Dulbecco's modified Eagle's medium, catalog no. 21063-029), FBS (Fetal Bovine Serum, Hyclone, catalog no. SH30910.03), 0.5% Trypsin-EDTA (ethylenediaminetetraacetic acid, catalog no. 15400054) and MitoTracker Deep Red FM (catalog no. M22426) were purchased from Thermo Fisher Scientific (Waltham, MA, USA). DMEM (catalog no. 08458-45), $Na_2HPO_4 \cdot 12H_2O$ (98%, catalog no. 31722-45) and KH_2PO_4 (99.5%, catalog no. 28721-55) were purchased from Nacalai Tesque (Kyoto, Japan). HeLa (human epithelial carcinoma) cells were obtained from ATCC (catalog no. CCL-2, Manassas, USA).

2.2. Preparation of Cationic Nanogels (NANOGEL-1 and NANOGEL-2) and Cationic Fluorescent Nanogel Thermometers (NANOGEL-3~6).

NIPAM, DBD-AA, MBAM, TMEDA and/or CTAC were dissolved in 19 mL of water (for the final concentration of each compound in a reaction mixture, see Table 1). Dry argon gas was bubbled through the solution at 70 °C for 30 min to remove the dissolved oxygen. ADIP [31] in water (1 mL) was added to initiate polymerization, and the mixture was stirred using a rod with a paddle at 250 rpm and 70 °C for 1 h under an argon atmosphere. The mixture was then poured into 400 mL of water, and the nanogels were precipitated by a salting-out technique. After purification by dialysis for at least one week, the nanogel dispersions were freeze-dried. The purity of NANOGEL-1~6 was confirmed by ^{1}H-NMR measurements with a Bruker AVANCE 400 spectrometer (Billerica, MA, USA) in D_2O (Figure S1). The yields are indicated in Table 1.

Table 1. Preparation of cationic nanogels (NANOGEL-1 and NANOGEL-2) and cationic fluorescent nanogel thermometers (NANOGEL-3~6).

Name	Final Concentrations in the Reaction Mixture						Yield (%)
	NIPAM (mM)	MBAM (mM)	DBD-AA (mM)	TMEDA (mM)	ADIP (mM)	CTAC (mM)	
NANOGEL-1	100	1		2.9	28	1.9	39
NANOGEL-2	100	1		2.9	28		14
NANOGEL-3	100	1	0.1	2.9	28	1.9	35
NANOGEL-4	100	1	0.2	2.9	28	1.9	47
NANOGEL-5	100	1	0.5	2.9	28	1.9	28
NANOGEL-6	100	1	1	2.9	28	1.9	46

2.3. Characterization of NANOGEL-1~6

TEM images were obtained with a Hitachi H-7100 transmission electron microscope (Hitachi High-Technologies, Tokyo, Japan). A drop of the nanogel solution in ethanol or water (0.01 w/v%, 5 µL) was placed on a formvar-coated copper grid. The specimen was air-dried at room temperature and then examined at an accelerating voltage of 75 kV. The hydrodynamic diameter and zeta potential were measured with a Malvern Instruments Zetasizer Nano ZS (currently Malvern Panalytical, Malvern, UK). The samples were equilibrated for 10 min at each temperature. The amount of fluorescent DBD-AA units in NANOGEL-3~6 was estimated from the comparison of the absorbance of their methanol solution with that of *N*,2-dimethyl-*N*-(2-{methyl[7-(dimethylsulfamoyl)-2,1,3-benzoxadiazol-4-yl]amino}ethyl)propenamide ($\varepsilon = 11,000\,M^{-1}\,cm^{-1}$ at 444 nm) as a model compound [33]. The fluorescence spectra of NANOGEL-3~6 were recorded in water and a 150 mM KCl solution at various temperatures using a JASCO FP-6500 spectrofluorometer (Tokyo, Japan) with a Hamamatsu R-7029 optional photomultiplier tube (Hamamatsu, Japan, operating range, 200–850 nm). The sample temperature was controlled using a JASCO ETC-273T temperature controller (Tokyo, Japan).

2.4. Introduction of Cationic Fluorescent Nanogel Thermometers into HeLa Cells

The HeLa cells were cultured on a 35-mm glass dish in high-glucose DMEM supplemented with FBS at 37 °C with 5% CO_2. The cationic fluorescent nanogel thermometers were introduced into the HeLa cells by two different methods: a standard method using cells adhered to a glass-bottom dish and a modified method using suspended cells. For the standard method, DMEM was removed from a glass-bottom dish containing HeLa cells at 30–50% confluency, and the cells were rinsed with 1 mL of 1×PBS (phosphate buffered saline, 10×PBS containing 28.9 g of $Na_2HPO_4 \cdot 12H_2O$, 2.0 g of KH_2PO_4, 80.0 g of NaCl and 2.0 g of KCl in a 1 liter solution). Then, PBS was replaced by 1 mL of cationic fluorescent nanogel thermometers in a 5 w/v% glucose solution (0.05 w/v%, 10 μL of a 5 w/v% stock solution in water diluted in 990 μL of a 5 w/v% glucose solution). A 5 w/v% stock solution in water was prepared and incubated at 4 °C at least overnight before the full solvation of nanogels. The dish was incubated at 25 or 37 °C for 5, 10 or 20 min without CO_2 supply. After incubation, the nanogel solution was removed, and the cells were rinsed with 1 mL of 1×PBS three times. Two milliliters of phenol red-free DMEM was added to the dish before imaging.

For the modified method using suspended cells, HeLa cells cultured in 100 mm culture dishes were rinsed with 1 mL of 1×PBS, then treated with 0.5 mL of a 0.05 w/v% trypsin-EDTA-1×PBS solution and incubated at 37 °C for 3–5 min. The detached cells were suspended in 9 mL of DMEM. One milliliter of the cell suspension was transferred to 1.5 mL tubes and centrifuged at 1200 rpm at 4 °C for 1 min. The collected cells were rinsed twice with 1 mL of 1×PBS, and then suspended in 1 mL of cationic fluorescent nanogel thermometers in a 5 w/v% glucose solution. After incubation at 25 °C for 20 min without CO_2 supply, the cells were again collected by centrifugation at 1200 rpm at 4 °C for 1 min and rinsed with 1 mL of 1×PBS three times before being suspended in 10 mL of DMEM. Two milliliters mL of cell suspension was added to a 35-mm glass-bottom dish and incubated at 37 °C with 5 % CO_2 for one or two nights before observation.

2.5. Fluorescence Imaging of HeLa Cells

Confocal fluorescence imaging was performed using a TCS-SP5 laser scanning confocal microscope equipped with an HCX PL APO Ibd.BL 63 × 1.4 N.A. oil objective (Leica Microsystems, Wetzlar, Germany). Cells loaded with cationic fluorescent nanogel thermometers were excited by a 458 nm argon laser, then the fluorescence images were acquired through bandpass 500–700 nm in a 1024 × 1024 pixel format, with zoom factors ranging from 1 to 10 and a scanning speed of 400 Hz. The contrast and brightness of the fluorescence images were enhanced using ImageJ with a constant signal intensity ratio. The incorporation efficiencies (%) of NANOGEL-**6** were determined using Equation (1),

$$\text{Incorporation efficiency (\%)} = \text{number of cells containing NANOGEL-}\mathbf{6} \\ /\text{number of cells} \times 100, \tag{1}$$

in which the total cell number was 183–401, and the cells that showed a fluorescence intensity higher than the threshold (equal to the maximum autofluorescence intensity) were counted as the "cells containing NANOGEL-**6**". For the co-visualization of NANOGEL-**6** and mitochondria, the HeLa cells were stained with 50 nM MitoTracker Deep Red FM (MitoTracker DR) in phenol red-free DMEM for 5 min at room temperature and then treated with NANOGEL-**6** by a standard method using adhered cells. A 458 nm argon laser was used to excite NANOGEL-**6**, and a 633 nm HeNe laser was used to excite MitoTracker DR. The fluorescence of NANOGEL-**6** was collected through bandpass 500–600 nm, and the fluorescence of MitoTracker DR was collected through bandpass 645–730 nm.

3. Results

3.1. Preparation of Cationic Nanogels (NANOGEL-1 and NANOGEL-2) and Cationic Fluorescent Nanogel Thermometers (NANOGEL-3~6)

The cationic nanogels (NANOGEL-1 and NANOGEL-2) and cationic fluorescent nanogel thermometers (NANOGEL-3~6) were synthesized by radical polymerization, due to the cationic radical initiator ADIP with the starting materials indicated in Table 1. Crude nanogels were purified by reprecipitation with a salting-out technique and subsequent dialysis for at least one week. The purity of NANOGEL-1~6 was confirmed by ^{1}H-NMR measurements (Figure S1 in Supplementary Materials). The moderate yields of NANOGEL-1~6 likely resulted from their substantial loss during the reprecipitation (due to high hydrophilicity of NANOGEL-1~6 with cationic surfaces) and the long-term dialysis. The size and surface charge were evaluated by TEM/DLS measurements and zeta potential measurements (Figure 2), respectively, and are summarized in Table 2. The temperature-dependent hydrodynamic diameters of NANOGEL-1~6 determined by DLS clearly show the thermosensitive characteristics of the polyNIPAM units (cf., volume phase transition temperature of the polyNIPAM gel is 32 °C [34]). Sufficient amounts of cationic terminals introduced by ADIP to increase the solubility of nanogels resulted in the isolation of nanogels at a high temperature (45 °C) without aggregation (Table 2). NANOGEL-1 without fluorescent DBD-AA units was obtained by referring to the concentrations of monomers adopted for the preparation of anionic NIPAM nanogels [15]. Interestingly, the cationic NANOGEL-2 could be obtained without CTAC, which is a rare example in the preparation of NIPAM nanogels without any surfactant molecules [35]. Then, the cationic fluorescent nanogel thermometers NANOGEL-3~6 were synthesized with various concentrations of an environment-sensitive fluorescent monomer, DBD-AA. The amounts of fluorescent DBD-AA units in NANOGEL-3~6 were converted to the corresponding concentrations when the nanogel concentration was 0.01 w/v% (Table 2).

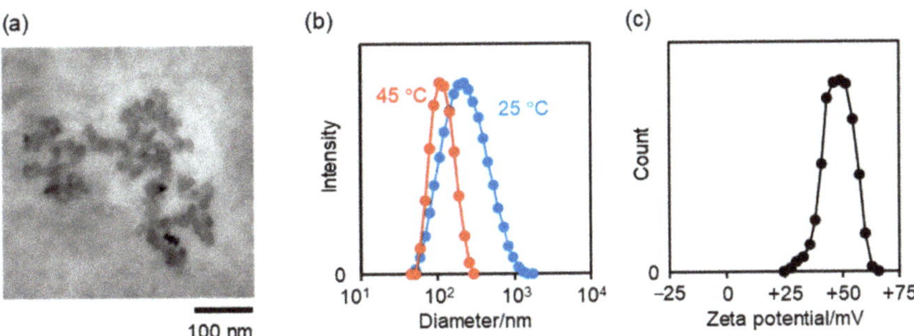

Figure 2. Characterization of NANOGEL-1 as a representative. (**a**) TEM image; (**b**) size distribution measured by DLS (0.001 w/v% in water at 25 and 45 °C); (**c**) zeta potential distribution (0.1 w/v% in water at 45 °C).

3.2. Fluorescence Responses of Cationic Fluorescent Nanogel Thermometers (NANOGEL-3~6) in Aqueous Solutions

The fluorescence spectra of NANOGEL-3~6 (0.01 w/v%) in water and 150 mM KCl solution were recorded with changing temperature (Figure 3). The fluorescence intensity ratio at 25 and 45 °C (defined as the fluorescence enhancement (FE) factor) and the maximum emission wavelengths at 25 and 45 °C are listed in Table 2. All samples showed fluorescence enhancements at approximately 32 °C, which is the lower critical solution temperature of PNIPAM nanogels [34]. The heat-induced fluorescence enhancement of NANOGEL-3~6 in 150 mM KCl solution was higher than that in water because of the salting out effects. It is known that salts, including KCl accelerate the dehydration of NIPAM units by hydrogen bonding with their amide groups and by increasing the surface tension of water in the hydration shell around the hydrophobic groups [36,37]. Nevertheless, the salting out effects by KCl on the fluorescence responses of NANOGEL-3~**6** were gradually saturated when the concentration of KCl exceeded 50 mM (Figure S2), which is preferred for the use under intracellular conditions (where $[K^+]$ is approximately 139 mM [38].) As indicated in Figure 3a and Table 2, the maximum emission wavelength at a high temperature (i.e., 45 °C) was remarkably shorter than that at a low temperature (i.e., 25 °C). This temperature-dependent maximum emission wavelength is due to the functional mechanism of NIPAM-based fluorescent polymeric thermometers [6], i.e., the drastic variation of the microenvironment near DBD-AA units with a heat-induced structural change in NIPAM units. As cationic fluorescent nanogel thermometers contain many fluorescent DBD-AA units, the heat-induced fluorescence enhancement became less with a small shift in the maximum emission wavelength between 25 and 45 °C (e.g., NANOGEL-**3** vs NANOGEL-**6** in Table 2). The decrease in sensitivity to the temperature variation is likely due to a structural disturbance of the thermoresponsive nanogels by the relatively bulky DBD-AA units.

Table 2. Physical and photophysical properties of cationic nanogels (NANOGEL-**1** and NANOGEL-**2**) and cationic fluorescent nanogel thermometers (NANOGEL-**3**~**6**).

Name	Diameter (nm) [1]		Zeta Potential (mV) [2]	[DBD-AA unit] (µM) [3]	FE [4] in Water	λ_{em} (nm)		FE [4] in 150 mM KCl sol.
	at 25 °C	at 45 °C				at 25 °C	at 45 °C	
NANOGEL-1	270 ± 31	123 ± 1.4	+47.8 ± 0.9	—	—	—	—	—
NANOGEL-2	284 ± 7.8	117 ± 2.7	+56.9 ± 0.7	—	—	—	—	—
NANOGEL-3	232 ± 25	90.2 ± 7.2	+37.3 ± 0.3	1.09	7.72	589	561	10.90
NANOGEL-4	311 ± 24	151 ± 11	+45.0 ± 1.0	2.42	6.03	585	563	8.60
NANOGEL-5	230 ± 32	133 ± 4.0	+48.8 ± 0.2	6.64	3.58	582	567	4.50
NANOGEL-6	314 ± 20	157 ± 27	+53.5 ± 0.2	12.0	1.95	577	570	2.30

[1] Determined by DLS. The av ± s.d. of peak values in triplicate measurements. [2] At 45 °C. The av ± s.d. of peak values of triplicate measurements. [3] When the nanogel concentration is 0.01 w/v%. [4] Fluorescence enhancement factor calculated as fluorescence intensity at λ_{em} at 45 °C divided by that at 25 °C.

3.3. Introduction of Cationic Fluorescent Nanogel Thermometers into Mammalian HeLa Cells

The method for introducing cationic fluorescent nanogel thermometers was optimized using the most strongly fluorescent NANOGEL-**6** and HeLa cells as model mammalian cells. One should note that weakly fluorescent nanogel thermometers with fewer DBD-AA units (e.g., NANOGEL-**4**) could not be detected inside the HeLa cells at a low temperature (e.g., 30 °C) with the same laser excitation power and detection sensitivity of the confocal laser scanning microscope used for the detection of NANOGEL-**6**. Figure 4 shows the effect of incubation time at 25 °C on the incorporation efficiency of NANOGEL-**6** when established standard conditions for cationic fluorescent polymeric thermometers (0.05 w/v% in a 5 % glucose solution [28]) were adopted. A high temperature (e.g., 37 °C) accelerated the introduction of cationic fluorescent nanogel thermometers into the HeLa cells, but induced unfavorable aggregation inside the cells. Therefore, we fixed the incubation time and temperature to be 20 min and 25 °C, respectively, in the subsequent experiments. Figure 5 displays the representative transmission and confocal fluorescence images of a HeLa cell containing NANOGEL-**6**, in which the mitochondria were additionally stained by MitoTracker DR. The cationic nanogel thermometer NANOGEL-**6** was detected in the form of dots in the HeLa cells and remained inside them once introduced. While a significant background noise was detected when cationic fluorescent nanogel thermometer NANOGEL-**6** was introduced into adherent HeLa cells, due to the attachment of NANOGEL-**6** on the surface of a glass-bottom dish, treatment of suspended HeLa cells with NANOGEL-**6** could significantly reduce this background noise (Figure 6). This alternative protocol, i.e., introducing NANOGEL-**6** into suspended HeLa cells, is considered for improving the sensitivity of intracellular thermometry by increasing the signal-to-noise ratio.

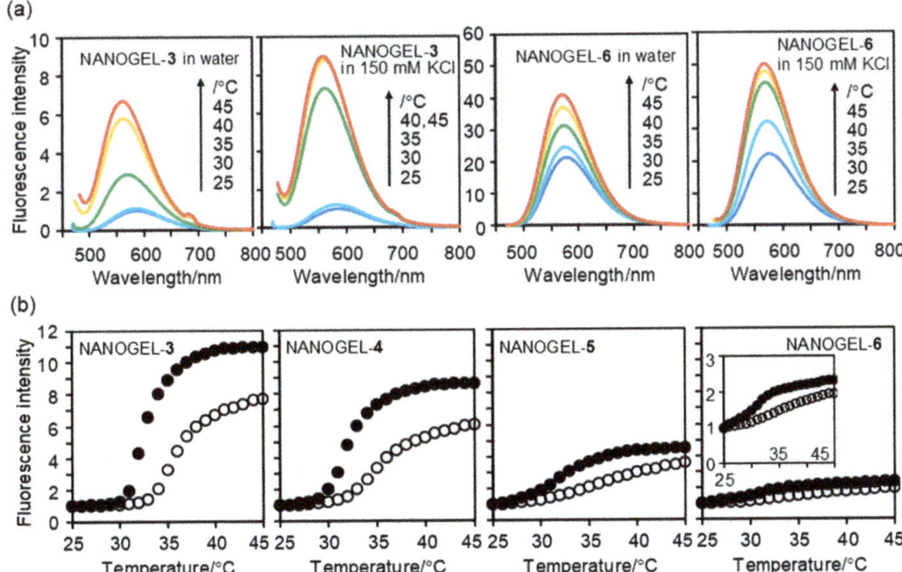

Figure 3. Fluorescence response of NANOGEL-3~6 to temperature variation. (**a**) Temperature-dependent fluorescence spectra of NANOGEL-3 and NANOGEL-6 in water and 150 mM KCl aqueous solution. The vertical units in (**a**) are identical.; (**b**) Fluorescence intensity of NANOGEL-3~6 in water (open circle) and 150 mM KCl aqueous solution (closed circle) at λ_{em} at 45 °C (see Table 2). The inset for NANOGEL-6 is vertically expanded. Concentration of NANOGEL-3~6: 0.01 w/v%; excitation: 456 nm. The shoulders at approximately 684 nm in the fluorescence spectra at 40 and 45 °C in panel (**a**) were due to the scattered excitation light.

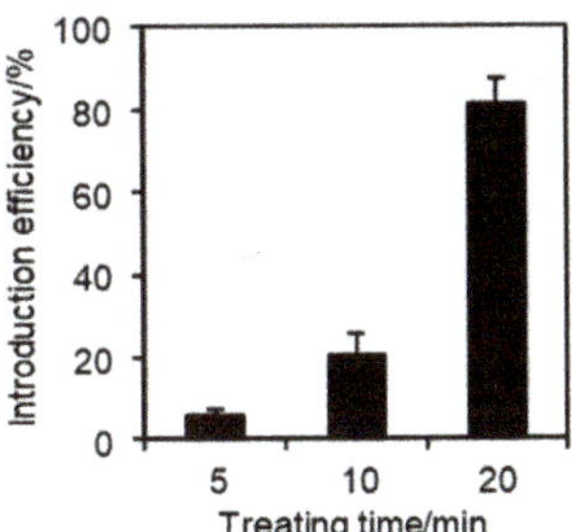

Figure 4. Effects of the incubation time on the incorporation efficiency. The HeLa cells were treated with NANOGEL-6 (0.05 w/v%) in 5 % glucose solution at 25 °C for the indicated durations. The vertical bars represent the s.d. evaluated by at least four independent experiments.

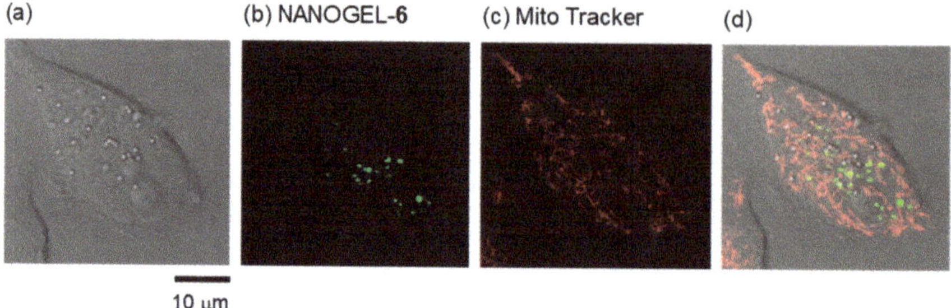

Figure 5. Cationic fluorescent nanogel thermometer NANOGEL-**6** in a live HeLa cell. (**a**) Transmitted light image; (**b**) confocal fluorescence image of NANOGEL-**6** (excitation: 458 nm; emission: 500–600 nm); (**c**) confocal fluorescence image of mitochondria visualized with Mito Tracker DR (excitation: 633 nm; emission: 645–730 nm); (**d**) merged image.

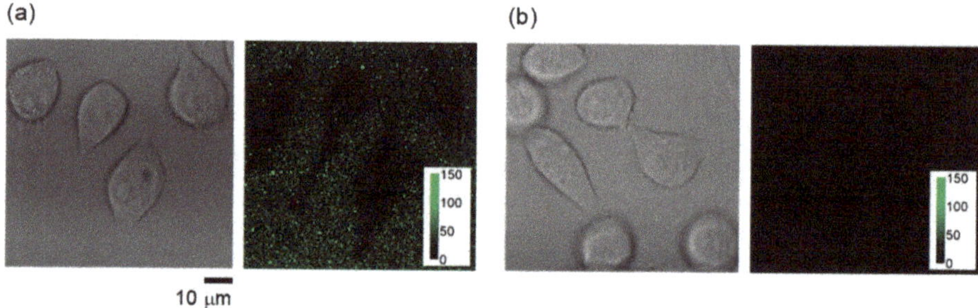

Figure 6. Reduction of background noise in the fluorescence image of HeLa cells with NANOGEL-**6** by the modified incorporation method using suspended cells. (**a**) Transmitted light image (left) and confocal fluorescence image (right, excitation: 458 nm; emission: 500–600 nm) at 37 °C when NANOGEL-**6** was introduced into adherent HeLa cells attached on a dish.; (**b**) Transmitted light image (left) and confocal fluorescence image (right, excitation: 458 nm; emission: 500–600 nm) at 37 °C when NANOGEL-**6** was introduced into suspended HeLa cells before scattering on the dish. The confocal fluorescence images were focused on the glass surface of the dish. Color bars in the fluorescence images indicate the fluorescence intensity (arbitrary unit).

3.4. Functions of Cationic Fluorescent Nanogel Thermometers inside HeLa Cells

The functions of NANOGEL-**6** in HeLa cells were examined as a representative cationic fluorescent nanogel thermometer. Similar to the response in aqueous solutions, the cationic fluorescent nanogel thermometer NANOGEL-**6** introduced into HeLa cells showed remarkable fluorescence enhancement when the temperature of the culture medium was increased (Figure 7).

Another unique property of cationic fluorescent nanogel thermometers prepared by ADIP is non-cytotoxicity. As demonstrated in Figure 8, the HeLa cells containing NANOGEL-**6** were capable of dividing in a similar manner to those without staining (i.e., control). The fluorescence response of NANOGEL-**6** to temperature variation was confirmed for the HeLa cells incubated for one day after the introduction of NANOGEL-**6** (Figure 8).

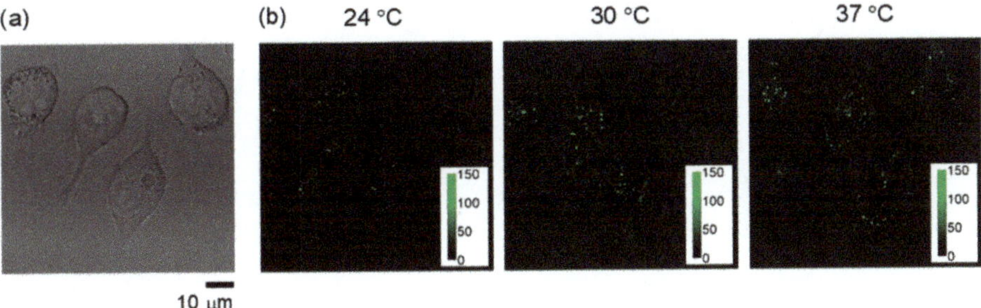

Figure 7. Fluorescence response of NANOGEL-6 in live HeLa cells. (**a**) Transmitted light image at 37 °C; (**b**) confocal fluorescence image (excitation: 458 nm; emission: 500–600 nm) at 24 °C (left), 30 °C (middle) and 37 °C (right). Color bars in the fluorescence images indicate the fluorescence intensity (arbitrary unit).

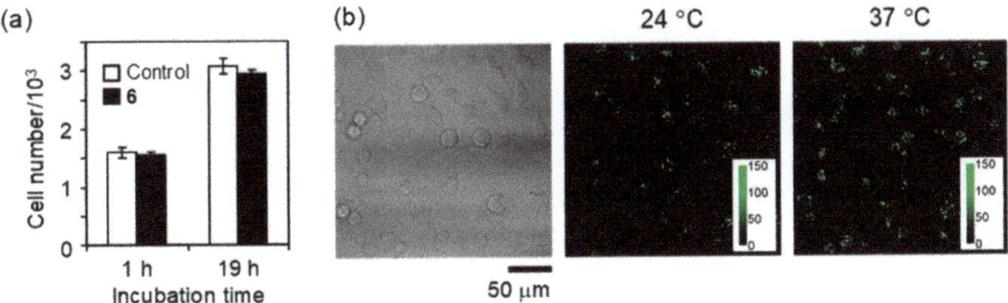

Figure 8. Non-cytotoxicity and long-term functionality of NANOGEL-6 in live HeLa cells. (**a**) Growth of HeLa cells containing NANOGEL-6 and none (control) (mean ± s.d.); (**b**) Transmitted light image at 37 °C (left) and confocal fluorescence image (excitation: 458 nm; emission: 500–600 nm) at 24 °C (middle) and 37 °C (right) at 24 h after the introduction of NANOGEL-6 into the cells. Color bars in the fluorescence images indicate the fluorescence intensity (arbitrary unit).

4. Discussion

In the present study, we demonstrated the synthesis of cationic nanogels (NANOGEL-1~6) and presented a functional assessment of cationic fluorescent nanogel thermometers (NANOGEL-3~6). One of the important advantages of fluorescent polymeric thermometers is the capability of functional integration by introducing additional units into a macromolecule. In the case of NANOGEL-3~6, the NIPAM units were assumed to be sensitive to temperature variations, whereas the DBD-AA units produced a fluorescence signal. The crosslinker MBAM units provided the robustness of nanoparticles. The positive charges on the surface, which originated from the radical initiator ADIP, enabled spontaneous entry into mammalian cells. Non-cytotoxicity was also a consequence of this cationic surface of the nanogel. Targeting to organelles by attaching a specific signal on the surface or signal normalization by introducing a second reference fluorophore [39] will be expected to improve in the future.

Table 3 summarizes both the advantages and disadvantages of the four types of fluorescent polymeric thermometers ever developed. Now, we can fill the last empty column in Table 3. In general, cationic fluorescent thermometers are more useful than anionic ones because the former can be introduced into mammalian cells under mild conditions without a microinjection technique so that a large number of samples can be treated. Fluorescent nanogel thermometers and fluorescent linear polymeric thermometers are complementary: fluorescent nanogel thermometers show low toxicity to live cells, but the spatial resolution in intracellular thermometry is low, while fluorescent linear thermometers show opposite characteristics. Recently, biological and even medical researchers have begun to utilize fluorescent polymeric thermometers in their own studies [40–42]. The cationic fluorescent nanogel thermometers developed in this study will contribute to the field of intracellular thermometry, due to their remarkable non-cytotoxicity, which will enable long-term observations.

Table 3. Comparison of fluorescent polymeric thermometers for intracellular thermometry.

Charge [1] / Morphology	Nanogel	Linear Polymer
Cationic	- moderate spatial resolution (single-cell level) [2] - spontaneously entering into cell - non-cytotoxicity	- high spatial resolution (200 nm) [3] - spontaneously entering into cell - relatively high cytotoxicity [4]
Anionic	- moderate spatial resolution (single-cell level) [2] - requiring microinjection - relatively low cytotoxicity	- high spatial resolution (200 nm) [3] - requiring microinjection - relatively high cytotoxicity [4]

[1] Neutral fluorescent polymeric thermometers cannot be utilized for intracellular thermometry, due to serious aggregation under the high ionic strength inside living cells. [2] The average temperature of a single cell can be monitored. [3] The temperature distribution inside a cell can be imaged. [4] Interfering with cell division.

Supplementary Materials: The following are available online at http://www.mdpi.com/2073-4360/11/8/1305/s1, Figure S1: 1H-NMR charts of NANOGEL-1~6, Figure S2: Effects of the KCl concentration on fluorescence responses of NANOGEL-3 as a representative. The fluorescence intensity is normalized at 25 °C. The samples were excited at 456 nm.

Author Contributions: Conceptualization, S.U.; methodology, S.U.; validation, T.H., K.K. and S.U.; formal analysis, T.H. and S.U.; investigation, T.H., K.K., N.I. and S.U.; resources, N.I. and S.U.; writing—original draft preparation, S.U.; writing—review and editing, T.H., N.I. and S.U.; visualization, T.H. and S.U.; project administration, N.I. and S.U.; funding acquisition, N.I. and S.U.

Funding: This research was funded by the JST (Development of Advanced Measurement and Analysis System) and JSPS (Grant-in-Aid for Scientific Research (B) (17H03075)).

Conflicts of Interest: The authors declare that they have no conflict of interest.

References

1. Uchiyama, S.; de Silva, A.P.; Iwai, K. Luminescent molecular thermometers. *J. Chem. Educ.* **2006**, *83*, 720–727. [CrossRef]
2. Jaque, D.; Vetrone, F. Luminescence nanothermometry. *Nanoscale* **2012**, *4*, 4301–4326. [CrossRef] [PubMed]
3. Brites, C.D.S.; Lima, P.P.; Silva, N.J.O.; Millán, A.; Amaral, V.S.; Palacio, F.; Carlos, L.D. Thermometry at the nanoscale. *Nanoscale* **2012**, *4*, 4799–4829. [CrossRef] [PubMed]
4. Wang, X.-D.; Wolfbeis, O.S.; Meier, R.J. Luminescent probes and sensors for temperature. *Chem. Soc. Rev.* **2013**, *42*, 7834–7869. [CrossRef] [PubMed]
5. Carlos, L.D.; Palacio, F. (Eds.) *Thermometry at the Nanoscale*; Royal Society of Chemistry: Cambridge, UK, 2016.
6. Uchiyama, S.; Matsumura, Y.; de Silva, A.P.; Iwai, K. Fluorescent molecular thermometers based on polymers showing temperature-induced phase transitions and labeled with polarity-responsive benzofurazans. *Anal. Chem.* **2003**, *75*, 5926–5935. [CrossRef] [PubMed]
7. Uchiyama, S.; Matsumura, Y.; de Silva, A.P.; Iwai, K. Modulation of the sensitive temperature range of fluorescent molecular thermometers based on thermoresponsive polymers. *Anal. Chem.* **2014**, *76*, 1793–1798. [CrossRef]
8. Iwai, K.; Matsumura, Y.; Uchiyama, S.; de Silva, A.P. Development of fluorescent microgel thermometers based on thermo-responsive polymers and their modulation of sensitivity range. *J. Mater. Chem.* **2005**, *15*, 2796–2800. [CrossRef]

9. Hu, J.; Liu, S. Responsive polymers for detection and sensing applications: Current status and future developments. *Macromolecules* **2010**, *43*, 8315–8330. [CrossRef]

10. Pietsch, C.; Schubert, U.S.; Hoogenboom, R. Aqueous polymeric sensors based on temperature-induced polymer phase transitions and solvatochromic dyes. *Chem. Commun.* **2011**, *47*, 8750–8765. [CrossRef]

11. Li, C.; Liu, S. Polymeric assemblies and nanoparticles with stimuli-responsive fluorescence emission characteristics. *Chem. Commun.* **2012**, *48*, 3262–3278. [CrossRef]

12. Uchiyama, S.; Gota, C.; Tsuji, T.; Inada, N. Intracellular temperature measurements with fluorescent polymeric thermometers. *Chem. Commun.* **2017**, *53*, 10976–10992. [CrossRef]

13. Graham, E.M.; Iwai, K.; Uchiyama, S.; de Silva, A.P.; Magennis, S.W.; Jones, A.C. Quantitative mapping of aqueous microfluidic temperature with sub-degree resolution using fluorescence lifetime imaging microscopy. *Lab Chip* **2010**, *10*, 1267–1273. [CrossRef]

14. Cellini, F.; Peteron, S.D.; Porfiri, M. Flow velocity and temperature sensing using thermosensitive fluorescent polymer seed particles in water. *Int. J. Smart Nano Mater.* **2017**, *8*, 232–252. [CrossRef]

15. Gota, C.; Okabe, K.; Funatsu, T.; Harada, Y.; Uchiyama, S. Hydrophilic fluorescent nanogel thermometer for intracellular thermometry. *J. Am. Chem. Soc.* **2009**, *131*, 2766–2767. [CrossRef]

16. Okabe, K.; Inada, N.; Gota, C.; Harada, Y.; Funatsu, T.; Uchiyama, S. Intracellular temperature mapping with a fluorescent polymeric thermometer and fluorescence lifetime imaging microscopy. *Nat. Commun.* **2012**, *3*, 705. [CrossRef]

17. Inada, N.; Uchiyama, S. Methods and benefits of imaging the temperature distribution inside living cells. *Imaging Med.* **2013**, *5*, 303–305. [CrossRef]

18. Qiao, J.; Chen, C.; Qi, L.; Liu, M.; Dong, P.; Jiang, Q.; Yang, X.; Mu, X.; Mao, L. Intracellular temperature sensing by a ratiometric fluorescent polymeric thermometer. *J. Mater. Chem. B* **2014**, *2*, 7544–7550. [CrossRef]

19. Hu, X.; Li, Y.; Liu, T.; Zhang, G.; Liu, S. Intracellular cascade FRET for temperature imaging of living cells with polymeric ratiometric fluorescent thermometers. *ACS Appl. Mater. Interfaces* **2015**, *7*, 15551–15560. [CrossRef]

20. Qiao, J.; Hwang, Y.-H.; Chen, C.-F.; Qi, L.; Dong, P.; Mu, X.-Y.; Kim, D.-P. Ratiometric fluorescent polymeric thermometer for thermogenesis investigation in living cells. *Anal. Chem.* **2015**, *87*, 10535–10541. [CrossRef]

21. Chen, Z.; Zhang, K.Y.; Tong, X.; Liu, Y.; Hu, C.; Liu, S.; Yu, Q.; Zhao, Q.; Huang, W. Phosphorescent polymeric thermometers for in vitro and in vivo temperature sensing with minimized background interference. *Adv. Funct. Mater.* **2016**, *26*, 4386–4396. [CrossRef]

22. Gong, D.; Cao, T.; Han, S.-C.; Zhu, X.; Iqbal, A.; Liu, W.; Qin, W.; Guo, H. Fluorescence enhancement thermoresponsive polymer luminescent sensors based on BODIPY for intracellular temperature. *Sens. Actuators B Chem.* **2017**, *252*, 577–583. [CrossRef]

23. Ding, Z.; Wang, C.; Feng, G.; Zhang, X. Thermo-responsive fluorescent polymers with diverse LCSTs for ratiometric temperature sensing through FRET. *Polymers* **2018**, *10*, 283. [CrossRef]

24. Zhang, H.; Jiang, J.; Gao, P.; Yang, T.; Zhang, K.Y.; Chen, Z.; Liu, S.; Huang, W.; Zhao, Q. Dual-emissive phosphorescent polymer probe for accurate temperature sensing in living cells and zebrafish using ratiometric and phosphorescence lifetime imaging microscopy. *ACS Appl. Mater. Interfaces* **2018**, *10*, 17542–17550. [CrossRef]

25. Qiao, J.; Chen, C.; Shangguan, D.; Mu, X.; Wang, S.; Jiang, L.; Qi, L. Simultaneous monitoring of mitochondrial temperature and ATP fluctuation using fluorescent probes in living cells. *Anal. Chem.* **2018**, *90*, 12553–12558. [CrossRef]

26. Morris, M.C.; Depollier, J.; Mery, J.; Heitz, F.; Divita, G. A peptide carrier for the delivery of biologically active proteins into mammalian cells. *Nat. Biotechnol.* **2001**, *19*, 1173–1176. [CrossRef]

27. Wadia, J.S.; Stan, R.V.; Dowdy, S.F. Transducible TAT-HA fusogenic peptide enhances escape of TAT-fusion proteins after lipid raft macropinocytosis. *Nat. Med.* **2004**, *10*, 310–315. [CrossRef]

28. Tsuji, T.; Yoshida, S.; Yoshida, A.; Uchiyama, S. Cationic fluorescent polymeric thermometers with the ability to enter yeast and mammalian cells for practical intracellular temperature measurements. *Anal. Chem.* **2013**, *85*, 9815–9823. [CrossRef]

29. Hayashi, T.; Fukuda, N.; Uchiyama, S.; Inada, N. A cell-permeable fluorescent polymeric thermometer for intracellular temperature mapping in mammalian cell lines. *PLoS ONE* **2015**, *10*, e0117677. [CrossRef]

30.   Inada, N.; Fukuda, N.; Hayashi, T.; Uchiyama, S. Temperature imaging using a cationic linear fluorescent polymeric thermometer and fluorescence lifetime imaging microscopy. *Nat. Protoc.* **2019**, *14*, 1293–1321. [CrossRef]

31.   Uchiyama, S.; Tsuji, T.; Kawamoto, K.; Okano, K.; Fukatsu, E.; Noro, T.; Ikado, K.; Yamada, S.; Shibata, Y.; Hayashi, T.; et al.   A cell-targeted non-cytotoxic fluorescent nanogel thermometer created with an imidazolium-containing cationic radical initiator. *Angew. Chem. Int. Ed.* **2018**, *57*, 5413–5417. [CrossRef]

32.   Gota, C.; Uchiyama, S.; Ohwada, T. Accurate fluorescent polymeric thermometers containing an ionic component. *Analyst* **2007**, *132*, 121–126. [CrossRef]

33.   Gota, C.; Uchiyama, S.; Yoshihara, T.; Tobita, S.; Ohwada, T. Temperature-dependent fluorescence lifetime of a fluorescent polymeric thermometer, poly(*N*-isopropylacrylamide), labeled by polarity and hydrogen bonding sensitive 4-sulfamoyl-7-aminobenzofurazan. *J. Phys. Chem. B* **2008**, *112*, 2829–2836. [CrossRef]

34.   Pelton, R. Temperature-sensitive aqueous microgels. *Adv. Colloid Interface Sci.* **2000**, *85*, 1–33. [CrossRef]

35.   Pelton, R.H.; Chibante, P. Preparation of aqueous latices with *N*-isopropylacrylamide. *Colloids Surf.* **1986**, *20*, 247–256. [CrossRef]

36.   Freitag, R.; Garret-Flaudy, F. Salt effects on the thermoprecipitation of poly(*N*-isopropylacrylamide) oligomers from aqueous solution. *Langmuir* **2002**, *18*, 3434–3440. [CrossRef]

37.   Burba, C.M.; Carter, S.M.; Meyer, K.J.; Rice, C.V. Salt effects on poly(*N*-isopropylacrylamide) phase transition thermodynamics from NMR spectroscopy. *J. Phys. Chem. B* **2008**, *112*, 10399–10404. [CrossRef]

38.   Lodish, H.; Berk, A.; Kaiser, C.A.; Krieger, M.; Bretscher, A.; Ploegh, H.; Amon, A.; Scott, M.P. *Molecular Cell Biology*, 7th ed.; W. H. Freeman: New York, NY, USA, 2013; p. 485.

39.   Uchiyama, S.; Tsuji, T.; Ikado, K.; Yoshida, A.; Kawamoto, K.; Hayashi, T.; Inada, N. A cationic fluorescent polymeric thermometer for the ratiometric sensing of intracellular temperature. *Analyst* **2015**, *140*, 4498–4506. [CrossRef]

40.   Tsuji, T.; Ikado, K.; Koizumi, H.; Uchiyama, S.; Kajimoto, K. Difference in intracellular temperature rise between matured and precursor brown adipocytes in response to uncoupler and β-adrenergic agonist stimuli. *Sci. Rep.* **2017**, *7*, 12889. [CrossRef]

41.   Kimura, H.; Nagoshi, T.; Yoshii, A.; Kashiwagi, Y.; Tanaka, Y.; Ito, K.; Yoshino, T.; Tanaka, T.D.; Yoshimura, M. The thermogenic actions of natriuretic peptide in brown adipocytes: The direct measurement of the intracellular temperature using a fluorescent thermoprobe. *Sci. Rep.* **2017**, *7*, 12978. [CrossRef]

42.   Hoshi, Y.; Okabe, K.; Shibasaki, K.; Funatsu, T.; Matsuki, N.; Ikegaya, Y.; Koyama, R. Ischemic brain injury leads to brain edema via hyperthermia-induced TRPV4 activation. *J. Neurosci.* **2018**, *38*, 5700–5709. [CrossRef]

Article

A Cu(II) Indicator Platform Based on Cu(II) Induced Swelling that Changes the Extent of Fluorescein Self-Quenching

Feifei Wang †, Roy P. Planalp and W. Rudolf Seitz *

Department of Chemistry, University of New Hampshire, Durham, NH 03824, USA;
fsq5@wildcats.unh.edu (F.W.); Roy.Planalp@unh.edu (R.P.P.)
* Correspondence: rudi.seitz@unh.edu; Tel.: +1-603-862-2408
† Current address: Department of Chemical and Materials Engineering, University of Alberta, Edmonton, AB
 T6G 1H9, Canada; fwang3@ualberta.ca; Guangzhou Medical University, Guangzhou 511436, China.

Received: 26 June 2019; Accepted: 13 November 2019; Published: 25 November 2019

Abstract: In this study, we established a new fluorescent indicator platform. The responsive element consists of poly(N-isopropylacrylamide) nanospheres that include small percentages of fluorescein and a ligand, anilinodiacetate (phenylIDA). Nanosphere diameters were determined to be in the range from 50 to 90 nm by scanning electron microscopy. They were entrapped in a polyacrylamide gel to prevent nanosphere aggregation. At pH 6, the ligand is negatively charged in the absence of metal ions. Charge-charge repulsion causes the nanosphere to swell. Dynamic light scattering measurements show that these nanospheres do not shrink and aggregate at high temperature. Cu(II) binding neutralizes the charge causing the particles to shrink. This brings fluoresceins closer together, increasing the degree of self-quenching. The intensity decreases by 30% as Cu(II) concentration increases. To rule out the possibility that the observed decrease in intensity was due to Cu(II) quenching of fluorescence, we also added Zn(II) and observed a decrease in intensity. This approach can be adapted to sense different metal ions and different concentrations of Cu(II) by changing the ligand.

Keywords: self-quenching; pNIPAM; cross-linked nanoparticles; copper; PA gel

---

1. Introduction

Cu(II) is an active producer of oxidative stress for both plants [1–3] and animals [4]. Human uptake of Cu is usually in the range of 0.6–1.6 mg per day [5]. Excess uptake of Cu in human beings is related to cancer and aging [5]. It is also reported to be related to diseases of the nervous system such as Alzheimer's, Menkes, and Wilson diseases [6,7]. Because of its biological effects, control of Cu contamination is an important aspect of environmental protection.

The biotic ligand model (BLM) considers the interactions of all parameters in a natural system to predict the bioavailability of metal ions [8,9]. Bioavailable Cu concentrations predicted by the BLM correlate well with measured Cu LC50s. Total Cu does not correlate well with actual toxicity [9]. However, the BLM is based on an indirect measurement of bioavailable Cu(II), that is, it is based on measurements of organic carbon, pH, other metal ions and several other parameters. At present there is no viable method for measuring bioavailable Cu(II) directly.

Several studies report ligands that change fluorescence when they bind Cu(II). These can potentially be used to measure bioavailable Cu(II). There are some fluorogenic ligands that have increased fluorescence when they bind Cu(II) [10–13]. However, some of them can only be applied in organic solvents such as THF [10] and acetonitrile [11,14], which are not appropriate for the detection of bioavailable Cu(II) in water systems. Low sensitivity, long response times, poor selectivity and ligands with inappropriate Cu(II)-complex formation constants are other problems that render reported ligands unsuitable for Cu(II) monitoring.

Many other fluorescent sensors have decreased or "turn off" fluorescence upon Cu(II) binding due to Cu quenching of the fluorogenic ligands [15–17]. The strategy of developing a fluorogenic ligand that is capable of measuring bioavailable Cu(II) has yet to succeed. Furthermore, even if successful, it would only be applicable to Cu(II).

We prefer to base detection on metal ion induced changes in a water-soluble polymer conformation detected via fluorescence. This approach separates the fluorophore from the metal, rendering it less subject to metal ion quenching, a frequent issue with Cu(II). Furthermore, the selectivity of this approach can be modified by changing the ligand while keeping the rest of the indicator platform. Du et al. synthesized a ratiometric fluorescent Cu(II) indicator platform [18]. Cu(II) binding neutralizes the charge on the ligand, which causes poly(N-isopropylacrylamide) (pNIPAM) to change conformation. This in turn affects the environment of a dansyl comonomer [18]. The indicator developed by Yao et al. [19] is based on fluorescence resonance energy transfer (FRET) [20]. Cu(II) binding introduces positive charge repulsion which separates copolymer strands disrupting FRET. However, neither of these systems has the required sensitivity for environmental Cu(II) measurements. In Du et al.'s indicator, the fluorophore utilized is not that efficient, and for Yao et al.'s indicator, the limit of detection is not low enough. Osambo et al. demonstrated an indicator platform based on changes in FRET accompanying metal ion induced nanoparticle swelling [21]. However, the excitation wavelength is too short to be practical. We also synthesized ratiometric indicators with both donor and acceptor fluorophores on the same polymer chain, but the signal changes with time due to slow polymer untangling. Therefore, our goal is to demonstrate an indicator platform that is both stable and sensitive, and involves wavelengths in the visible spectrum.

The indicator discussed in this paper is based on cross-linked pNIPAM nanoparticles. A negatively charged ligand is used to make the nanoparticle swell in the absence of metal ions. Addition of metal ions neutralizes the negative charge causing the nanoparticle to shrink. This results in a change in fluorescein concentration per unit volume. The fluorescence signal of fluorescein decreases with increasing concentration due to self-quenching when the concentration is above a critical concentration [22]. Our approach is illustrated schematically in Figure 1. However, nanoparticles alone can undergo self-agglomeration, which affects the volume change, and may also block the Cu(II) binding sites. In order to avoid agglomeration, the nanoparticles were embedded in a polyacrylamide gel. The PA gel increases the stability of the single nanoparticles. This approach makes it possible to synthesize particles with a wider range of sizes.

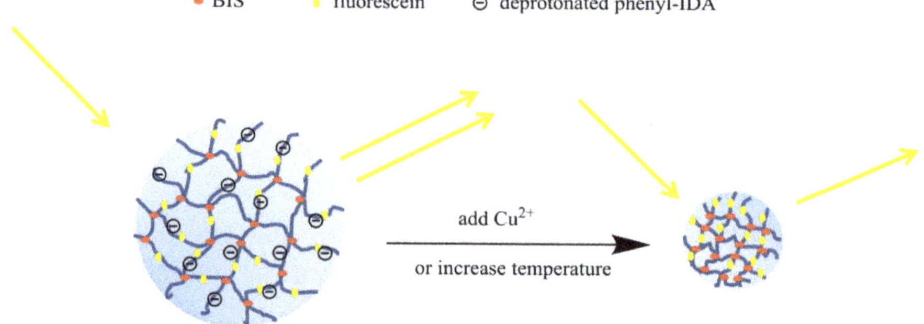

Figure 1. Sensing mechanism of self-quenching poly(N-isopropylacrylamide) (pNIPAM) nanoparticles. Because of the negative charges on the ligand, the nanoparticles swell. When Cu(II) bind to the ligand, the charge neutralization results in less swelling of the nanoparticles, hence the fluorescence intensity also decreases.

The data we show here are for Cu(II). However, the approach is general because binding of other metal ions will also change the charge on the polymer backbone leading to swelling or shrinking depending on whether the charge increases or decreases. Thus, the indicator platform we demonstrate here is applicable to other metal ions depending on the particular ligand that is incorporated into the polymer.

2. Experimental Materials

Materials: Sodium dodecyl sulfate (SDS), *N*-isopropyl acrylamide (NIPAM), *N,N′*-Methylenebisacrylamide (BIS), fluorescein *o*-acrylate, potassium persulfate (KPS), acrylamide, ammonium persulfate (APS), *N,N,N′,N′*-Tetramethylethylenediamine (TEMED), Copper (II) nitrate trihydrate, and Zinc (II) nitrate hexahydrate were purchased from Sigma-Aldrich. Aqueous solutions were prepared from doubly distilled water from a Corning Mega-Pure distillation apparatus. Dialysis tubing with a molecular weight cut-off (MWCO) of 3.5–5 kDa was purchased from Spectrum Labs.

Equipment: Fluorescence responses were measured using the scan mode on a Varian Cary Eclipse fluorometer equipped with a Peltier thermostatted single cell holder. Scanning Electron Microscopy (SEM) was performed on a Tescan Lyra3 GMU Focused Ion Beam (FIB) SEM. A Branson model 1800 sonicator was used for reagent dissolution and sonication. A Buchi RE111 Rotavapor was used to evaporate solvents. Separation of precipitated polymer from the solution was performed on a Beckman GP centrifuge (8000 rpm) or an Eppendorf centrifuge 5415 C (14,000 rpm). A FreeZone Plus 2.5 Liter Cascade Benchtop Freeze Dry System was used to lyophilize samples.

3. Procedures

3.1. The Synthesis of Anilinodiacetic Acid Ligand (Phenyl-IDA) in Ester Form (Tert-Butoxycarbonyl Methyl-(3-Vinyl-Phenyl)-Amino) Acetic Acid Tert-Butyl Ester

The esterified ligand was synthesized according to previous literature [21]. The structure is shown in Figure 2. The esterified ligand (compound **a**) was copolymerized into the indicator platform and then hydrolyzed to the acid form (compound **b**). It was deprotonated (compound **c**) in pH 6 buffer.

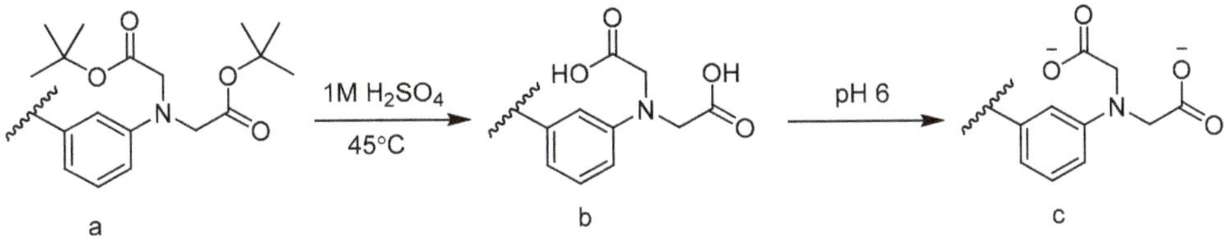

Figure 2. Hydrolysis and deprotonating of phenyl-IDA ligand: (**a**) Phenyl-IDA ester; (**b**) Phenyl-IDA; (**c**) Deprotonating phenyl-IDA.

3.2. Emulsion Polymerization of Self-Quenching pNIPAM Nanoparticles

Surfactant, sodium dodecyl sulfate (SDS) (0.14 g, 0.5 mmol) was added to a round bottom flask containing 45 mL deionized water. 1.4 g (12.3 mmol) NIPAM, 0.038 g (0.246 mmol) *N,N′*-Methylenebisacrylamide (BIS), 0.005 g fluorescein *o*-acrylate and 0.095 g (0.2 mmol) phenyl-IDA ligand in ester form were added to the mixture under stirring. The solution was stirred and degassed with N_2 for 30 min. Then the flask was placed in an oil bath at 70 °C. 0.05 g (0.2 mmol) potassium persulfate (KPS) was dissolved in 5 mL DI water, degassed for 5 min and injected into the heated reaction mixture with a syringe. After 6 h, polymerization was quenched by exposure to air. The mixture was dialyzed against deionized water using dialysis tubing with a 3.5–5 kDa molecular weight cut-off with stirring. The water was changed twice daily. After 7 days, the mixture was lyophilized to obtain a pale yellow powder.

3.3. Removal of the Ester to Produce the Ligand

The lyophilized nanoparticles with phenyl-IDA were suspended in 50 mL 1 M H_2SO_4 solution in a round bottom flask. Acidification (Figure 2) was conducted in an oil bath at 50 °C with stirring for 8 h. Then the mixture was filtered through a glass frit and rinsed with water several times. Another round of acidification was conducted to make sure all of the ligand was hydrolyzed to the acid form. The product was lyophilized and then characterized using a fluorometer and SEM.

3.4. Self-Quenching Cross-Linked Nanoparticles Embedded in the PA Gel

Acrylamide (0.475 g), BIS (0.025 g), 250 μL 20× Tris/Borate/EDTA (TBE) buffer and dry pNIPAM particles obtained from the above procedure were added to the 20 mL vial. DI water was added to the vial to bring the volume to 3.83 mL. The solution was degassed for 15 min. Then 20 μL 10% (w/v) APSand 4 μL TEMED were added to the mixture and the solution was gently but thoroughly swirled. The gel solution was quickly and gently introduced into the mold and covered to minimize exposure to oxygen. Gel polymerization proceeded for 2 h. The obtained PA thin film (1 mm) was placed in DI water for 2 days and rinsed with DI water several times in order to remove unreacted monomer and TBE buffer.

3.5. Fluorescence Measurement of Nanoparticles Alone

Fluorescence was measured in a 3 mL polystyrene cuvette with 0.1 M pH 6 3-(*N*-morpholino) propanesulfonic acid (MOPS) buffer. This pH keeps the phenyl-IDA ligand in its deprotonated form (compound c in Figure 2) since the pKa_1 of *N*-Phenyliminodiacetic acid is 2.41 and the pKa_2 is 5.05 [23]. This pH also prevents $Cu(OH)_2$ formation. The particles were suspended in the buffer. The concentration of phenyl-IDA ligand is 10^{-5} M in the cuvette based on the calculations using feed amounts. Cu(II) ions were added from a $Cu(NO_3)_2$ stock solution to the cuvette with a micro pipette in μL, to avoid a significant volume change. Cu(II) concentrations were increased from pCu 7 to pCu 4. Zn(II) ions were added from a $Zn(NO_3)_2$ stock solution to the cuvette with a micro pipette in μL. Since Zn(II) has a lower formation constant with phenyl-IDA, the concentration of Zn(II) was directly brought up to 10^{-4} M to see the response. Both excitation and emission slit widths were 10 nm. The sample was excited at 450 nm. The particles show fluorescein emission at 514 nm.

3.6. Fluorescence Measurement of Nanoparticles Embedded in the PA Gel

The PA gel formed with self-quenching pNIPAM nanoparticles was immersed in DI water and then rinsed several times in order to remove particles that are not immobilized in the gel. The thin film (1 mm) of PA gel sample was fixed using a clean polytetrafluoroethylene (PTFE) holder with a hole so that the gel can completely cover the hole. The location of the hole was adjusted to let incident light go through the gel sample in the fluorometer (Figure 3). MOPS buffer of 0.1 M pH 6 was added to the cuvette. The excitation wavelength was 450 nm. The slit widths were 5 nm. The sample emitted at 514 nm. The theoretical concentration of ligand phenyl-IDA in the cuvette is 10^{-5} M.

Figure 3. PTFE holder with gel sample at the hole. The height of the hole can be adjusted in order to let incident light go through it and let the detector receive the fluorescence.

4. Results and Discussion

4.1. Morphology of the Nanoparticles

Scanning electron microscopy (SEM) was used to study the morphological features of self-quenching pNIPAM particles. The sample was taken from the stock suspension of 0.1 g/L prepared with the lyophilized powder. A platinum sputtering layer was coated onto the sample stub after the sample was dried.

The size of the dry particles ranges from 40 to 90 nm (Figure 4). The surface of the particles is not that smooth, and there is some deformation in shape. Some of this is the result of two particles merging to form a larger particle. There is not much agglomeration of nanoparticles, probably because of the dilute solution.

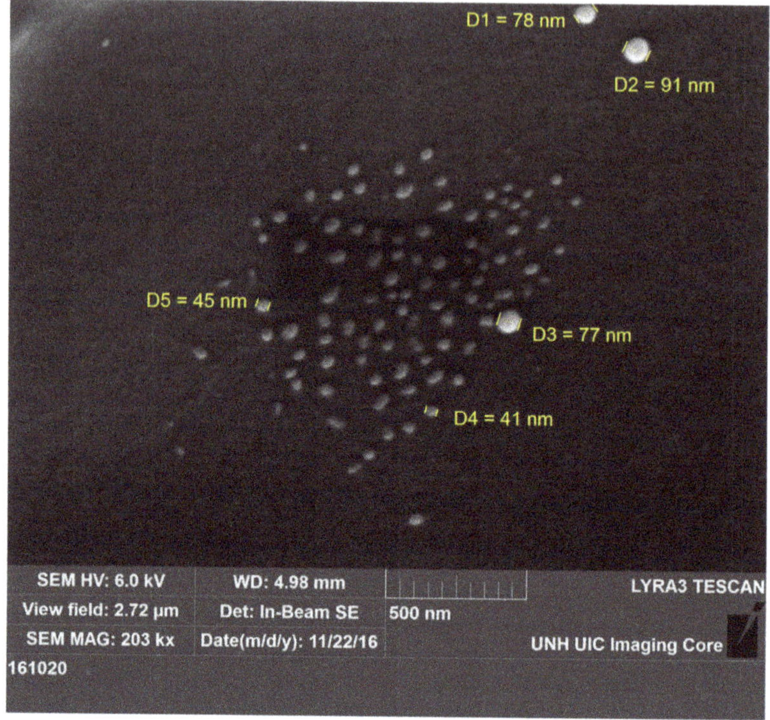

Figure 4. SEM image of self-quenching pNIPAM nanoparticles. The sizes of the particles range from 40 to 90 nm.

4.2. Fluoresence Study of the Self-Quenching pNIPAM Nanoparticles Alone

4.2.1. Thermal Response

PNIPAM has reverse solubility upon heating. The thermal response is characterized by a lower critical solution temperature (LCST), where the pNIPAM abruptly transitions from hydrophilic to hydrophobic. This occurs because hydrogen bonding gets weaker with increasing temperature, reaching a point where it is no longer able to prevent hydrophobic collapse [24]. The LCST of pNIPAM is in the range of 30–35 °C [25]. When the polymer chains are cross-linked, the polymer is swollen with water below the LCST and collapses, excluding water above the LCST. Our pNIPAM nanoparticles are prepared with 2 mol% BIS, 2 mol% phenyl-IDA, 0.2% (w/v) fluorescein *o*-acrylate and NIPAM, by emulsion polymerization in order to control the diameter of the particles. The concentration of fluorescein in the particles was set at 0.2 g per 100 g of nanoparticles, which is the critical concentration for self-quenching [22]. This is based on the assumptions that the density of the nanoparticles is 1 g/mL. Therefore we can get high fluorescence intensity and significant self-quenching simultaneously. The fluorescence signal decreases with increasing temperature from 25 °C to 46 °C (Figure 5). This thermal response could be from both thermal quenching and particle shrinking that leads to more self-quenching. When the temperature increases, pNIPAM shrinks, causing the fluorescein to be closer to each other, which leads to self-quenching. Our data show only a small decrease in fluorescence with temperature with no sign of a large change that would be indicative of the thermal phase transition. From this we conclude that the presence of phenyl IDA in the polymer chain is preventing hydrophobic collapse of the cross-linked nanoparticles. This is consistent with our expectation that we have removed the t-butyl groups leaving carboxylates that are deprotonated at the pH of this measurement. The literature pKa_1 of N-Phenyliminodiacetic acid is 2.41 while pKa_2 is 5.05 [23], so in the pH 6 MOPS buffer the phenyl IDA ligand is deprotonated to produce negative charges. Based on the acid-base equilibrium, the fraction of the deprotonated form of ligand can be estimated using Equation (1):

$$A = [L^{2-}]/c_L = Ka_1 Ka_2/([H^+]^2 + Ka_1[H^+] + Ka_1 Ka_2) \tag{1}$$

where α is the fraction of deprotonated form of phenyl IDA ligand that has two charges, c_L is the total concentration of ligand. The fraction α of phenyl IDA with two charges is 89.9% in pH 6 buffers. That means the rest of the ligand should have one charge. This calculation assumes that the published solution pKa values for phenyl IDA apply to phenyl IDA that has been copolymerized with NIPAM. In practice, this is unlikely. However, given the small value of pKa_1, we feel safe in assuming that essentially all the immobilized ligand is charged.

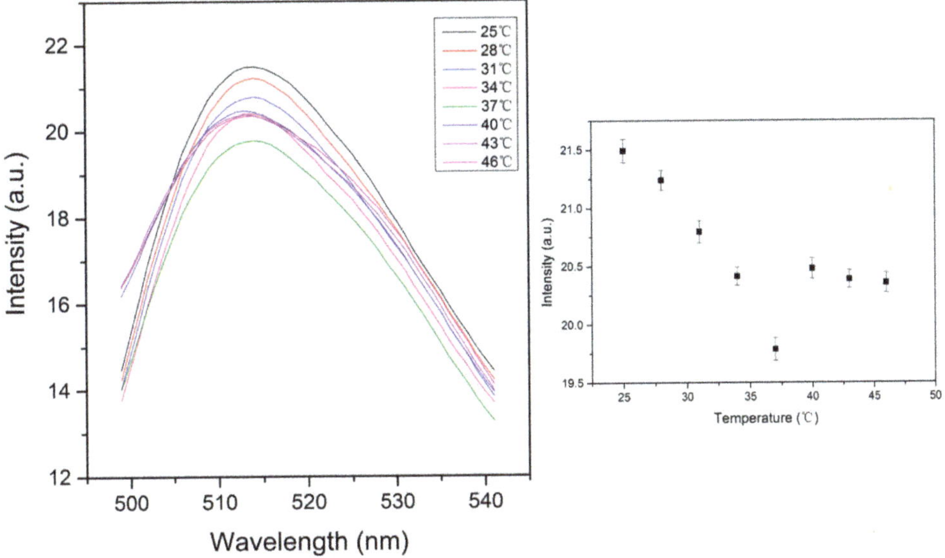

Figure 5. Thermal response of self-quenching pNIPAM nanoparticles from 25 °C to 46 °C. The intensity values in the right figure were taken from the peak intensity in the left figure. (ex: 450 nm, slit widths: 10 nm).

4.2.2. Response to Cu(II)

Cu(II) quenches fluorescence when bound to a fluorophore because it is paramagnetic [26]. This system avoids the problem by exploiting a polymer conformational change that leads to a change in the emission from the fluorophore. The synthesized self-quenching pNIPAM nanoparticles exhibit decreased fluorescence with increasing Cu(II) concentration, below and above the LCST (25 °C and 46 °C) (Figure 6). This occurs when Cu(II) binding neutralizes the negative charges, causing particles to shrink and the self-quenching of fluorescein to increase. There are big changes around pCu = 5.3. This occurs when Cu(II) concentration is close to the ligand concentration, which is estimated to be 10^{-5} M in the buffer based on initial amounts in the polymerization. This is sufficient to neutralize all the charges on the polymer. The log Kf for Cu(II)-phenyl-IDA is 6.37, large enough so that essentially all the Cu(II) is bound to ligand when the concentrations of both are close to 10^{-5} M. When the ligand is on the polymer, water has less access to the charges on the ligand to stabilize them. Because of this we expect the log Kf for Cu(II) phenyl-IDA to be somewhat larger than 6.37 for the polymer bound ligand.

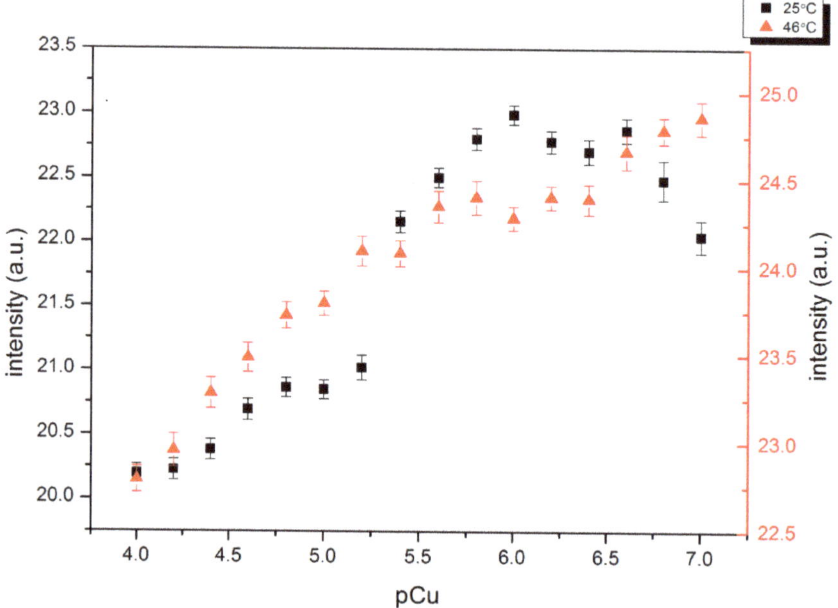

Figure 6. Cu(II) response of self-quenching pNIPAM nanoparticles alone at 25 °C and 46 °C. pCu = −log [Cu^{2+}]. Smaller pCu represents higher Cu(II) concentration.

The concentration of the nanoparticles in the cuvette was 0.01 g/L in order to get an easily measurable signal while still avoiding particle aggregation. The measurements of the response to Cu(II) were taken within 10 min, and at either temperature, there was no visible aggregation. This relative stability can be ascribed to the negative charges on the ligands. It is expected that the uncharged nanoparticles may aggregate when the temperature is above the LCST.

4.3. Fluorescence Study of the Self-Quenching pNIPAM Nanoparticles Embedded in the PA Gel

The intensity change shown in Figure 6 is not as large as we expected. The possible cause is that there is minor aggregation, which inhibits the volume change upon Cu(II) binding. In order to prevent aggregation, an embedding medium was developed. Polyacrylamide (PA) is a cross-linked gel that is commonly used for polyacrylamide gel electrophoresis. PA gel is transparent, relatively chemically inert and the pore size can be controlled [27]. These features make it not only a good support in electrophoresis, but also a potential embedding medium for biological functional units like enzymes, antibodies, and synthetic agents or particles [28,29]. The self-quenching pNIPAM nanoparticles were embedded in a PA gel in order to prevent possible particle aggregation and increase the signal change.

The pore size of the gel was controlled to 3.4–34 nm by choosing the appropriate total monomer concentration and weight percentage of cross-linker [30]. The pNIPAM nanoparticles were mixed with the gel solution before polymerization and were trapped in the gel after gel formation.

The fluorescence intensity of the embedded nanoparticles continuously decreases with increasing temperature (Figure 7). The higher signal is due to less aggregation. This change is much larger than that of pNIPAM particles alone (Figure 5). This is due to the higher stability of particles in the PA gel where they cannot aggregate, which affects the signal. The thermal phase transition of pNIPAM can be observed as the slight slope change around 37 °C in the graph of fluorescence intensity vs. temperature (Figure 7). This is consistent with the response of pNIPAM particles alone (Figure 5). The decrease in the fluorescence intensity is very large, from 51 to 36 a.u as the temperature increases from 25 to 46 °C. In a separate experiment we determined that the fluorescence of fluorescein decreases by approximately 1.1% per degree C for fluorescein in pH 6 buffer. This means that the decrease observed in Figure 7 is greater than the temperature effect on fluorescein and, therefore, presumably involves a degree of increased selfquenching due to particle shrinking. The change is 30%, much larger than 5%, the change for particles only, as shown in Figure 5.

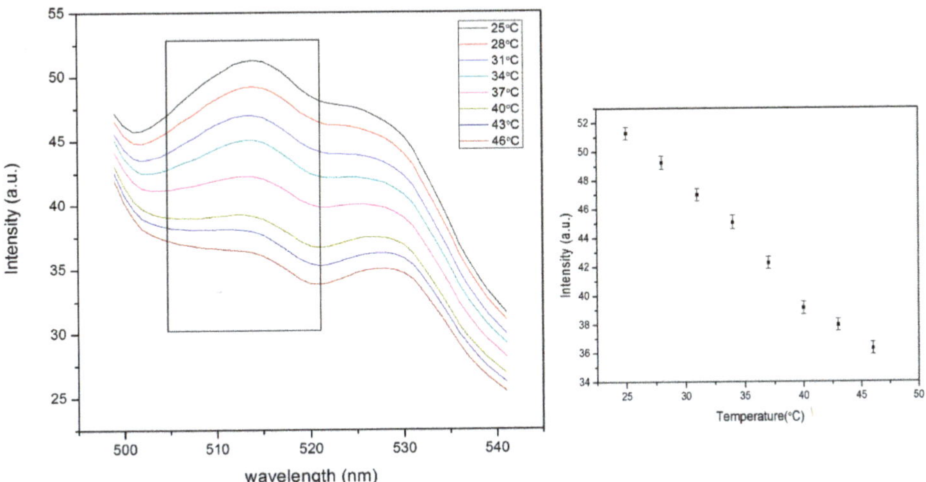

Figure 7. Thermal response of self-quenching pNIPAM nanoparticles embedded in the polyacrylamide (PA) gel from 25 °C to 46 °C. The intensity values in the figure on the right were taken from the peak intensity at 515 nm in the figure on the left. The tailing before 500 nm is from the background scattering of the PA gel. The peak at 530 nm is from the fluorescence of PTFE holder. (ex: 450 nm, slit widths: 5 nm).

The Cu(II) induced response of the embedded pNIPAM nanoparticles at 25 °C is also much larger than that of the pNIPAM particles alone. The fluorescence intensity drops from 44 to 33 a.u. This is due to the neutralization of the negative charges of phenyl-IDA by Cu(II) binding (Figure 8). Charge neutralization allows the particles to shrink, thus causing more fluorescein self-quenching.

In a control experiment, we determined that Cu(II) concentrations as high as 0.001 M did not quench fluorescein fluorescence when Cu(II) was added to solutions of fluorescein in pH 6 buffer. This rules out the possibility of quenching by solution phase Cu(II) as an explanation for the observed intensity decrease.

The shapes of the fluorescence spectra of PA gel supported nanoparticles (Figures 7 and 8) are different from those of nanoparticles alone (Figure 5). This is due to some background scattering from the gel. The excitation wavelength was fixed at 450 nm in order to avoid the overlapping of the scattering peak from water with the fluorescein peak.

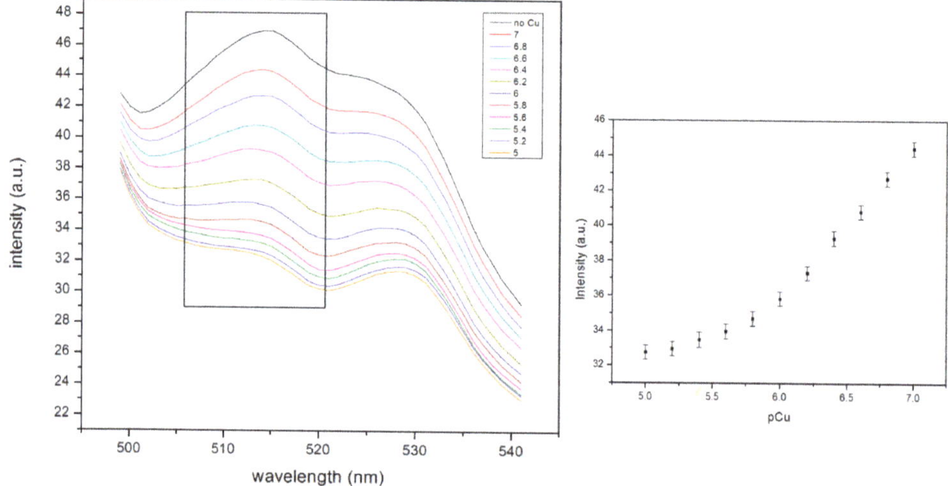

Figure 8. Cu(II) response of self-quenching pNIPAM nanoparticles in the PA gel at 25 °C. The intensity values in the figure on the right is taken from the peak intensity at 515 nm in the figure on the left (ex: 450 nm, slit widths: 5 nm).

The metal ion response is slow because diffusion of metal ions into a PA gel takes more time than binding to the pNIPAM particle alone. Because the pore size (3.4~34 nm) of the PA gel is much larger than Cu (II), and PA has excellent hydrophilicity, the absorption of Cu(II) should be rapid. If the pINPAM particles in the gel are evenly distributed, the metal ions are absorbed into the gel and bind to the ligand. The binding of Cu(II) to the ligand may also be slowed down by the interaction with the gel. The time it takes to reach equilibrium is hard to estimate. Therefore the data were collected when the signal stabilized. The time it took to stabilize was about 10 min for each set of data.

The high signal intensity in Figure 8 is due to the stability provided by the PA gel. The nanoparticles stay apart, leading to less self-quenching from adjacent particles. The concentration of the particles in the gel can be decreased to a much lower level than 0.01 g/L with increased slit width. This is important for the application to environmental monitoring since the concentration of the indicator needs be lowered so that the presence of ligand does not perturb the natural system.

4.4. Zn(II) Responses of Self-Quenching pNIPAM Nanoparticles Alone and Particles Embedded in the PA Gel

To confirm that the decrease in fluorescence intensity is mainly the result of the volume change rather than Cu(II) quenching, fluorescence measurements with Zn(II) addition were conducted. Unlike Cu(II), Zn(II) does not quench fluorescence. Zn(II) ions were added from a $Zn(NO_3)_2$ stock solution to the cuvette. Both the pNIPAM nanoparticles alone and the embedded pNIPAM nanoparticles show decreased fluorescence signal upon Zn(II) addition (Figure 9). This response confirms that the metal ions added to the particle cause a volume change, while some of the observed response to Cu(II) may be due to quenching. We also see a decrease in intensity due to increased fluorescein self-quenching.

The literature formation constant for Cu(II)-phenyl IDA is log K_f = 6.37 while that for Zn(II)-phenyl IDA is log K_f = 3.53 [31]. The ligand phenyl-IDA has much lower affinity towards Zn(II). With the same level of indicator present, the Zn(II) addition may not have the same effect as Cu(II) addition because the Zn(II) does not completely bind to the ligand. The decrease in intensity with Cu(II) (Figures 6 and 7) is larger than with Zn(II) (Figure 9).

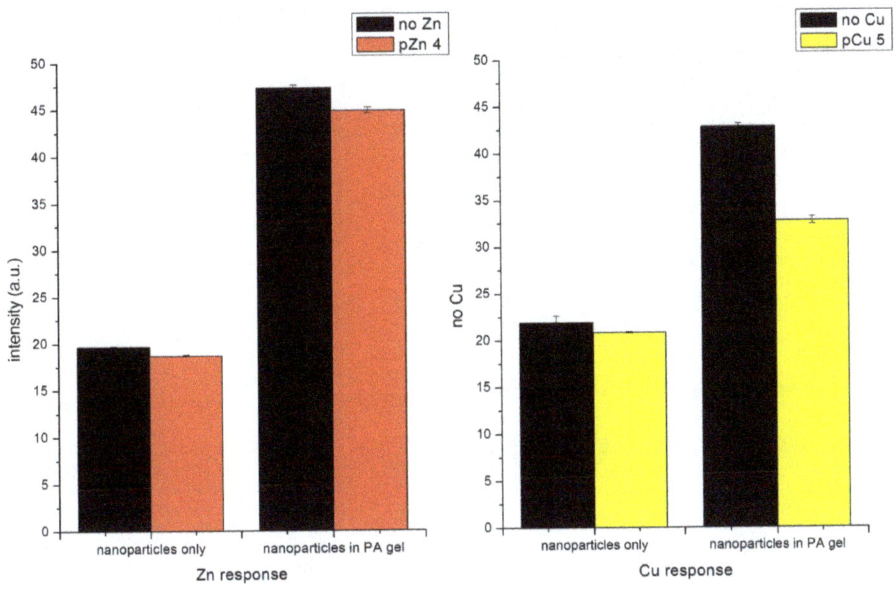

Figure 9. Comparison of Zn(II) response and Cu(II) response of self-quenching pNIPAM nanoparticles alone and particles embedded in the PA gel.

5. Conclusions

A fluorescent metal ion indicator based on cross-linked pNIPAM nanoparticles was synthesized by copolymerizing fluorescein, ligand phenyl-IDA and NIPAM and cross-linker. The negative charges on the ligand make the nanoparticle swell. When Cu(II) ions are added to the system, they bind to the ligand and neutralize the negative charges, decreasing the swelling extent. The shrinkage of the particles leads to a shorter distance between adjacent fluoresceins, thus increasing self-quenching. The fluorescence intensity decreases with increasing Cu(II) concentration. Embedding the nanoparticles in the PA gel causes a larger change in the quenching due to shrinkage.

This indicator platform has several advantages over other platforms: (a) This indicator responds to Cu(II), which normally quenches fluorescence. The separation of ligand and fluorophores makes the binding site of Cu(II) separate from the fluorophore, thus decreasing the paramagnetic quenching effect on fluorescence. (b) The sensitivity and selectivity of the indicator can be modified by utilizing different ligands without changing the excitation wavelength. Theoretically, the indicator platform responds to all metal ions with appropriate ligands. (c) It was reported that a cross-linked structure improves the thermal stability of polymers [32], and in this indicator platform, it also helps to solve the problem of the stability of free-floating or end-grafted polymer chains, since the untangling of the polymer chains is not an issue. The nanoparticles are also easy to purify and recycle by centrifugation. (d) The self-quenching pNIPAM nanoparticles were embedded in a PA gel in order to prevent possible particle aggregation and increase the signal change.

The ultimate purpose of the indicator is to measure bioavailable metal ions in the environment. Future work may involve the application of a reference fluorophore in the gel to obtain ratiometric measurements, which can reduce error due to instrumental drift and simplify calibration. Meanwhile, we also plan to try an array with a donor fluorophore on one end of the pNIPAM chain and an acceptor fluorophore on the other end, to get a much larger signal change. The sensitivity can also be improved by using fluorophores with high quantum yield, and the limit of detection can be improved with high Cu(II) affinity ligands, which enables accurate readouts with low indicator concentration. This is beneficial to environmental monitoring since the equilibria in the environment would not be disturbed by the indicator. Moreover, the indicator is expected to selectively respond to target metal ions without the interference of other metals when metal-selective ligand systems are incorporated into the polymer.

Author Contributions: F.W. performed most of the experiments described in this manuscript. The manuscript was largely written by W.R.S. with input from both coauthors. R.P.P. consulted on the research and developed the ligand synthesis.

Funding: This research received no external funding.

Acknowledgments: The authors thank Tianyu Ren for measuring the temperature coefficient for fluorescein in buffer and for confirming that 0.001 M Cu(II) does not quench the fluorescence of fluorescein in pH 6 buffer.

Conflicts of Interest: The authors declare no conflict of interest.

References

1. Gallego, S.M.; Benavídes, M.P.; Tomaro, M.L. Effect of heavy metal ion excess on sunflower leaves: Evidence for involvement of oxidative stress. *Plant Sci.* **1996**, *121*, 151–159. [CrossRef]
2. Mahmood, T.; Islam, K.R. Response of Rice Seedlings to Copper Toxicity and Acidity. *J. Plant Nutr.* **2006**, *29*, 943–957. [CrossRef]
3. Assche, F.; Clijsters, H. Effects of metals on enzyme activity in plants. *Plant Cell Environ.* **1990**, *13*, 195–206. [CrossRef]
4. Zimmer, A.M.; Barcarolli, I.F.; Wood, C.M.; Bianchini, A. Waterborne copper exposure inhibits ammonia excretion and branchial carbonic anhydrase activity in euryhaline guppies acclimated to both fresh water and sea water. *Aquat. Toxicol.* **2012**, *122–123*, 172–180. [CrossRef] [PubMed]
5. Peña, M.M.O.; Lee, J.; Thiele, D.J. A Delicate Balance: Homeostatic Control of Copper Uptake and Distribution. *J. Nutr.* **1999**, *129*, 1251–1260. [CrossRef]
6. Gaetke, L.M.; Chow, C.K. Copper toxicity, oxidative stress, and antioxidant nutrients. *Toxicology* **2003**, *189*, 147–163. [CrossRef]
7. Strausak, D.; Mercer, J.F.B.; Dieter, H.H.; Stremmel, W.; Multhaup, G. Copper in disorders with neurological symptoms: Alzheimer's, Menkes, and Wilson diseases. *Brain Res. Bull.* **2001**, *55*, 175–185. [CrossRef]
8. Paquin, P.R.; Gorsuch, J.W.; Apte, S.; Batley, G.E.; Bowles, K.C.; Campbell, P.G.C.; Delos, C.G.; Di Toro, D.M.; Dwyer, R.L.; Galvez, F.; et al. The biotic ligand model: A historical overview. *Comp. Biochem. Physiol. C Toxicol. Pharmacol.* **2002**, *133*, 3–35. [CrossRef]
9. Di Toro, D.M.; Allen, H.E.; Bergman, H.L.; Meyer, J.S.; Paquin, P.R.; Santore, R.C. Biotic ligand model of the acute toxicity of metals. 1. Technical Basis. *Environ. Toxicol. Chem.* **2001**, *20*, 2383–2396. [CrossRef]
10. Ghosh, P.; Bharadwaj, P.K.; Mandal, S.; Ghosh, S. Ni(II), Cu(II), and Zn(II) cryptate-enhanced fluorescence of a trianthrylcryptand: A potential molecular photonic or operator. *J. Am. Chem. Soc.* **1996**, *118*, 1553–1554. [CrossRef]
11. Banthia, S.; Samanta, A. Photophysical and Transition-Metal Ion Signaling Behavior of a Three-Component System Comprising a Cryptand Moiety as the Receptor. *J. Phys. Chem. B* **2002**, *106*, 5572–5577. [CrossRef]
12. Xiang, Y.; Tong, A.; Jin, P.; Ju, Y. New fluorescent rhodamine hydrazone chemosensor for Cu(II) with high selectivity and sensitivity. *Org. Lett.* **2006**, *8*, 2863–2866. [CrossRef] [PubMed]
13. Zhang, J.F.; Zhou, Y.; Yoon, J.; Kim, Y.; Kim, S.J.; Kim, J.S. Naphthalimide Modified Rhodamine Derivative: Ratiometric and Selective Fluorescent Sensor for Cu^{2+} Based on Two Different Approaches. *Org. Lett.* **2010**, *12*, 3852–3855. [CrossRef] [PubMed]
14. Zhou, Y.; Wang, F.; Kim, Y.; Kim, S.-J.; Yoon, J. Cu^{2+}-Selective Ratiometric and "Off-On" Sensor Based on the Rhodamine Derivative Bearing Pyrene Group. *Org. Lett.* **2009**, *11*, 4442–4445. [CrossRef] [PubMed]
15. Wu, S.-P.; Liu, S.-R. A new water-soluble fluorescent Cu(II) chemosensor based on tetrapeptide histidyl-glycyl-glycyl-glycine (HGGG). *Sens. Actuators B* **2009**, *141*, 187–191. [CrossRef]
16. Sumner, J.P.; Westerberg, N.M.; Stoddard, A.K.; Fierke, C.A.; Kopelman, R. Cu^+- and Cu^{2+}-sensitive PEBBLE fluorescent nanosensors using DsRed as the recognition element. *Sens. Actuators B Chem.* **2006**, *113*, 760–767. [CrossRef]

17. Jung, H.S.; Kwon, P.S.; Lee, J.W.; Kim, J.I.; Hong, C.S.; Kim, J.W.; Yan, S.; Lee, J.Y.; Lee, J.H.; Joo, T.; et al. Coumarin-Derived Cu^{2+}-Selective Fluorescence Sensor: Synthesis, Mechanisms, and Applications in Living Cells. *J. Am. Chem. Soc.* **2009**, *131*, 2008–2012. [CrossRef]
18. Du, J.; Yao, S.; Seitz, W.R.; Bencivenga, N.E.; Massing, J.O.; Planalp, R.P.; Jackson, R.K.; Kennedy, D.P.; Burdette, S.C. A ratiometric fluorescent metal ion indicator based on dansyl labeled poly (N-isopropylacrylamide) responds to a quenching metal ion. *Analyst* **2011**, *136*, 5006–5011. [CrossRef]
19. Yao, S.; Jones, A.M.; Du, J.; Jackson, R.K.; Massing, J.O.; Kennedy, D.P.; Bencivenga, N.E.; Planalp, R.P.; Burdette, S.C.; Seitz, W.R. Intermolecular approach to metal ion indicators based on polymer phase transitions coupled to fluorescence resonance energy transfer. *Analyst* **2012**, *137*, 4734–4741. [CrossRef]
20. Zadran, S.; Standley, S.; Wong, K.; Otiniano, E.; Amighi, A.; Baudry, M. Fluorescence resonance energy transfer (FRET)-based biosensors: Visualizing cellular dynamics and bioenergetics. *Appl. Microbiol. Biotechnol.* **2012**, *96*, 895–902. [CrossRef]
21. Osambo, J.; Seitz, W.; Kennedy, D.; Planalp, R.; Jones, A.; Jackson, R.; Burdette, S. Fluorescent Ratiometric Indicators Based on Cu(II)-Induced Changes in Poly (NIPAM) Microparticle Volume. *Sensors* **2013**, *13*, 1341–1352. [CrossRef] [PubMed]
22. Nichols, J.J.; King-Smith, P.E.; Hinel, E.A.; Thangavelu, M.; Nichols, K.K. The Use of Fluorescent Quenching in Studying the Contribution of Evaporation to Tear Thinning. *Investig. Ophthalmol. Vis. Sci.* **2012**, *53*, 5426–5432. [CrossRef] [PubMed]
23. Lin, H.-K.; Gu, Z.-X.; Chen, Y.-T. Linear thermodynamic function relationships in coordination chemistry: Calorimetric study on nickel(II)—, cobalt(II)—, zinc(II)—N-(p-substituted phenyl)iminodiacetic acid binary systems and nickel(II)—, cobalt(II)—, zinc(II)—nitrilotriacetic acid—N-(p-substituted phenyl)iminodiacetic acid ternary systems. *Thermochim. Acta* **1994**, *242*, 51–64. [CrossRef]
24. Schild, H.G. Poly (N-isopropylacrylamide): Experiment, theory and application. *Prog. Polym. Sci.* **1992**, *17*, 163–249. [CrossRef]
25. Schild, H.G.; Tirrell, D.A. Microcalorimetric detection of lower critical solution temperatures in aqueous polymer solutions. *J. Phys. Chem.* **1990**, *94*, 4352–4356. [CrossRef]
26. Yang, W.; Chen, X.; Su, H.; Fang, W.; Zhang, Y. The fluorescence regulation mechanism of the paramagnetic metal in a biological HNO sensor. *Chem. Commun.* **2015**, *51*, 9616–9619. [CrossRef] [PubMed]
27. Rüchel, R.; Steere, R.L.; Erbe, E.F. Transmission-electron microscopic observations of freeze-etched polyacrylamide gels. *J. Chromatogr. A* **1978**, *166*, 563–575. [CrossRef]
28. Bilal, M.; Rasheed, T.; Iqbal, H.M.N.; Hu, H.; Wang, W.; Zhang, X. Horseradish peroxidase immobilization by copolymerization into cross-linked polyacrylamide gel and its dye degradation and detoxification potential. *Int. J. Biol. Macromol.* **2018**, *113*, 983–990. [CrossRef]
29. Safronov, A.P.; Shankar, A.; Mikhnevich, E.A.; Beketov, I.V. Influence of the particle size on the properties of polyacrylamide ferrogels with embedded micron-sized and nano-sized metallic iron particles. *J. Magn. Magn. Mater.* **2018**, *459*, 125–130. [CrossRef]
30. Barril, P.; Nates, S. Introduction to Agarose and Polyacrylamide Gel Electrophoresis Matrices with Respect to Their Detection Sensitivities. In *Gel Electrophoresis-Principles and Basics*; Magdeldin, S., Ed.; InTechOpen: London, UK, 2012. [CrossRef]
31. Lin, H.-K.; Liu, Z.-J.; Tang, X.-H.; Chen, R.-T. Further Study of Linear Free Energy Relationships in Complex Compounds. *Acta Phys. Chim. Sin.* **1993**, *9*, 565–568. [CrossRef]
32. Mane, S.; Ponrathnam, S.; Chavan, N. Effect of Chemical Cross-linking on Properties of Polymer Microbeads: A Review. *Can. Chem. Trans.* **2016**, *3*, 473–485. [CrossRef]

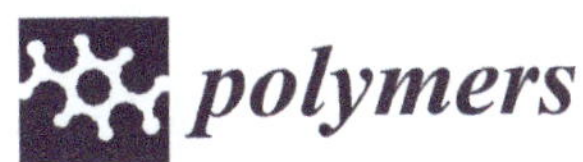 

Article

Exploring NIR Aza-BODIPY-Based Polarity Sensitive Probes with ON-and-OFF Fluorescence Switching in Pluronic Nanoparticles

Bahar Saremi [1,2], Venugopal Bandi [3,†], Shahrzad Kazemi [3], Yi Hong [2,4], Francis D'Souza [3,*] and Baohong Yuan [1,2,*]

1   Ultrasound and Optical Imaging Laboratory, Department of Bioengineering, The University of Texas at Arlington, Arlington, TX 76019, USA; bahar.saremi@uta.edu
2   Joint Biomedical Engineering Program, The University of Texas at Arlington and The University of Texas Southwestern Medical Center at Dallas, Dallas, TX 75390, USA; yihong@uta.edu
3   Department of Chemistry, University of North Texas, Denton, TX 76203, USA; venu235@gmail.com (V.B.); Shahrzadkazemi@my.unt.edu (S.K.)
4   Department of Bioengineering, The University of Texas at Arlington, Arlington, TX 76019, USA
*   Correspondence: francis.dsouza@unt.edu (F.D.); baohong@uta.edu (B.Y.);
    Tel.: +1-940-369-8832 (F.D.); +1-817-272-2917 (B.Y.)
†   Current Address: Department of Surgery, University of Pennsylvania Perelman School of Medicine, Philadelphia, PA 19104, USA.

Received: 18 December 2019; Accepted: 20 February 2020; Published: 2 March 2020

Abstract: Because of their deep penetration capability in tissue, red or near infrared (NIR) fluorophores attract much attention in bio-optical imaging. Among these fluorophores, the ones that respond to the immediate microenvironment (i.e., temperature, polarity, pH, viscosity, hypoxia, etc.) are highly desirable. We studied the response of six NIR aza-BODIPY-based and structurally similar fluorophores to polarity and viscosity for incorporation inside Pluronic nanoparticles as switchable fluorescent probes (SFPs). Based on our results, all of these fluorophores were moderately to strongly sensitive to the polarity of the microenvironment. We concluded that attaching amine groups to the fluorophore is not necessary for having strong polarity sensitive probes. We further studied the response of the fluorophores when embedded inside Pluronic nanoparticles and found that four of them qualified as SFPs. We also found that the switching ratio of the fluorophore-encapsulated Pluronic nanoparticles (I_{ON}-to-I_{OFF}) is related to the length of the hydrophobic chain of the Pluronic tri-block copolymers. As such, the highest switching ratio pertained to F-68 with the lowest hydrophobic block poly (propylene oxide) (PPO chain of only 30 units).

Keywords: aza-BODIPY; environment-sensitive; polarity-sensitive; near-infrared fluorescence imaging; temperature-sensitive; ultrasound switchable fluorescence probe; Pluronic; F-127; F-96; F-68; thermosensitive

---

1. Introduction

Fluorescence imaging is a sensitive and noninvasive method for investigating physiological and biomolecular processes in vitro and in vivo. Tuning the excitation light to 650–900 nm (red/near infrared (NIR)) has made fluorescence imaging in centimeter-deep tissue possible [1]. The red/NIR region is suitable for deep in vivo imaging as water and hemoglobin have their lowest absorption, while tissue autofluorescence and scattering are relatively low [1–4]. Despite the advantages that red/NIR imaging confers, there are a limited number of fluorescent dyes with excitation and emission in this region. Indocyanine green (ICG), one of the cyanine family dyes, remains the only FDA-approved NIR dye currently administered for clinical applications [1].

Environment-sensitive fluorescent probes are capable of responding to changes in the immediate microenvironment. By changing emission characteristics in response to stimuli or cellular conditions such as polarity, pH, viscosity, hypoxia, ions, etc. [5,6], these probes often forewarn of a severe disease (e.g., higher blood viscosity in diabetic patients [5,7] and lower pH and hypoxia in the tumor tissue of cancer patients [8]).

Among the environmental stimuli, polarity is an important stimulus associated with hydrophobicity of proteins and consequently a broad range of diseases such as Alzheimer's, in which elevation of hydrophobicity is concurrent with the increase in aggregation-prone proteins [9]. In addition to conveying vital information about the degree of aggregation of proteins, polarity-sensitive fluorescent probes have major applications in the synthesis of thermosensitive switches in conjunction with thermosensitive polymers (Figure 1a).

Although temperature-sensitive fluorescent probes have been reported for numerous applications, they suffer from low sensitivity. Rhodamine-B, a widely used probe for thermometry, has a mild thermosensitivity of 2.3% per degree Kelvin [10]. Multicolor methods, on the other hand, have not attained sensitivities of more than 10% [10]. By contrast, polarity-sensitive fluorescent probes, in conjunction with thermoresponsive polymers, show a dramatic change in fluorescence intensity in response to the change in polarity and, consequently, temperature.

Recently, a new polarity-sensitive fluorophore was introduced, and its characteristics and application as a switchable fluorescent probe (SFP), when incorporated inside polymeric nanoparticles, were studied [11]. In this system, the heat generated from the focused ultrasound would increase the temperature of the tissue at the focus of the ultrasound. Increasing the temperature to above the lower critical solution temperature (LCST) of the probes would elicit a response from the thermoresponsive polymer. The shrinkage of the nanoparticle would affect its water content, and consequently decrease the polarity of the immediate microenvironment of the fluorophores. This would trigger the fluorophore molecules to switch from an "OFF" or dark state, at which they emit weakly, to an "ON" state, at which they emit strongly.

In pursuit of more versatile and stronger environment-sensitive probes, herein, a set of aza-BODIPY-based and structurally related fluorophores were investigated, and their polarity and/or viscosity sensitivity, as well as the structural characteristics leading to such behavior, were explored. The response of the fluorophores was also tested when embedded inside Pluronic nanoparticles (Figure 1a,b).

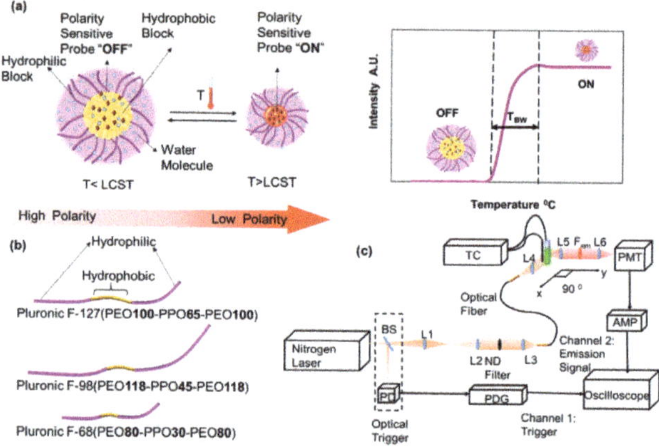

Figure 1. (**a**) Nanoparticles made of thermoresponsive Pluronic block copolymers, embedded with polarity-sensitive fluorophores, respond to the increase in temperature. By shrinking and excreting the water from the core of the nanoparticles, polarity is significantly decreased and polarity-sensitive fluorophores switch to "ON." (**b**) Pluronic F-127, F-98, and F-68 triblock copolymers with hydrophobic and hydrophilic chains [12,13]. (**c**) A schematic of the optical measurement system.

2. Materials and Methods

2.1. Fluorophore Characteristics

The optical properties of six aza-BODIPY-based and structurally related fluorophores were investigated in regard to their structure as provided in Figure 2. All fluorophores were synthesized by Dr. D'Souza's team. The synthesis methods for fluoprophores 1, 3, 4, 5 and 6 have been reported previously [2,14–16]. The synthesis method for fluorophore 2 is provided in Supplementary Materials S4. The fluorophores' names, excitation wavelength (λ_{ex}), and detection emission wavelength (λ_{em}) are summarized in Table 1. The fluorescence spectra are provided in Supplementary Materials S1.

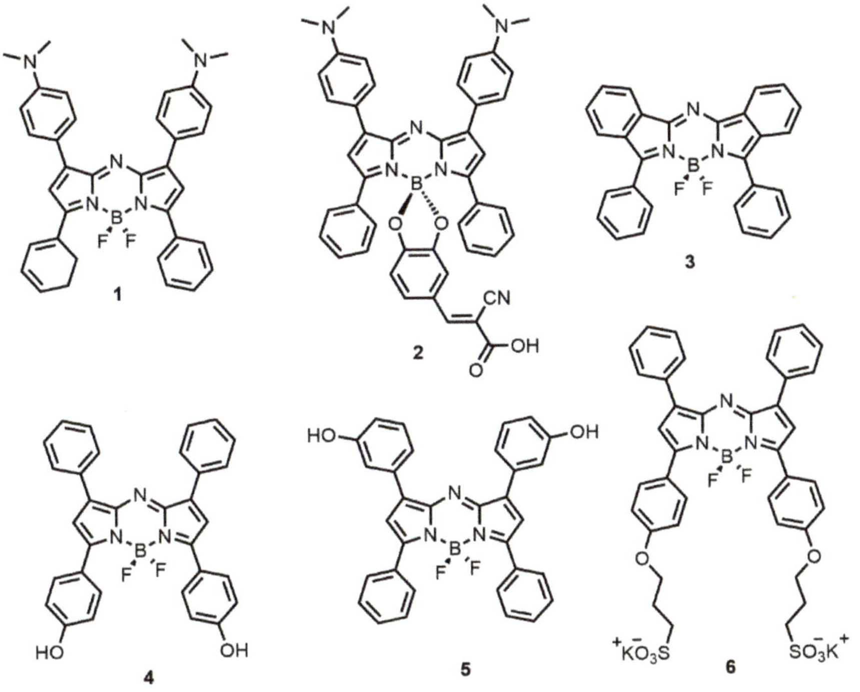

Figure 2. Structure of the aza-BODIPY-based and structurally related fluorophores (synthesis methods have been previously reported for fluorophores 1, 3, 4, 5, 6 [2,14–16] and for fluorophore 2 in Supplementary Materials S4).

Table 1. Summary of the fluorophores' names, excitation, and emission detection wavelengths adopted in the experiments [2,14–16] and Supplementary Materials S4.

	Fluorophore Name	Abbrev. Name	λ_{ex} (nm)	λ_{em} (nm)
1	Top Dimethyl amine ADPF2	TOP DMAADP	644	785/62 BP
2	Top Dimethyl amine ADPCNCA	TOP DMAADPCA	644	808 LP
3	Benzannulated ADPF2	Fused ADPF$_2$	710	785/62 BP
4	ADP (OH)$_2$ (Bottom)	ADP(OH)$_2$(Bottom)	655	711/25 BP
5	Top (OH)$_2$ ADP	Top (OH)$_2$ ADP	655	711/25 BP
6	ADP-Di Sulphonic acid	ADP DI(SA)	644	711/25 BP

2.2. Fluorescence Measurement System

The fluorescence measurement system utilized for lifetime and intensity measurements was discussed in previous work [17,18], with minor changes to accommodate the temperature control system and corresponding filters for different fluorescent dyes. Briefly, a combined laser system from Optical Building Blocks Corporation (Birmingham, NJ, USA) generated an 800 ps pulse at the excitation wavelength of each dye. All lenses used in the system were purchased from Thorlabs Inc. (Newton, NJ, USA), if not otherwise stated. The generated fluorescence light passed through a converging lens, and the corresponding emission band-pass or long-pass filter (BP, FF01-711/25-25, FF01-785/62-25, or BLP01-808R-25 from Semrock, Rochester, NY, USA). The emitted fluorescent light was detected by a photomultiplier tube (PMT, H10721-20, Hamamatsu, Japan). A pulse delay generator (PDG, DG645, Stanford Research Systems, Sunnyvale, CA, USA) triggered the 2.5 GHz oscilloscope (DPO 7254, Tektronix, Beaverton, OR, USA). The output of the PMT was converted to a voltage signal and was amplified by a broadband preamplifier (C5594, bandwidth from 50 kHz to 1.5 GHz, Hamamatsu, Japan). The signal was ultimately acquired by the multichannel and broadband oscilloscope. Each emission decay pulse recorded from the oscilloscope, used for calculating fluorescence lifetime, was averaged 100 times. A temperature controller (PTC10, Stanford Research System, Sunnyvale, CA, USA) was used to control the temperature (Figure 1c).

2.3. Fluorescence Emission Intensity and Lifetime Measurement

Since the fluorescence signal is acquired by an oscilloscope, the peaks of the decay curves indicate the voltage, and are proportional to the fluorescence intensity. For lifetime calculations, the acquired signal and the impulse response function were deconvolved and the decay curve was fitted to a monoexponential decay function [18]. Calculations were done using MATLAB (Natick, MA), with an iterative method to find the best fit with lowest residue. For fluorescence intensity calculations in polarity experiment, after averaging the emission curves 100 times, a moving average filter (n = 5) was applied, and the peak height of the decay curve was obtained in each solvent for all fluorophores. Fluorescence intensity of each fluorophore in water was normalized to 1. For fluorescence intensity calculations in fluorophore-encapsulated Pluronic nanoparticles, the same protocol was followed without a moving average filter. Herein, the peak voltage of the decay curve at each temperature was obtained (mV), and the switching ratio was calculated as discussed in the text. For comparisons regarding the Pluronic nanoparticles, a two-tail *t*-test (equal variance) was conducted. Pearson linear correlation coefficient was calculated with R program.

2.4. Solvents with Different Polarities

Fluorescence lifetime and emission intensity of the fluorophores were measured in different solvents. Solvents were obtained from Sigma-Aldrich Corporate (St. Louis, MO, USA), if not otherwise stated. Characteristics of the solvents used are provided in Table 2. The final concentration of fluorophores was kept at 50 nM.

Table 2. Solvent properties [19–21].

Solvent	Solvent Type	$E_T(30)$ (kcal/mol)	Dielectric Constant ε
Water	Polar protic	62.8	80.1
DMSO	Dipolar aprotic	45.1	46.7
Dichloromethane	Polar aprotic	40.7	9
Benzene	Nonpolar	34.3	2.3
Toluene	Nonpolar	33.9	2.4
Glycerol	Polar protic	57	42.5
Ethylene glycol	Polar protic	53.8	31.8

2.5. Solvents with Different Viscosities

To cover a range of viscosities, different mixtures of ethylene glycol (EG) and glycerol (Gl) were prepared. Solutions were prepared by mixing Gl and EG at different volume ratios: Gl/EG (*v/v*) % of (0/100)%, (16/84)%, (50/50)%, (75/25)%, (92/8)%, and (100/0)%. The viscosity of the mixtures (Table 3) were measured with an A&D SV-10 Viscomter (Tokyo, Japan).

Table 3. Viscosities pertaining to Gl/EG (*v/v*) % solutions.

Glycerol %	0	16	50	75	92	100
Viscosity (m Pa.s)	12.7	20.8	53.5	117.0	273.5	634.5

2.6. Preparation of Fluorophore-Encapsulated Pluronic Nanoparticles

Nanoparticles were synthesized based on a revised protocol from Pluronic block copolymers consisting of poly(ethylene oxide) (PEO) and hydrophobic poly(propylene oxide) (PPO) [22]. Pluronic F-127 (PEO100-PPO65-PEO100) [12] obtained from Sigma-Aldrich (St. Louis, MO, USA), as well as Pluronic F-98 (PEO118-PPO45-PEO118) [13] and Pluronic F-68 (PEO80-PPO30-PEO80) [13] obtained from BASF (Florham Park, NJ, USA), were dissolved in deionized water (D.I. water) at pH 8.5 with a 5% (*w/v*) ratio with stirring. The fluorophores and tetrabutylammonium iodide (TBAI) from Sigma-Aldrich (St. Louis, MO, USA) were dissolved with a molar ratio of 1:8 in 6 mL chloroform (Fisher Scientific, Pittsburgh, PA, USA) by sonication using a bath ultrasonic cleaner (Branson 1510, Branson Ultrasonic Corporation, Danbury, CT, USA) and added drop-wise to 15 mL of the Pluronic solution while stirring at 1200 rpm. The sample was then sonicated with an XL-2020 probe-sonicator (Misonix, Farmingdale, NY, USA), while the probe intensity was kept at ~5–5.5. The sample was then moved to a beaker covered by aluminum foil with generated holes and stirred at 200–300 rpm overnight to evaporate the chloroform (final concentration of fluorophores was kept at about 50 µM). Nanoparticles were then filtered with (10,000 MWCO) Amicon ultra centrifugation filters (Merck Millipore, Billerica, MA, USA) and large particles and/or impurities were filtered by 0.45 µm membrane filters (Fisher Scientific, Pittsburgh, PA, USA). The effect of filtration and dilution on LCST has been extensively studied in the Supplementary Materials S3. For filtration, samples were diluted 5 times (for facilitating the process), centrifuged at 4500 G with a Legend X1 centrifuge (Sorvall™ Legend™ X1, Thermo, Marietta, OH, USA), reduced, and then brought back to the initial volume with D.I. water. All measurements were done with filtered 1% samples, except for when the effect of filtration was studied, where 1% and 0.2% samples, with and without filtering, were prepared.

3. Results and Discussion

3.1. Response to Polarity

To investigate the effect of polarity, the fluorescence intensity of the fluorophores, were measured in different solvents. By changing the solvents, from water to toluene, the polarity of the microenvironment changed 28.9 units of polarity index, E_T (30). This polarity index was defined as the molar electronic transition energy of the negatively solvatochromic pyridinium N-phenolate betaine dye as probe molecule, measured in kilocalories per mole (kcal mol^{-1}) at room temperature (25 °C), and normal pressure (1 bar) [23]. Normalized fluorescence intensity vs polarity index is plotted in Figure 3. The decay curves are provided in Supplementary Materials S2. The maximum emission peak for each fluorophore among solvents is referred to as I_{max}, and the emission peak in water is denoted as I_{water}.

As demonstrated in Figure 3, based on the normalized fluorescence intensity values, ADP(OH)2 Bottom (i.e., fluorophore 4) had the strongest dependency on the polarity with I_{max}-to-I_{water} ratio of ~478, followed by Top (OH)$_2$ ADP (i.e., fluorophore 5) with a ratio of ~9.58. Top Dimethyl amine ADPCNCA (TOP DMAADPCA, i.e., fluorophore 2) and Top Dimethyl amine ADPF2 (TOP DMA ADP, i.e., fluorophore 1) had a ratio of 3.77 and 5.04 times, respectively. ADP Di Sulphonic acid (ADP Di

(SA), i.e., fluorophore 6) had a ratio of 3.8 times, followed by Benzannulated ADPF2 (i.e., fluorophore 3) that had the lowest ratio of 2.16 times. A previous publication reported that Top Dimethyl amine ADPF2 (TOP DMA ADP, i.e., fluorophore 1) has polarity sensitivity between 677 and 692 nm [14]. Because of the particular interest in the red/NIR region, we measured the polarity sensitivity between 723 and 847 nm, in this study.

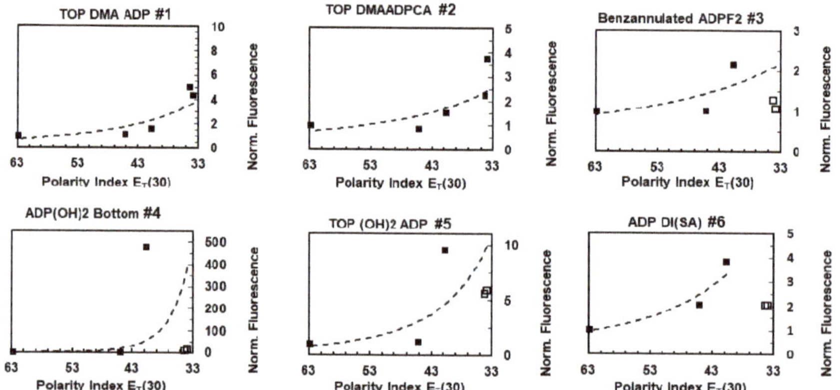

Figure 3. Peak fluorescence emission signal intensity vs solvent polarity. Power function fitted to the points depicted with black squares. The fluorescence intensity in water is normalized to 1.

The significant decrease in the fluorescence intensity from the highest polarity value, observed for fluorophores 3, 4, 5 and 6 in benzene and toluene (the last two squares) might be due to lower solubility in highly nonpolar solvents. Since we are only interested in the response to polarity, the points in which the fluorescence intensity was affected by solubility were shown with hollow squares and not included in fitting. To limit the effect of solubility, very low concentrations (50 nM) were used, and no aggregation was observed. In general, the fluorescence emission intensity of all the fluorophores increased with the decrease in polarity. The existence of amine groups attached to the fluorophore system seems unnecessary for generating a strong polarity-sensitive probe. As such, fluorophore 4 is a highly sensitive probe, despite not having an amine group.

In the case of fluorophores with substitutions of electron-donating groups to the aryl rings attached to the core of the fluorophore, polarity sensitivity was most prominent when the hydroxyl groups were attached to the para position of aryls at 3- and 5- (fluorophore 4), and to a lesser degree, when hydroxyl groups were attached to the meta position of the aryls at 1- and 7- (fluorophore 5), or when the sulfonate groups were attached to the aryls at 3- and 5- positions (fluorophore 6). More conclusive results were obtained by synthesizing Pluronic nanoparticles embedded with the fluorophores.

3.2. Response to Viscosity

The fluorescence intensity and lifetime of the fluorophores were measured in regard to the viscosity of the environment, while polarity was kept relatively constant (below 3.2 units of E_T (30)). To quantitatively investigate the relationship between the fluorescence intensity with the viscosity, the Förster–Hoffmann equation was used [24,25]:

$$\log (I) = C + X \log(\eta) \tag{1}$$

Here, I is the peak emission intensity, η is the solvent viscosity, C is a constant related to temperature and concentration, and X is a constant related to fluorophore properties. For each fluorophore, the logarithm of signal peak in EG (lowest viscosity) was normalized to 1. As evident from the graphs (Figure 4), the fluorescence intensity of fluorophores 2, 4, 5, and 6 was relatively insensitive to viscosity.

The fluorescence intensity of ADP BF$_2$ (fluorophore 3) dropped in the viscous medium, while the fluorescence intensity of fluorophore 1 increased about 14 times at the beginning, and then decreased. The decrease in fluorescence intensity could be due to lack of a response to viscosity (fluorophore 3), or a response (fluorophore 1) compounded by the decrease in fluorescent emission, caused by the change in polarity in this experiment. It should be noted that by changing the composition of the mixture, polarity changed 3.2 units of E_T (30), and inevitably the response to viscosity was convolved with the response to polarity. Although our result does not refute the fact that loss of energy due to rotational motion happens, it implies that viscosity sensitivity is not a major contributor to the phenomenon under study. Our data is consistent with previous studies reported in the literature that phenyl substitutions at 3- and 5- positions cannot rotate freely as a result of the steric hindrance caused by interaction of the hydrogen in the C-H bond with the fluorine atom from the BF2 [26,27].

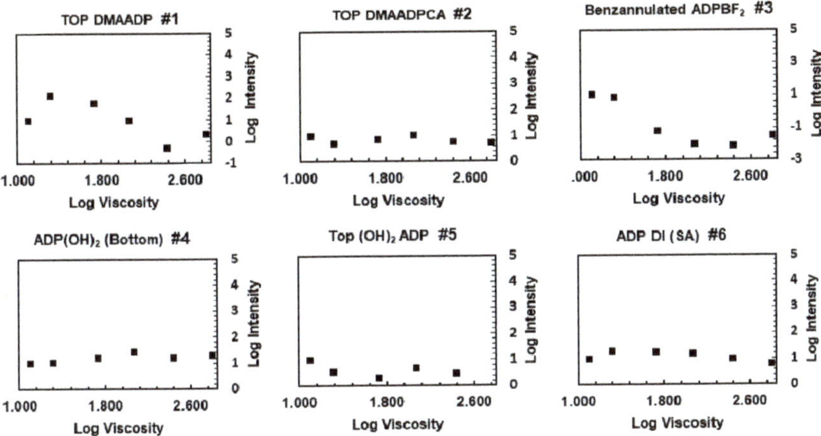

Figure 4. Fluorescence intensity in response to the change in the viscosity.

3.3. Fluorophore Encapsulation in Pluronic Nanoparticles

Based on the polarity sensitivity results, while all fluorophores responded to the change in polarity, it was concluded that the attachment of amine groups to the fluorophore was not necessary for having strong polarity-sensitive probes. For the next step, the fluorophores were encapsulated into Pluronic F-127 nanoparticles, and their response to temperature was measured with increasing the temperature (Figure 5). Fluorophore 4 was also encapsulated in Pluronic F-98 and F-68 nanoparticles, to investigate the effect of hydrophobic and hydrophilic chain lengths of Pluronic nanoparticles.

As shown in Figure 5, when the temperature is below the LCST (lower critical solution temperature) of the nanoparticles (black arrow), the fluorescence emission is very weak because the fluorophores in the nanoparticles are exposed to a microenvironment in which water molecules are rich, and therefore, the polarity is high and thus, the fluorophore has a low emission efficiency. When the temperature is above the LCST, the fluorescence emission increases rapidly. Eventually, the fluorescence intensity reaches a plateau (red arrow). This can be understood as follows. When the temperature is above the LCST, the fluorophores in the nanoparticles are exposed to a lack-of-water microenvironment in which the polarity is low, and thus the fluorophore has a high emission efficiency. Herein, the temperature range between the first local maximum and the LCST is the temperature transition bandwidth (T_{bw}). If the T_{bw} is as narrow as a few degrees, the Pluronic-fluorophore system can act as a switch, in which the fluorophore is considered to be at an "OFF" state when temperature is below the LCST, and at an "ON" state when temperature is above the first local maximum.

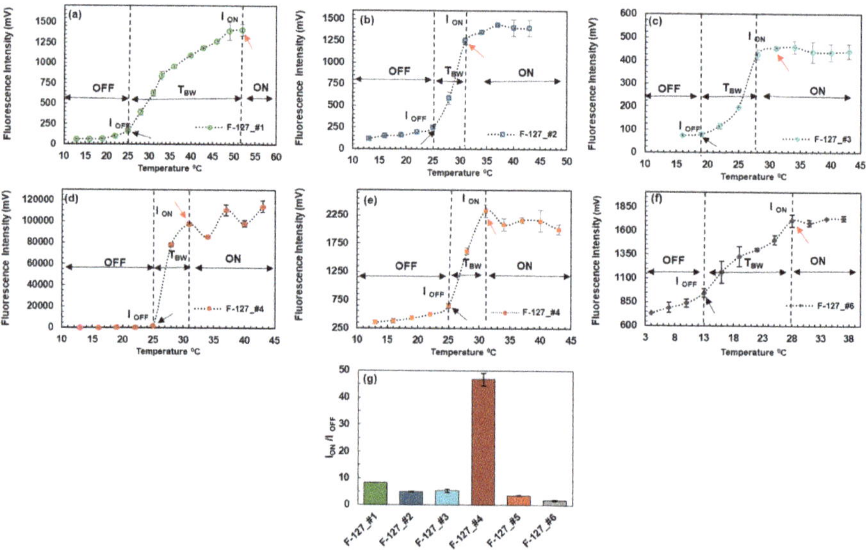

Figure 5. (**a–f**) Response of Pluronic F-127 loaded with fluorophores to the change in temperature. Black arrows show the LCST and red arrows the first local maximum (major slope change in the case of a plateau). (**g**) Pluronic F-127 nanoparticles loaded with fluorophore 4 acted as a very strong switch. Both F-127 nanoparticles loaded with fluorophore 2, 3 and 5, act as moderate switches. F-127 nanoparticles loaded with fluorophore 1, and 6 did not qualify as a switch due to having a large T_{BW} ($\geq$15 °C). (Values are based on Edge method).

For calculating the switching ratio (I_{ON}-to-I_{OFF}), two methods are proposed: (1) The edge method, in which the amplitude of the voltage at the first local maximum (or in the case of a plateau, the first point at major slope change) is divided by the amplitude of the voltage at the LCST, and (2) the vicinity method, in which the amplitude of the voltage at the first local maximum is divided by the averaged amplitude of the voltage of the points at the vicinity of the LCST (the temperature of choice below LCST depends on the application, and 3 °C was chosen here).

As shown in Figure 5 and Table 4, Pluronic F-127 with fluorophore 4 had the highest ratio of switching among all other fluorophores (46.74 times). Since the T_{BW} for this system is 6 °C, this polymer-fluorophore system qualified as a very strong switch.

Despite the (I_{ON}-to-I_{OFF}) ratio of Pluronic F-127 with fluorophore 1 (8.37 times), this system doesn't behave as a switch. This is due to the very large T_{BW} of more than 27 °C. Pluronic F-127 with fluorophores 2, 3 and 5 with an (I_{ON}-to-I_{OFF}) of 4.97, 5.4 and 3.59 times, respectively, act as moderate switches. F-127 nanoparticle system with fluorophore 6, has an (I_{ON}-to-I_{OFF}) of (1.81 times) over a wide bandwidth of 15 degrees and didn't behave as a switch.

Compared to F-127, Pluronic F-68 with fluorophore 4 elicits a higher value of I_{ON}-to-I_{OFF} of 72.74 times (Figure 6). Based on the Pearson linear correlation coefficient, there was a strong inverse correlation between the switch ratio and the length of the hydrophobic chain of the Pluronic tri-block copolymer. As such, the shorter length of the hydrophobic chain of the Pluronic tri-block copolymer would yield a higher switching ratio (30 PPO units of F-68 compared to the 65 PPO units of F-127). Although future research is needed to delineate this phenomenon, we suspect that the fluorophores are embedded in the hydrophobic region of the micelles, and the shorter length of the hydrophobic chain would produce a less polar environment upon being heated.

Table 4. Summary of switching properties of the fluorophore-encapsulated Pluronic nanoparticles.

Pluronic Nanoparticles	I_{ON}-to-I_{OFF} Edge Method	I_{ON}-to-I_{OFF} Vicinity Method	LCST * (°C) Nanoparticles	T_{BW} (°C)	Lifetime (ns) (OFF to ON Edge Method)
F-127_#1	8.37 ± 0.11	13.74 ± 1.39	25	27	0.41 ± 0.01 to 1.09 ± 0.01
F-127_#2	4.97 ± 0.4	6.36 ± 0.17	25	6	0.27 ± 0.1 to 0.75 ± 0.06
F-127_#3	5.40 ± 0.76	5.75 ± 0.34	19	9	1.42 ± 0.45 to 1.81 ± 0.05
F-127_#4	46.74 ± 3.35	109.1 ± 18.9	25	6	2.0 ± 0.14 to 3.02 ± 0.03
F-127_#5	3.59 ± 0.08	4.65 ± 0.14	25	6	2.55 ± 0.07 to 3.02 ± 0.17
F-127_#6	1.81 ± 0.15	2.04 ± 0.18	13	15	2.96 ± 0.23 to 3.23 ± 0.08
F-98_#4	62.26 ± 9.9	137.23 ± 0.15	34	6	1.5 ± 0.1 to 1.65 ± 0.05
F-68_#4	72.74 ± 0.81	102.07 ± 3.72	61	9	1.14 0.3 to 2.33 ± 0.04

* Lower critical solution temperature.

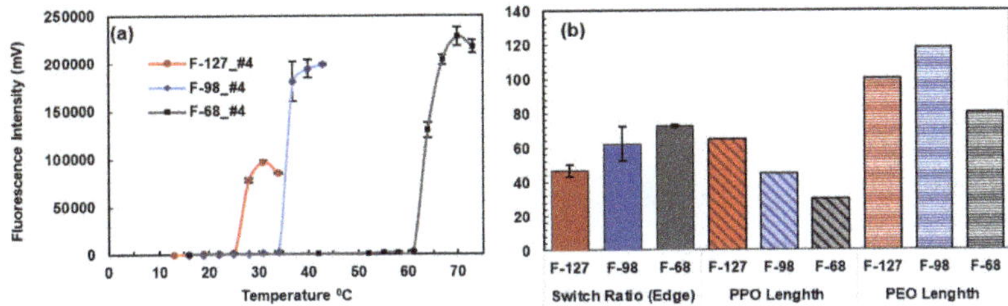

Figure 6. (**a**) Response of Pluronic F-127, F-98, and F-68 nanoparticles loaded with fluorophore 4 to the change in temperature. (**b**) Pluronic F-68 nanoparticles loaded with fluorophore 4 showed a stronger switching ratio compared to Pluronic F-127 nanoparticles (*p*-value of 0.009). The switch ratio is inversely correlated to length of hydrophobic (PPO) chain ($\rho = -0.99$, *p*-value = 0.018).

The observed significant polarity sensitivity of the fluorophores, considering their interconnected π-system, might be due to specific solvent-fluorophore interactions such as "charge transfer pathways," which mainly include photo-induced electron transfer (PeT) and internal charge transfer (ICT).

4. Conclusions

Six members of the aza-BODIPY-based and structurally related fluorophores with excitation and emission in the red/NIR region were investigated. It was shown that fluorophore 4 was a very strong polarity-sensitive probe, followed by fluorophores 1, 2, 5, 6, and 3. It was also shown that attaching amine groups to the fluorophore is not necessary for having strong polarity-sensitive probes. After encapsulating fluorophores into thermosensitive Pluronic nanoparticles, it was found that fluorophores 4, 2, 3, and 5 can be used as ON-and-OFF fluorescence switches because of the dramatic change in peak emission fluorescence intensity over a narrow range of temperatures (fluorophore 4 had the strongest switching response). It was also shown that the switching ratio of the fluorophore-encapsulated Pluronic nanoparticles (I_{ON}-to-I_{OFF}) increased with decreasing the number of hydrophobic chains. As such, Pluronic F-68 with a PPO length of 30 had the highest switching ratio, while the Pluronic F-127 nanoparticles with the PPO length of 65 had the lowest.

Supplementary Materials: The following are available online at http://www.mdpi.com/2073-4360/12/3/540/s1, Supplementary Materials S1: Absorbance and Emission Spectra, Supplementary Materials S2: Fluorescence Lifetime Decay Curves, Supplementary Materials S3: Effect of Filtration and Dilution on LCST, Supplementary Materials S4: Compound 2 (Top DMAADPCA) Synthesis Procedure.

Author Contributions: Conceptualization, B.Y.; methodology, B.Y., B.S., V.B., S.K.; software, B.S.; validation, B.S.; formal analysis, B.S.; investigation, B.S.; resources, B.Y., F.D., Y.H.; data curation, B.S.; writing—original draft preparation, B.S. and B.Y.; writing—review and editing, B.S., B.Y., V.B., Y.H., F.D.; visualization, B.S., V.B.; supervision, B.Y., Y.H., F.D.; project administration, B.Y.; funding acquisition, B.Y. All authors have read and agreed to the published version of the manuscript.

Funding: This work was supported in part by funding from the CPRIT RP170564 (BY) and NSF CBET-1253199 (BY).

Acknowledgments: We appreciate David Manivanh's help in measuring the viscosity of our samples. Our gratitude goes out to Guido Verbeck and the Laboratory for Imaging Mass Spectrometry at the University of North Texas for MALDI-Orbitrap Mass Spectrometry data.

Conflicts of Interest: The authors declare no conflict of interest.

References

1. Weissleder, R. A clearer vision for in vivo imaging. *Nat. Biotechnol.* **2001**, *19*, 316–317. [CrossRef] [PubMed]
2. Tasior, M.; Murtagh, J.; Frimannsson, D.O.; McDonnell, S.O.; O'Shea, D.F. Water-solubilised BF2-chelated tetraarylazadipyrromethenes. *Org. Biomol. Chem.* **2010**, *8*, 522–525. [CrossRef] [PubMed]
3. Frangioni, J.V. In vivo near-infrared fluorescence imaging. *Curr. Opin. Chem. Boil.* **2003**, *7*, 626–634. [CrossRef] [PubMed]
4. Lu, H.; Mack, J.; Yang, Y.; Shen, Z. Structural modification strategies for the rational design of red/NIR region BODIPYs. *Chem. Soc. Rev.* **2014**, *43*, 4778–4823. [CrossRef]
5. Yang, Z.; Cao, J.; He, Y.; Yang, J.H.; Kim, T.; Peng, X.; Kim, J.S. Macro-/micro-environment-sensitive chemosensing and biological imaging. *Chem. Soc. Rev.* **2014**, *43*, 4563–4601. [CrossRef] [PubMed]
6. Klymchenko, A. Solvatochromic and Fluorogenic Dyes as Environment-Sensitive Probes: Design and Biological Applications. *Accounts Chem. Res.* **2017**, *50*, 366–375. [CrossRef]
7. Irace, C.; Carallo, C.; Scavelli, F.; De Franceschi, M.S.; Esposito, T.; Gnasso, A. Blood Viscosity in Subjects With Normoglycemia and Prediabetes. *Diabetes Care* **2013**, *37*, 488–492. [CrossRef]
8. Kato, Y.; Ozawa, S.; Miyamoto, C.; Maehata, Y.; Suzuki, A.; Maeda, T.; Baba, Y. Acidic extracellular microenvironment and cancer. *Cancer Cell Int.* **2013**, *13*, 89. [CrossRef]
9. Mitchell, M.J.; King, M.R. Increased Protein Hydrophobicity in Response to Aging and Alzheimer's disease. *NIH Public Access* **2014**, *48*, 1–23. [CrossRef]
10. Estrada-Pérez, C.E.; Hassan, Y.A.; Tan, S. Experimental characterization of temperature sensitive dyes for laser induced fluorescence thermometry. *Rev. Sci. Instrum.* **2011**, *82*, 74901. [CrossRef]
11. Cheng, B.; Bandi, V.; Yu, S.; D'Souza, F.; Nguyen, K.T.; Hong, Y.; Tang, L.; Yuan, B. The Mechanisms and Biomedical Applications of an NIR BODIPY-Based Switchable Fluorescent Probe. *Int. J. Mol. Sci.* **2017**, *18*, 384. [CrossRef] [PubMed]
12. Pitto-Barry, A.; Barry, N.P.E. Pluronic® block-copolymers in medicine: from chemical and biological versatility to rationalisation and clinical advances. *Polym. Chem.* **2014**, *5*, 3291–3297. [CrossRef]
13. Lee, C.-F.; Tseng, H.-W.; Bahadur, P.; Chen, L.-J. Synergistic Effect of Binary Mixed-Pluronic Systems on Temperature Dependent Self-assembly Process and Drug Solubility. *Polymers* **2018**, *10*, 105. [CrossRef] [PubMed]
14. Killoran, J.; McDonnell, S.O.; Gallagher, J.F.; O'Shea, D.F. A substituted BF2-chelated tetraarylazadipyrromethene as an intrinsic dual chemosensor in the 650–850 nm spectral range. *New J. Chem.* **2008**, *32*, 483–489. [CrossRef]
15. Gresser, R.; Hummert, M.; Hartmann, H.; Leo, K.; Riede, M.K. Synthesis and Characterization of Near-Infrared Absorbing Benzannulated Aza-BODIPY Dyes. *Chem. A Eur. J.* **2011**, *17*, 2939–2947. [CrossRef] [PubMed]
16. Bandi, V.; El-Khouly, M.E.; Nesterov, V.N.; Karr, P.A.; Fukuzumi, S.; D'Souza, F. Self-Assembled via Metal-Ligand Coordination AzaBODIPY-Zinc Phthalocyanine and AzaBODIPY-Zinc Naphthalocyanine Conjugates: Synthesis, Structure, and Photoinduced Electron Transfer. *J. Phys. Chem. C* **2013**, *117*, 5638–5649. [CrossRef]
17. Cheng, B.; Wei, M.; Liu, Y.; Pitta, H.; Xie, Z.; Hong, Y.; Nguyen, K.T.; Yuan, B. Development of Ultrasound-Switchable Fluorescence Imaging Contrast Agents Based on Thermosensitive Polymers and Nanoparticles. *IEEE J. Sel. Top. Quantum Electron.* **2013**, *20*, 67–80. [CrossRef]
18. Saremi, B.; Wei, M.-Y.; Liu, Y.; Cheng, B.; Yuan, B. Re-evaluation of biotin-streptavidin conjugation in Förster resonance energy transfer applications. *J. Biomed. Opt.* **2014**, *19*, 85008. [CrossRef]
19. Haidekker, M.A.; Brady, T.; Lichlyter, D.; Theodorakis, E. Effects of solvent polarity and solvent viscosity on the fluorescent properties of molecular rotors and related probes. *Bioorganic Chem.* **2005**, *33*, 415–425. [CrossRef]
20. Web Services at the European Bioinformatics Institute. Available online: https://www.ebi.ac.uk/chebi/ (accessed on 14 November 2019).

21.   Murov, S.L. Properties of Solvents Used in Organic Chemistry. Available online: http://murov.info/orgsolvents.htm (accessed on 14 November 2019).

22.   Kim, T.H.; Chen, Y.; Mount, C.; Gombotz, W.R.; Li, X.; Pun, S.H. Evaluation of Temperature-Sensitive, Indocyanine Green-Encapsulating Micelles for Noninvasive Near-Infrared Tumor Imaging. *Pharm. Res.* **2010**, *27*, 1900–1913. [CrossRef]

23.   Reichardt, C. Solvatochromic Dyes as Solvent Polarity Indicators. *Chem. Rev.* **1994**, *94*, 2319–2358. [CrossRef]

24.   Su, D.; Teoh, C.L.; Gao, N.; Xu, Q.-H.; Chang, Y.-T. A Simple BODIPY-Based Viscosity Probe for Imaging of Cellular Viscosity in Live Cells. *Sensors* **2016**, *16*, 1397. [CrossRef] [PubMed]

25.   Haidekker, M.A.; Theodorakis, E.A. Environment-sensitive behavior of fluorescent molecular rotors. *J. Boil. Eng.* **2010**, *4*, 11. [CrossRef] [PubMed]

26.   Zhang, X.; Yu, H.; Xiao, Y. Replacing Phenyl Ring with Thiophene: An Approach to Longer Wavelength Aza-dipyrromethene Boron Difluoride (Aza-BODIPY) Dyes. *J. Org. Chem.* **2011**, *77*, 669–673. [CrossRef]

27.   Chen, J.; Reibenspies, J.; Derecskei-Kovacs, A.; Burgess, K. Through-space 13C–19F coupling can reveal conformations of modified BODIPY dyes†. *Chem. Commun.* **1999**, *24*, 2501–2502. [CrossRef]

MDPI

St. Alban-Anlage 66

4052 Basel

Switzerland

Tel. +41 61 683 77 34

Fax +41 61 302 89 18

www.mdpi.com

Polymers Editorial Office

E-mail: polymers@mdpi.com

www.mdpi.com/journal/polymers

www.ingramcontent.com/pod-product-compliance
Lightning Source LLC
LaVergne TN
LVHW071014170726
843515LV00006B/1133